Feng · Shi   Mathematical Theory of Elastic Structures

Springer-Verlag Berlin Heidelberg GmbH

Feng Kang   Shi Zhong-Ci

Mathematical Theory of Elastic Structures

With 74 Illustrations

 Springer

Feng Kang †
Shi Zhong-Ci

Institute of Computational Mathematics,
Chinese Academy of Sciences,
P.O. Box 2719,
Beijing 100080,
The People's Republic of China

Revised edition of the original Chinese edition published by Science Press Beijing 1981 as the 6th volume in the Series in Pure and Applied Mathematics.

Distribution rights throughout the world, excluding the People's Republic of China, granted to Springer-Verlag Berlin Heidelberg New York London Paris Tokyo Hong Kong Barcelona Budapest

Mathematics Subject Classification (1991): 73C20, 73K03, 73K05, 73K10, 73K20, 73V05, 73V25, 65N30

```
          Library of Congress Cataloging-in-Publication Data

Feng, K' ang.
    Mathematical theory of elastic structures / Feng Kang, Shi
  Zhongci.
        p.    cm.
    Includes bibliographical references and index.
    ISBN 978-3-662-03288-6     ISBN 978-3-662-03286-2 (eBook)
    DOI 10.1007/978-3-662-03286-2
    1. Elastic analysis (Engineering)  2. Structural analysis
  (Engineering)   I. Shih, Chung-tz'u.   II. Title.
  TA653.F46   1995
  624.1'7--dc20
                                                    95-39142
                                                        CIP
```

ISBN 978-3-662-03288-6

© Springer-Verlag Berlin Heidelberg 1996
Originally published by Springer-Verlag Berlin Heidelberg New York in 1996
Softcover reprint of the hardcover 1st edition 1996

Typesetting: Science Press Beijing. The People's Republic of China

SPIN: 10013542   41/3143-543210 – Printed on acid-free paper.

Preface

Elasticity theory is a classical discipline. The mathematical theory of elasticity in mechanics, especially the linearized theory, is quite mature, and is one of the foundations of several engineering sciences. In the last twenty years, there has been significant progress in several areas closely related to this classical field, this applies in particular to the following two areas.

First, progress has been made in numerical methods, especially the development of the finite element method. The finite element method, which was independently created and developed in different ways by scientists both in China and in the West, is a kind of systematic and modern numerical method for solving partial differential equations, especially elliptic equations. Experience has shown that the finite element method is efficient enough to solve problems in an extremely wide range of applications of elastic mechanics. In particular, the finite element method is very suitable for highly complicated problems. One of the authors (Feng) of this book had the good fortune to participate in the work of creating and establishing the theoretical basis of the finite element method. He thought in the early sixties that the method could be used to solve computational problems of solid mechanics by computers. Later practice justified and still continues to justify this point of view. The authors believe that it is now time to include the finite element method as an important part of the content of a textbook of modern elastic mechanics.

The second area is the development of composite elastic structural mechanics. In modern engineering practice we face not only the geometrically simple, elastic body, but also, more importantly, bodies composed of several elastic members, including those with different dimensions and with different properties, i.e., composite structures, such as aerospace structures, reactor structures, tall building structures, off-shore platform structures, underground structures.

The development of composite elastic mechanics has great significance in both practice and theory.

The development of composite elastic structural mechanics and the development of the finite element method, influence each other. To this date, however, the mathematical basis of composite structure theory is still not sufficiently rigorous and complete. At the same time, the theory of elliptic equations on composite manifolds developed in recent years by one of the authors (Feng), can be applied to place composite structure theory in a comparatively strict theoretical framework, aiding the further development of both theory and application. Thus, it also makes it possible to include a mathematical theory of composite structures in this book on modern elastic structural mechanics.

In view of these considerations this book covers the following main topics: 1. The classical theory of linear elasticity 2. The mathematical theory of the composite elastic structures. Here the mathematical method proposed by the authors is presented. 3. The numerical method for solving elastic structural problems—the finite element method. The authors try to treat these three topics within the framework f a unified theory. It seems likely that the material covered in the book has not been published before. To this end, the whole book carries on a theoretical discussion on the mathematical basis of the principle of minimum potential energy, using displacement as the fundamental variable. The emphasis is on the accuracy and completeness of the mathematical formulation of elastic structural problems. This is another unique feature of this book. As is well known, the variational principle is one of the mathematical formulations of elasticity.

The variational principle can be used to deal with almost any elastic problem, although it is not the only possible method. As is also well known, although the principle of minimum potential energy based on displacement is not the only form of the variational principle of elasticity, it is sufficient for our needs and has the greatest generality. It is especially suitable for the problems with high complexity. We should point out that the finite element method, based on the mathematical form of the variational principle, especially the form of the variational principle based on displacement, has had great success in practice. Proceeding according to this guiding idea, we hope to achieve our goal rather economically.

The authors express their heartfelt thanks to Lin Qun, Wang Jin–xian, Yan Chang–zhou, Fu Zi–zhi and others for their enthusiastic support and help during the process of writing, lecturing on and revising the book.

Feng Kang

Shi Zhong-Ci

April 1, 1994

Contents

Chapter 1

Simple Modes of Elastic Deformation

§1   Simple Stretching and Compression of Springs

1.1   Deformation mode

Consider a spring of length L. One end of the spring, O, is fixed. The spring is stretched or compressed by an external force which is applied at the other end of the spring, A.

Suppose the elongation of the spring, i.e., the displacement of point A, is u under the action of the external force f (Fig. 1). We define that: $u > 0$ means stretching of the spring, and $u < 0$ compression. It has been shown by experiment that, within certain range, the applied external force f is proportional to the elongation u of the spring:

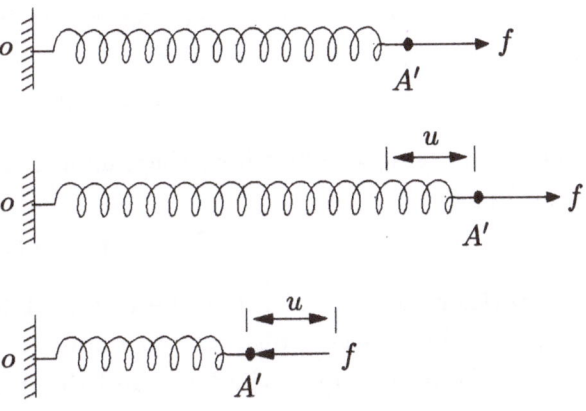

Fig. 1

$$f = cu, \tag{1.1}$$

where $c > 0$ is the spring constant depending on the material property and the geometric shape of the spring.

A so-called elastic reaction force is induced in the spring interior when one of its ends is loaded. According to Newton's third law, the elastic reaction force induced by the elongation u of the spring is

$$R = -cu, \tag{1.2}$$

where R is the internal force of the spring and the direction of the internal force is opposite to that of the external force oppsitely. This is Hooke's law.

Hence, when the elongation of the spring is u under the action of the external force f, the resultant force is

$$F = R + f = -cu + f.$$

In particular, when the resultant force $F = 0$, i.e.,

$$-cu + f = 0, \tag{1.3}$$

the spring is in equilibrium. This is the equilibrium equation of the mass point A. Hence we obtain the displacement in the state of the equilibrium configuration,

$$u = f/c,$$

and then, using Hooke's law, the elastic reaction force at the point A is found to be

$$R = -cu = -f.$$

Furthermore, we consider the resultant force F of the spring. When a mass point moves a distance du under the action of the external force F and the directions of the force and the displacement are the same, the work done is

$$dw = F du.$$

The potential energy dJ of the mass point, customarily expressed as the negative of the work, is

$$dJ = -dw = -F du,$$

or

$$\frac{dJ}{du} = -F. \tag{1.4}$$

Integrating over du, we obtain

$$J = -\int F du + J_0, \tag{1.5}$$

where J_0 is a constant of integration denoting the reference potential energy, which may as well be taken as zero. Therefore, if the potential energy

J is known, then the force F can be obtained by differentiating the potential energy, then changing its sign. Conversely, if the force F is known, then the potential energy J can be obtained by integrating the force, then changing its sign.

Now the resultant force of the spring is

$$F = F(u) = -cu + f,$$

and its direction coincides with that of the stretching and compression, so the potential energy is

$$J(u) = -\int F du = \frac{1}{2}cu^2 - fu, \tag{1.6}$$

This is the potential energy possessed by the spring when the elongation of the spring is u (whether the spring is in the equilibrium configuration or not) under the action of the external force f.

The resultant force F of the spring is composed of two parts. One is the internal force F_{in}, the other is the external force F_{ex}, and

$$F_{in} = R = -cu, \qquad F_{ex} = f,$$

$$F = F_{in} + F_{ex}.$$

Correspondingly, the potential energy J can also be divided into two parts

$$J = J_{in} + J_{ex},$$

$$J_{in} = \frac{1}{2}cu^2, \qquad J_{ex} = -fu,$$

and

$$F_{in} = -\frac{dJ_{in}}{du} = -cu, \qquad F_{ex} = -\frac{dJ_{ex}}{du} = f.$$

J_{in} is the elastic energy in the interior of the spring, which is a quadratic function of the displacement u and positive definite, i.e.,

$$J_{in} \geq 0,$$

$$J_{in} = 0, \quad \text{if and only if} \quad u = 0.$$

It is noted that there is a factor of $1/2$ in the expression of J_{in}. J_{ex} is called the potential energy of the external work, which is a linear function of u and J_{ex} has a negative sign.

1.2   Variational principles and equilibrium equation

In the section of 1.1, the equilibrium equation is derived from the intuitive equilibrium principle of forces. It is important that the equilibrium equation in elasticity can also be derived from a completely different way, i.e., from the method based on the principle of minimum potential energy.

In fact, according to (1.4), the relation between the potential energy and the force is

$$F = -\frac{dJ}{du}.$$

Hence the equilibrium equation $F = 0$ is equivalent to $\frac{dJ}{du} = 0$. Since the second derivative of J is:

$$\frac{d^2 J}{du^2} = c > 0,$$

the displacement u of the equilibrium configuration makes the potential energy J a minimum. Conversely, the state that makes the potential energy a minimum must be the equilibrium configuration. That is the principle of minimum potential energy.

Thus, the equilibrium problem in mechanics can be reduced to an extremum problem in mathematics, i.e., a variational problem

$$J(u) = \frac{1}{2}cu^2 - fu = \text{Min.} \tag{1.7}$$

The principle of minimum potential energy may still have an alternative equivalent form, the so-called principle of virtual work. Suppose the displacement u of the equilibrium configuration is incremented by a virtual displacement v, and becomes $u + v$. Then the potential energy of the spring becomes $J(u+v)$ from $J(u)$. Since $J(u)$ is a quadratic function of u,

$$J(u + v) = J(u) + J'(u)v + \frac{1}{2}J''(u)v^2$$

$$= J(u) + J'(u)v + \frac{1}{2}cv^2.$$

Obviously, the necessary and sufficient condition for making J a minimum is

$$J'(u)v = 0,$$

i.e., for any virtual displacement v we have

$$cuv - fv = 0, \tag{1.8}$$

The mechanical meaning is that the displacement of the equilibrium configuration makes the total virtual work vanish. Hence it is also called the principle of virtual work.

To sum up, the solutions of the following three problems are equivalent:

1. The principle of minimum potential energy:

$$J(u) = \frac{1}{2}cu^2 - fu = \text{Min};$$

2. The principle of virtual work:

$$cuv - fv = 0, \text{ for any virtual displacement } v;$$

3. The equation of equilibrium:

$$cu - f = 0.$$

For the simplest equilibrium problem of the stretching and compression of a spring with one degree of freedom, the equivalence of the three kinds of mathematical formulations above mentioned is almost tautological, without any actual difference in contents. However, a similar equivalence generally holds also for all other problems of elastic equilibrium which will discuss in the book. It will be shown that, different mathematical formulations will lead to different methods of solution, and their practical effects are very different. In other words, the equivalence in mathematics does not mean equal effects in practice.

§2   Stretching and Compression of Uniform Rods

2.1   Deformation modes

Suppose a slender rod Ω with length L and cross sectional area A, which is uniform along the x-axis, is fixed at one end and is stretched by uniformly distributed longitudinal loads exerted on the other end. Let the resultant force of the load be f, and the elongation of the rod be δL (Fig. 2). As in the case of the stretching and compression of springs, a form of Hooke's law holds, i.e., the force exerted per unit area of the rod f/A is proportional to the relative elongation $\delta L/L$:

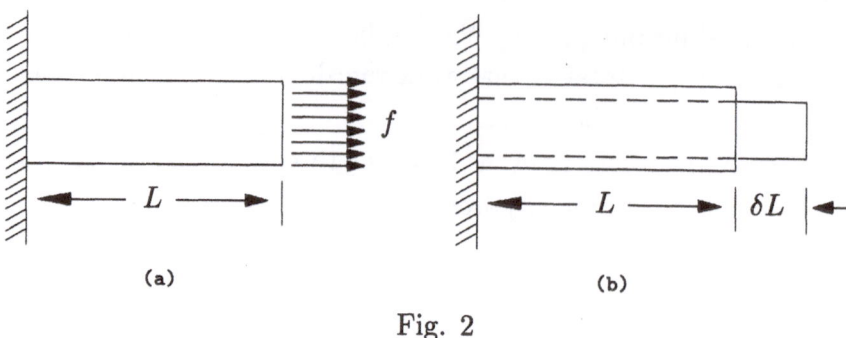

Fig. 2

$$\frac{f}{A} = E\frac{\delta L}{L},\tag{2.1}$$

where E is a constant called elastic modulus or Young's modulus, which depends only on the material but independent of the geometry of the rod. Since $\frac{\delta L}{L}$ is dimensionless, the dimension of the elastic modulus E is $\frac{f}{A}$, i.e., force/area.

Under the action of the load, the rod deforms elastically. The interior of the rod is in strained state, and yields internal forces. Taking an arbitrary cross section S, orthogonal to the longitudinal x-axis of the rod, the rod Ω is cut into two parts Ω^+ and Ω^- (Fig. 3a). There will be forces acting on each other across S. The force per unit area across the cross section S is called stress. Let the stress exerted on the negative side Ω^- by the positive side Ω^+ across S be σ. Assuming that the loads on the end section are uniformly distributed and the rod is slender, the stress σ will also be uniformly distributed over S. Therefore, the resultant force of the stress exerted on the negative side by the positive side over the whole cross section, i.e., the internal force, is

$$Q = \sigma A.\tag{2.2}$$

The internal force exerted on the positive side by the negative side over the cross section S is $-Q$, therefore $Q > 0$ means that the force is tensile and $Q < 0$ means that the force is compressive.

Arbitrarily take two cross sections S and S', with coordinates be x and x', and with corresponding internal forces be $Q(x)$ and $Q(x')$, respectively. Now consider the equilibrium of the block between these two cross sections (Fig. 3c). Since there is no load between S and S',

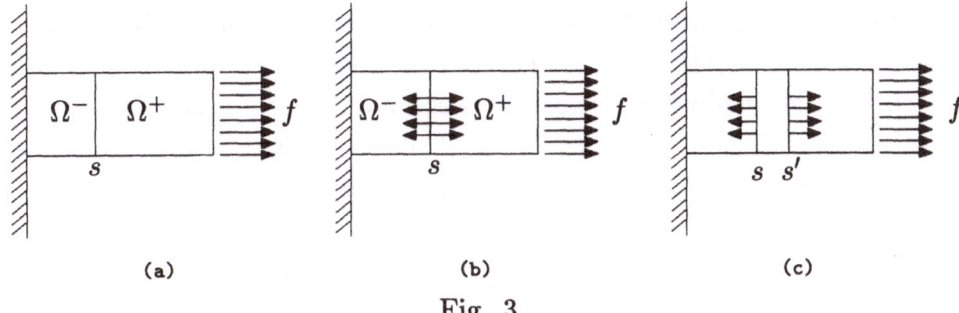

Fig. 3

$$-Q(x) + Q(x') = 0,$$

so that

$$\sigma(x) = \sigma(x'),$$

i.e., $\sigma(x)$ is independent of the coordinate x of the cross section:

$$\sigma(x) \equiv \sigma = \text{const}.$$

Then considering the equilibrium of the segment Ω^+ (Fig. 3b), we obtain

$$-Q + f = 0,$$

i.e.,

$$-A\sigma + f = 0, \tag{2.3}$$

this is the equilibrium equation expressed in terms of stresses, from which the stress can be found immediately as

$$\sigma = \frac{f}{A}.$$

The elastic deformation of the rod is measured in terms of the relative elongation, called the strain, and denoted by ε. Then Hooke's law can be expressed in terms of stress and strain as

$$\sigma = E\varepsilon, \qquad \varepsilon = \frac{\delta L}{L}, \tag{2.4}$$

where $\varepsilon > 0$ corresponds to the tensile deformation and $\varepsilon < 0$ corresponds to the compressive deformation. Under the condition of the above mentioned uniformly loaded slender rod, the strain ε, like the stress σ, is also uniform.

If the stress σ inside the rod is nonuniform, then the strain ε is also nonuniform. It is the limit of the relative elongation. Suppose the rod deforms under loading, so that the mass point located at coordinate x is displaced by $u(x)$, and the displacement of the point $x' = x + \Delta x$ near x is $u(x') = u(x + \Delta x)$, then the limit of the relative elongation is

$$\lim_{\Delta x \to 0} \frac{u(x + \Delta x) - u(x)}{\Delta x} = u'(x).$$

Thus, at each point x, the strain is given by the derivative $u'(x)$ of the displacement $u(x)$, i.e.,

$$\varepsilon = u'. \tag{2.5}$$

When the strain is uniform, the displacement u is a linear function of x

$$u = u(x) = u_1 + \frac{u_2 - u_1}{L}x, \tag{2.6}$$

where

$$u_1 = u(0) = 0, \qquad u_2 = u(L) = \delta L.$$

From (2.5), (2.6) and Hooke's law (2.4), we can rewrite the equilibrium equation (2.3), representing stresses as

$$-\frac{EA}{L}u_2 + f = 0, \tag{2.7}$$

The coefficient $\dfrac{EA}{L}$ is called the tensile rigidity of the rod. Hence the displacement can be found immediately as

$$u_2 = \frac{Lf}{EA}.$$

One can take the displacement u in the equilibrium equation (2.7) as the unknown, and solve for it first, then solve for the strain and the stress by (2.5) and Hooke's law. The method of solving the equilibrium equation by solving first for the displacement is called the displacement method. Conversely, one can take the stress σ as unknown in the equilibrium equation (2.3), and solve for it first, then solve for the strain and the displacement by Hooke's law and (2.5). The method of solving the equilibrium equation in this way is called the force method. These two methods of solving the equilibrium equation are the ones commonly used in elasticity. The book will mainly discuss the displacement method.

Just the elastic modulus E is not enough to describe all elastic properties of the material. Experiments show that there are transverse compressive (tensile) deformations associated with the longitudinal tensile (compressive) deformation caused by the longitudinal load (Fig. 2b). In fact, u, ε, and σ defined above only refer to the longitudinal x-direction. Analogous displacements v and w, the strains $\varepsilon_y = \dfrac{\partial v}{\partial y}, \varepsilon_z = \dfrac{\partial w}{\partial z}$, and as well as the stresses σ_y and σ_x can also be defined in the other two transverse directions, i.e., in the y − and z − directions. For the above mentioned longitudinal stretching under longitudinal load, Hooke's law is written as

$$\sigma_x = E\varepsilon_x,$$

while there exist the transverse compressions proportional to the longitudinal stretching

$$\varepsilon_y = \varepsilon_z = -v\nu_x.$$

Here, the dimensionless coefficient ν is a constant dependent only on the material, called Poisson's ratio. The negative sign in the expression means that, under the action of the longitudinal load, the longitudinal and the transverse strains of opposite sign. One can prove that, the elastic property of an isotropic (i.e., it is invariant under arbitrary rotation) solid material can be completely described by the elastic modulus E and Poisson's ratio ν. However, this is not true for the antistrophic materials. For example, 3 material constants are needed for cubic crystals and 21 material constants for general triclinic crystals. We will prove later that Poisson's ratio ν satisfies $0 < \nu < \dfrac{1}{2}$.

Table 1

Material	Elastic modulus (Dyne/cm^2)	Poisson's ratio
Steel	20.0×10^{11}	0.28
Copper	11.0×10^{11}	0.34
Concrete	2.7×10^{11}	0.10
Rubber	0.50×10^{11}	0.48

The elastic moduli and Poisson's ratios of some engineering materials are listed in Table 1. It can be seen that, E is a very large value. That means, the elastic deformation is very small under the condition of an ordinary load. In fact, Hooke's law (i.e., the so-called elastic law), which describes the linear relationship between the stress and strain, holds only for small deformations. When the stress has increased to a certain extent, the material starts to yield, great deformation may be caused by small increase of stress, and the material becomes plastic. Finally, it will be damaged as the stress successively increases. For some engineering materials, the capacities for resistance to stretching and compression are different, e.g., concrete can endure greater compression but less tension. Furthermore, the stress-strain curve of some engineering material goes along a path in the $\varepsilon - \sigma$ plane during loading, and another different path during unloading. Such materials remain more or less residual deformation and can only partially return to their original state after the load has been removed. All of such cases deviate from the ideal linear elastic law and will not be discussed in the book, though they are often encountered in practice.

2.2   Variational principles and equilibrium equations

Now, starting with energy, we use the variational principles to further analyze the tensile and compressive deformations of uniform rods.

The work is done by the internal force during the deformation of the rod and the strain energy is stored.

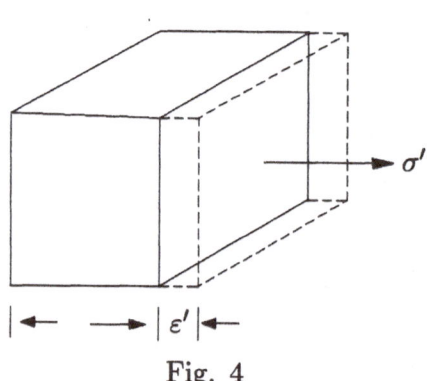

Fig. 4

Arbitrarily choose a unit cross sectional area and of unit length inside the rod as a volume element. Suppose the stress σ' inside the body, i.e., the force per unit area, changes from 0 to σ. The corresponding strain ε', i.e., the elongation per unit length, changes from 0 to ε (Fig. 4). Letting

$$\sigma' = t\sigma, \qquad \varepsilon' = t\varepsilon,$$

t changes from 0 to 1. Therefore, the strain energy stored in the unit volume (called the volume density of strain energy) is

$$W = \int_0^\varepsilon \sigma' d\varepsilon' = \int_0^1 t\sigma\varepsilon dt = \sigma\varepsilon \int_0^1 t dt = \frac{1}{2}\sigma\varepsilon. \qquad (2.8)$$

Note the factor $\frac{1}{2}$ in the expression. Because no work has been done by the longitudinal load in causing the simultaneous transverse deformations, it contributes nothing to the strain energy.

From (2.8), the strain energy per unit length of the rod (called the line density of strain energy) is

$$\overline{W} = WA = \frac{1}{2}A\sigma\varepsilon = \frac{1}{2}EA\varepsilon^2,$$

so the total strain energy of the rod (supposing that the left end is fixed, i.e., $u_1 = 0$) is

$$P = \overline{W}L = \frac{1}{2}EAL\varepsilon^2 = \frac{1}{2}\frac{EA}{L}(u_2 - u_1)^2 = \frac{1}{2}cu_2^2,$$

where
$$C = \frac{EA}{L}. \tag{2.9}$$

On the other hand, work fu_2 is done by the right end load f through the local displacement u_2, so the potential energy of the external work is $-fu_2$. Then the total potential energy of the rod is

$$J(u_2) = \frac{1}{2}cu_2^2 - fu_2. \tag{2.10}$$

This is a problem with are degree of freedom and the variable u_2 is its unknown . Its form is exactly the same as (1.6) for the potential energy of springs during stretching and compression in §1, hence there are still three equivalent mathematical formulations (write $u_2 = u$ and $v_2 = v$ for brevity):

1. The principle of minimum potential energy:

$$J(u) = \frac{1}{2}cu^2 - fu = \text{Min},$$

2. The principle of virtual work:

$$cuv - fv = 0, \quad \text{for any virtual displacement } v,$$

3. The equilibrium equation: $cu = f$.

In the following, we consider the case of a rod of which both ends are loaded by loads f_1 and f_2, respectively. Both end displacements u_1 and u_2 are unrestrained, and the strain energy is

$$P(\boldsymbol{u}) = P(u_1, u_2) = \frac{1}{2}c(u_2 - u_1)^2 = \frac{1}{2}(C\boldsymbol{u}, \boldsymbol{u}), \tag{2.1}$$

where

$$C = \begin{bmatrix} c & -c \\ -c & c \end{bmatrix} = c \begin{bmatrix} 1 & -1 \\ -1 & 1 \end{bmatrix}, u = \begin{bmatrix} u_1 \\ u_2 \end{bmatrix}.$$

where (u, v) denotes the inner product of the vectors u and v. The matrix C of the quadratic form (2.11) for the strain energy is called the stiffness matrix. It is symmetric, and is positive semi-definite because the strain energy $P(u)$ is non negative.

The potential energy of the external work is

$$-F(u) = (f_1 u_1 + f_2 u_2) = -(f, v), f = \begin{bmatrix} f_1 \\ f_2 \end{bmatrix}, \tag{2.12}$$

and the total potential energy of the rod is

$$J(u) = J(u_1, u_2) = \frac{1}{2}(Cu, u) - F(u). \tag{2.13}$$

$J(u)$ is a quadratic function of the variable elements u_1 and u_2. $J(u)$ has two first partial derivatives and four second partial derivatives:

$$J' = \begin{bmatrix} \dfrac{\partial J}{\partial u_1} \\ \dfrac{\partial J}{\partial u_2} \end{bmatrix} = \begin{bmatrix} cu_1 - cu_2 - f_1 \\ -cu_1 + cu_2 - f_2 \end{bmatrix},$$

$$J'' = \begin{bmatrix} \dfrac{\partial^2 J}{\partial u_1^2} & \dfrac{\partial^2 J}{\partial u_2 \partial u_2} \\ \dfrac{\partial^2 J}{\partial u_2 \partial u_1} & \dfrac{\partial^2 J}{\partial u_2^2} \end{bmatrix} = \begin{bmatrix} c & -c \\ -c & c \end{bmatrix} = C,$$

$J(u)$ has the Taylor expansion

$$J(u_1 + v_1, u_2 + v_2) = J(u_1, u_2) + \Big(\frac{\partial J}{\partial u_1} v_1 + \frac{\partial J}{\partial u_2} v_2 \Big)$$
$$+ \int_{i,j=1}^{2} \frac{\partial^2 J}{\partial u_i \partial u_j} v_i v_j.$$

Since the matrix of second derivatives $J'' = C$ is positive semi-definite, it can be seen from the above expansion that a necessary and sufficient condition for making the total potential energy J a minimum is $J' = 0$,

i.e.,

$$\begin{cases} \dfrac{\partial J}{\partial u_1} = cu_1 - cu_2 - f_1 = 0, \\[2mm] \dfrac{\partial J}{\partial u_2} = -cu_1 + cu_2 - f_2 = 0. \end{cases}$$

Written in matrix form, this becomes

$$\boldsymbol{Cu} = \boldsymbol{f}. \tag{2.14}$$

The vanishing of first derivatives of J is also equivalent to the principle of virtual work.

$$dJ = \frac{\partial J}{\partial u_1} v_1 + \frac{\partial J}{\partial u_2} v_2 = (cu_1 - cu_2 - f_1)v_1 + (-cu_1$$

$$+cu_2 - f_2)v_2 = 0, \qquad \text{for any } v_1, v_2.$$

Written in inner product form,

$$(\boldsymbol{Cu}, \boldsymbol{v}) - (\boldsymbol{f}, \boldsymbol{v}) = 0, \quad \text{for any } \boldsymbol{v} = \begin{bmatrix} v_1 \\ v_2 \end{bmatrix}. \tag{2.15}$$

Equation (2.14) is just the equilibrium equation of the rod unrestrained at both ends. If we arbitrarily take a cross section S cutting the rod into two parts Ω^- and Ω^+ (Fig. 3a), we can establish the equilibrium for each part:

$$\Omega^- : f_1 + \sigma A = 0,$$

$$\Omega^+ : -\sigma A + f_2 = 0,$$

respectively. By substituting

$$\sigma = E\varepsilon = E\frac{u_2 - u_1}{L}$$

into the above expressions, we get (2.14).

The coefficient matrix C of (2.14) is the coefficient matrix of the quadratic form of strain energy.

It is singular because its determinant

$$\det C = c^2 - c^2 = 0.$$

Hence, equation (2.14) may have no solutions. The corresponding system of homogeneous equations

$$
\begin{cases}
-c(v_2 - v_i) = 0 \\
c(v_2 - v_1) = 0
\end{cases}
$$

has a general solution

$$
v = a \begin{bmatrix} 1 \\ 1 \end{bmatrix} = a v^{(1)}, \quad v^{(1)} = \begin{bmatrix} 1 \\ 1 \end{bmatrix},
$$

where a is an arbitrary constant.

Obviously, a necessary and sufficient condition for the existence of solutions of the system of nonhomogeneous equation (2.14) is that the load vector f and the vector v are orthogonal:

$$
(f, v) = a(f, v^{(1)}) = a(f_1 + f_2) = 0,
$$

i.e.,

$$
f_1 + f_2 = 0. \tag{2.16}
$$

This is the self-equilibrating condition of the two external end loads, which is the prerequisite for equilibrium of a unrestrained rod, called the compatibility condition. Under this condition, there exists a solution of the equilibrium equation. A particular solution is

$$
u^* = \begin{bmatrix} 0 \\ f_2/c \end{bmatrix},
$$

while the general solution is

$$
u = v + u^* = \begin{bmatrix} a \\ a + f_2/c \end{bmatrix}.
$$

Hence the solution is unique only up to a rigid translation by $av^{(1)}$, which corresponds to a rigid translation. However, the stress solution is unique:

$$
\sigma = f_2/A = -f_1/A.
$$

We see that, the stress σ is uniform over the rod, hence we still have a problem of single degree of freedom, viewed in terms of the stress. However, the problem is one of two degrees of freedom u_1 and u_2, viewed in terms of the displacement.

Now let the left end load be changed into an elastic support. Suppose the elastic reaction force exerted on the end by the support is proportional to the deviation of the displacements

$$-c_1(u_1 - \bar{u}_1) = -c_1 u_1 + f_1, \quad f_1 = c_1 \bar{u}_1,$$

where $c_1 > 0$ is the spring constant of the support, $\bar{u}_1$ is a reference displacement, and f_1 is known. Hence the supporting force can be considered as the sum of the elastic reaction force $-c_1 u_1$ and the load $f_1 = c_1 \bar{u}_1$ which is independent of the displacement u_1. So the strain energy $\frac{1}{2} c_1 u_1^2$ of the support should be added to the total strain energy of the system in addition to the strain energy $\frac{1}{2} c(u_2 - u_1)^2$ of the rod, i.e.,

$$P(\boldsymbol{u}) = P(u_1, u_2) = \frac{1}{2} c(u_2 - u_1)^2 + \frac{1}{2} c_1 u_1^2$$

$$= \frac{1}{2}(C\boldsymbol{u}, \boldsymbol{u}),$$

where the stiffness matrix

$$C = \begin{bmatrix} c + c_1 & -c \\ -c & c \end{bmatrix} \tag{2.17}$$

is obviously symmetric and positive definite.

The potential energy of external work is still

$$-F(\boldsymbol{u}) = -(\boldsymbol{f}, \boldsymbol{u}).$$

The total potential energy of the system is

$$J(\boldsymbol{u}) = J(u_1, u_2) = \frac{1}{2}(C\boldsymbol{u}, \boldsymbol{u}) - (\boldsymbol{f}, \boldsymbol{u}).$$

As in the two cases discussed above, there are three mathematical formulations:

1. $J(\boldsymbol{u}) = \frac{1}{2}(C\boldsymbol{u}, \boldsymbol{u}) - (\boldsymbol{f}, \boldsymbol{u}) = \text{Min},$
2. $(C\boldsymbol{u}, \boldsymbol{v}) - (\boldsymbol{f}, \boldsymbol{v}) = 0, \quad \text{for any } \boldsymbol{v},$ (2.18)
3. $C\boldsymbol{u} = \boldsymbol{f}.$

In order to verify that (2.18), derived from the variational principle, is just the equilibrium equations, one can simply separate the rod into two parts Ω^- and Ω^+, and write out the equilibrium equations respectively,

$$\Omega^- : (-c_1 u_1 + f_1) + \sigma A = 0,$$

$$\Omega^+ : -\sigma A + f_2 = 0,$$

and then substitute $E(u_2 - u_1)/L$ for the stress σ.

The coefficient matrix C of the equilibrium equations (2.18) is just the stiffness matrix (2.17), which is positive definite. Hence there exists a unique solution

$$u_1 = \frac{f_1 + f_2}{c_1}, \quad u_2 = \frac{f_1 + f_2}{c_1} + \frac{f_2}{c}. \tag{2.19}$$

The corresponding stress solution is

$$\sigma = f_2/A,$$

so the stress in the rod is still uniform.

Two points can be seen from the above discussion: 1. The stiffness matrix is always symmetric and positive definite or positive semi-definite, unclear–I think it should be: 2. The stiffness matrix, i.e., the quadratic form of strain energy, and acts both as the coefficient matrix of the quadratic form of strain energy, and as the coefficient matrix of the equilibrium equations is positive definite whenever a geometrical constraint or elastic support is imposed on the end point.

The foregoing analysis of the stretching and compression of uniform rods can be immediately generalized to the piecewise uniform rod and to the plane system of rods in tension. These two cases will be discussed in the following two sections, respectively.

2.3   Piecewise uniform rods

Suppose a rod $x_1 \leq x \leq x_N$ is divided into $N - 1$ segments: $[x_1, x_2], [x_2, x_3], \cdots, [x_{N-1}, x_N]$. Each segment $[x_{i-1}, x_i]$ possesses a uniform material constant E_i, a uniform sectional area A_i, and a length $E_i = x_i - x_{i-1}$. There are loads $f_1, \cdots, f_N$ exerted on the interface sections $x = x_1, \cdots, x_N$, respectively, but no further loads are exerted on the interior of each segment. This guarantees that both the stress and the strain are uniform:

$$\sigma = \sigma_i, \qquad \varepsilon = \varepsilon_i,$$

while the displacement u is linear within each segment $[x_{i-1}, x_i]$. Hence, as a whole, both $\sigma(x)$ and $\varepsilon(x)$ are piecewise constant, and $u(x)$ is piecewise linear, i.e., a broken-line distribution determined by the nodal displacements $u_1, \cdots, u_N$ (Fig. 5).

The strain energy of this composite rod equals the sum of the strain energies of the individual segments. Since the strain energy of the segment $[x_{i-1}, x_i]$, based on the (2.11), is

$$\frac{1}{2}c_i(u_i - u_{i-1})^2, \quad c_i = \frac{E_i A_i}{L_i}, \qquad i = 2, \cdots, N,$$

the total strain energy is

$$P(\boldsymbol{u}) = P(u_1, \cdots, u_N) = \frac{1}{2}\int_{i=2}^{N} c_i(u_i - u_{i-1})^2$$

$$= \frac{1}{2}(C\boldsymbol{u}, \boldsymbol{u}). \qquad (2.20)$$

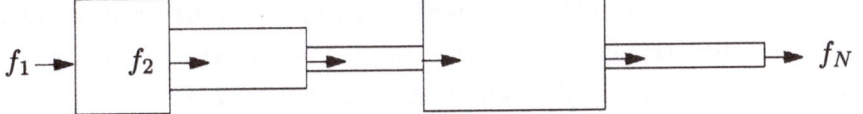

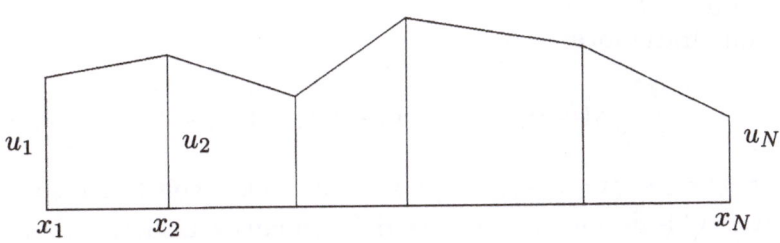

Fig. 5

The displacement vector is $\boldsymbol{u} = (u_1, \cdots, u_N)^T$, where the superscript T always denotes the transpose of a vector or a matrix, and the matrix

$$C = [c_{ij}]_{i,j=1,\cdots,N} = \begin{bmatrix} c_2 & -c_2 & & 0 \\ -c_2 & c_2 + c_3 & -c_3 & \\ & \ddots & \ddots & \ddots \\ 0 & & -c_N & c_N \end{bmatrix} \qquad (2.21)$$

is tridiagonal symmetric. C is called the stiffness matrix of the composite rod. Because $P(\boldsymbol{u}) \geq 0$ for any u, C is positive semi-definite.

The potential energy of the external work of the system is $-F(\boldsymbol{u})$,

$$F(\boldsymbol{u}) = F(u_1, \cdots, u_N) = \int_{i=1}^{N} f_i u_i = (\boldsymbol{f}, \boldsymbol{u}), \qquad (2.22)$$

so the total potential energy is

$$J(\boldsymbol{u}) = P(\boldsymbol{u}) - F(\boldsymbol{u}) = \frac{1}{2}(C\boldsymbol{u}, \boldsymbol{u}) - f(\boldsymbol{f}, \boldsymbol{u}). \qquad (2.23)$$

By the principle of minimum potential energy, the displacement of the equilibrium configuration minimizes the total potential energy a minimum:

$$J(\boldsymbol{u}) = \text{Min}.$$

On the other hand, since the stiffness matrix C is symmetric positive semi-definite, it follows from the extremum theorem of multivariable functions in calculus that the following three problems are equivalent, i.e., either they all have the same solution, or none of them has a solution.

1. $J(\boldsymbol{u}) = \dfrac{1}{2}(C\boldsymbol{u}, \boldsymbol{u}) - (\boldsymbol{f}, \boldsymbol{u}) = \text{Min}$, $\qquad\qquad$ (2.24)

2. $(C\boldsymbol{v}, \boldsymbol{v}) - (\boldsymbol{f}, \boldsymbol{v}) = 0$,   for any $\boldsymbol{v} = (v_1, \cdots, v_N)^T$, $\qquad$ (2.25)

3. $\dfrac{\partial J}{\partial u_i} = \displaystyle\int_{j=1}^{N} c_{ij} u_j - f_i = 0$, $i = 1, \cdots, N$. $\qquad$ (2.26)

The quadratic form

$$(C\boldsymbol{v}, \boldsymbol{v}) = \frac{1}{2}\int_{i=2}^{N} c_i (v_i - v_{i-1})^2 = 0 \Longleftrightarrow v_i - v_{i-1} = 0, \quad i = 2, \cdots, N,$$

i.e., $v_1 = v_2 = \cdots = v_N = a$ (italic is an arbitrary constant), so the stiffness matrix C is degenerate and the determinant vanishes. The nontrivial general solution of the homogeneous equations

$$\int_{j=1}^{N} c_{ij} v_j = 0, \quad i = 1, \cdots, N$$

is

$$v_1 = v_2 = \cdots = v_N = a$$

which corresponds to the rigid translation.

By the fundamental theorem of linear algebra, a necessary and sufficient condition for the existence of solutions to the degenerate equations (2.26) is that the right hand vector $\boldsymbol{f}$ and the general solution $\boldsymbol{v} = (a, \cdots, a)^T$ of the system of homogeneous equations be orthogonal:

$$(\boldsymbol{f}, \boldsymbol{v}) = a(f_1 + \cdots + f_N) = 0,$$

i.e.,

$$f_1 + \cdots + f_N = 0. \tag{2.27}$$

Such a condition means the self-equilibration of the external loads. Because of the equivalence of the above three problems, it is also the necessary and sufficient condition for the equilibrium of unrestrained, piecewise, and uniform rods. we have already mentioned this in the discussion of the uniform rod.

If the condition (2.27) is satisfied, then there exists a solution of the system of equations (2.26), but the solution is only unique up to a constant (rigid translation):

$$u_i = u_i^* + a, \quad i = 1, \cdots, N,$$

where u^* is an arbitrary particular solution of the system of equations. Although the displacement solution is not unique, the solutions for the stress σ_i or the strain ε_i are still unique because the rigid translation contributes nothing to the stress or strain. When a constraint condition

$$\int_{i=1}^{N} u_i = 0$$

is imposed on the solution u, the displacement solution becomes unique as well.

One can easily see that, the system of extremum equations (2.26) in problem 3 is just the equilibrium equation. For let $x_{i-\frac{1}{2}}$ be the midpoint of the segment $[x_{i-1}, x_i]$. Writing out the equilibrium equations of forces on each segment $[x_{i-\frac{1}{2}}, x_{i+\frac{1}{2}}]$ and $[x_1, x_{\frac{3}{2}}], [x_{N-\frac{1}{2}}, x_N]$, where $i = 2, \cdots, N-1$, we obtain

$$\begin{cases} \sigma_2 A_2 + f_1 = 0, \\ -\sigma_2 A_2 + \sigma_3 A_3 + f_2 = 0, \\ \cdots\cdots\cdots \\ -\sigma_N A_N + f_N = 0. \end{cases} \tag{2.28}$$

Using the stress-strain relation and the strain-displacement relation, the system of equations (2.28), where stresses are used as the unknowns, can be transformed into the system of equations (2.26), where displacements are used as unknowns.

Hence, as in the case of uniform rods, the stiffness matrix C is symmetric positive semi-definite, and is the coefficient matrix not only of the quadratic form of the strain energy but also of the equilibrium equations.

If the left end of the rod is fixed, $u_1 = 0$, then the strain energy is

$$P(\boldsymbol{u}) = P(u_2, \cdots, u_N) = \frac{1}{2}c_2 u_2^2 + \frac{1}{2}\int_{i=3}^{N} c_i(u_i - u_{i-1})^2$$

$$= \frac{1}{2}(C\boldsymbol{u}, \boldsymbol{u}),$$

$$\boldsymbol{u} = (u_2, \cdots, u_N)^T,$$

the stiffness matrix becomes symmetric positive definite,

$$C = \begin{bmatrix} c_2 + c_3 & -c_3 & & 0 \\ -c_3 & c_3 + c_4 & -c_4 & \\ & \ddots & \ddots & \ddots \\ 0 & & -c_N & c_N \end{bmatrix}. \tag{2.29}$$

This is because

$$(C\boldsymbol{v}, \boldsymbol{v}) = 0 \iff v_2 = 0, \ v_i - v_{i-1} = 0, \ i = 3, \cdots, N,$$

i.e.,

$$v_2 = v_3 = \cdots = v_N = 0.$$

The potential energy of the external work is

$$-F(\boldsymbol{u}) = -(f_2 u_2 + \cdots + f_N u_N) = -(\boldsymbol{f}, \boldsymbol{u}),$$

$$\boldsymbol{f} = (f_2, \cdots, f_N)^T.$$

The total potential energy is

$$J(\boldsymbol{u}) = P(\boldsymbol{u}) - F(\boldsymbol{u}) = \frac{1}{2}(C\boldsymbol{u}, \boldsymbol{u}) - (\boldsymbol{f}, \boldsymbol{u}).$$

Since the matrix C is symmetric positive definite, the three equivalent problems which are analogous to (2.24)–(2.26) have the same unique solution.

If the left end is changed into an elastic support, and the supporting force is $-c_1 u_1 + f_1$, then the elastic energy $\frac{1}{2} c_1 u_1^2$ should be added to the strain energy of the system. We obtain

$$P(\boldsymbol{u}) = P(u_1, \cdots, u_N) = \frac{1}{2} c_1 u_1^2 + \frac{1}{2} \sum_{i=2}^{N} c_i (u_i - u_{i-1})^2$$

$$= \frac{1}{2} (C\boldsymbol{u}, \boldsymbol{u}),$$

$$\boldsymbol{u} = (u_1, \cdots, u_N)^T,$$

where

$$C = \begin{bmatrix} c_1 + c_2 & -c_2 & & 0 \\ -c_2 & c_2 + c_3 & -c_3 & \\ & \ddots & \ddots & \ddots \\ 0 & & -c_N & c_N \end{bmatrix}. \tag{2.30}$$

Since $(C\boldsymbol{v}, \boldsymbol{v}) = 0 \iff v_1 = 0, v_i - v_{i-1} = 0, i = 2, \cdots, N$, i.e.,

$$v_1 = v_2 = \cdots = v_N = 0,$$

the matrix C is symmetric positive definite. The potential energy of the external work is

$$-F(\boldsymbol{u}) = -(f_1 u_1 + \cdots + f_N u_N) = -(\boldsymbol{f}, \boldsymbol{u}),$$

$$\boldsymbol{f} = (f_1, \cdots, f_N)^T.$$

The total potential energy is

$$J(\boldsymbol{u}) = P(\boldsymbol{u}) - F(\boldsymbol{u}) = \frac{1}{2}(C\boldsymbol{u}, \boldsymbol{u}) - f(\boldsymbol{f}, \boldsymbol{u}).$$

Hence, as in the case where the left end is fixed, there exists unique solutions of all three equivalent problems. The only differences are that coefficient matrices C are slightly different and the range of subscripts should be taken $i, j = 1, \cdots, N$.

We see from the above discussion that the stiffness matrix is positive semi-definite when both ends are unrestrained, and that the stiffness matrix becomes positive definite if one of the ends is fixed or elastically supported, i.e., the stiffness matrix can be made positive definite by adding geometric constraints or elastic support.

2.4 Systems of rods in plane tension

In the foregoing, we have discussed the tensile and compressive deformations of piecewise uniform rods. A peculiarity of composite rods of this
kind is that the axial directions of the individual rods are coincident. With
modifications, the analysis above can also be used to discuss so-called systems of rods in plane tension, which are composed of rods in tension lying
in the same plane, but without a common axial direction.

Suppose each rod of the system of rods is uniform, the loads are all
exerted at the ends of the rods (called nodal points for brevity), and
all connections among the rods are pin-connected. Axial stretching and
compression, but no flexural deformation normal to the axial direction, is
considered for each rod.

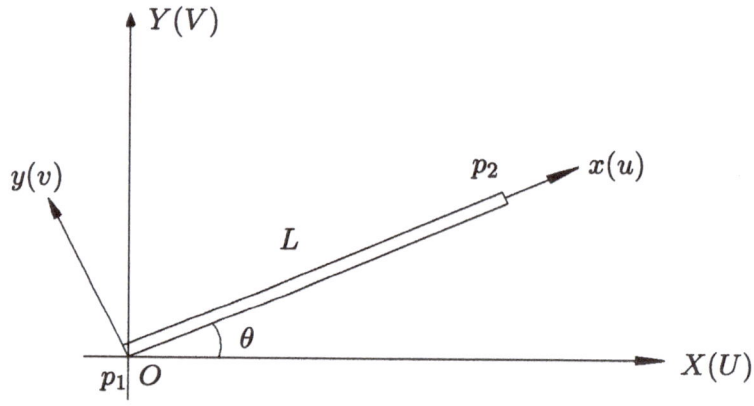

Fig. 6

Since the axial directions of the rods are not completely coincident, for
convenience of treatment, we introduce a system of coordinates (xy) for
each rod, called the local coordinates, with the convention that the axial
direction of the rod is taken as the $x-$ direction. The displacement vector
in the local coordinates is denoted by (u, v). The $y-$directional component of the displacement vector $v = 0$ because no transverse bending is
considered. Besides the local coordinates of the rods we also need global
coordinates (X, Y) in the plane as reference coordinates for the whole system of rods. The displacement vector in the global coordinates is denoted
by (U, V).

Suppose there is a rod L with two end points denoted by P_1 and

P_2. The local and the global coordinates are shown in Fig. 6. The transformation relations between these two sets of coordinates are

$$\begin{cases} x = X\cos\theta + Y\sin\theta, \\ y = -X\sin\theta + Y\cos\theta. \end{cases}$$

Correspondingly, the transformation relations among the local and global components of the displacement vector are

$$\begin{cases} u = U\cos\theta + V\sin\theta, \\ v = -U\sin\theta + V\cos\theta. \end{cases} \tag{2.31}$$

We first consider this case where the ends are not fixed. Hence suppose that the stiffness matrix in the local coordinates of the rod, which is unrestrained or elastically supported at both ends, is C. Its concrete form can be seen from the relevant expressions in Section 2.2. Then the strain energy of the rod is

$$P(\boldsymbol{u}) = P(u_1, u_2) = \frac{1}{2}(C\boldsymbol{u}, \boldsymbol{u}),$$

the potential energy of the external work is

$$-F(\boldsymbol{u}) = -(f_1 u_1 + f_2 u_2) = -(\boldsymbol{f}, \boldsymbol{u}),$$

where f_1 and f_2 are the axial loads exerted on the end points P_1 and P_2 of the rod, respectively.

Now we discuss the transformation from the local to the global displacement vector. From the transformation relations (2.31),

$$\boldsymbol{u} = \begin{bmatrix} u_1 \\ u_2 \end{bmatrix} = \begin{bmatrix} \cos\theta & \sin\theta & 0 & 0 \\ 0 & 0 & \cos\theta & \sin\theta \end{bmatrix} \begin{bmatrix} U_1 \\ V_1 \\ U_2 \\ V_2 \end{bmatrix} = R\boldsymbol{U}, \tag{2.32}$$

the transformation matrix is

$$R = \begin{bmatrix} \cos\theta & \sin\theta & 0 & 0 \\ 0 & 0 & \cos\theta & \sin\theta \end{bmatrix},$$

$$U = \begin{bmatrix} U_1 \\ V_1 \\ U_2 \\ V_2 \end{bmatrix}. \tag{2.33}$$

Hence in the global coordinates the strain energy of the rod is

$$P(\boldsymbol{U}) = \frac{1}{2}(CR\boldsymbol{U}, R\boldsymbol{U}) = \frac{1}{2}(R^T CR\boldsymbol{U}, \boldsymbol{U})$$

$$= \frac{1}{2}(K\boldsymbol{U}, \boldsymbol{U}). \tag{2.34}$$

The potential energy of the external work is

$$-F(\boldsymbol{U}) = -(\boldsymbol{f}, R\boldsymbol{U}) = -(R^T \boldsymbol{f}, \boldsymbol{U}) = -(\boldsymbol{F}, \boldsymbol{U}). \tag{2.35}$$

The matrix $K = R^T CR$ is just the stiffness matrix of the rod in global coordinates, and is obviously still symmetric positive semi-definite; $\boldsymbol{F} = R^T \boldsymbol{f}$ is the corresponding load column matrix.

Assuming in global coordinates the strain energy of the i-th rod is

$$P(\boldsymbol{U}^{(i)}) = \frac{1}{2}(K^{(i)}\boldsymbol{U}^{(i)}, \boldsymbol{U}^{(i)}),$$

the potential energy of the external work is

$$-F(\boldsymbol{U}^{(i)}) = -(\boldsymbol{F}^{(i)}, U^{(i)}).$$

Summing up the strain energies and the potential energies of external work of all the rods, respectively, the total strain energy and the total potential energy of external work of the system of rods can be obtained:

$$\begin{cases} P(\boldsymbol{U}) = \displaystyle\int_i P(\boldsymbol{U}^{(i)}) = \frac{1}{2}\int_i (K^{(i)}\boldsymbol{U}^{(i)}, \boldsymbol{U}^{(i)}) = \frac{1}{2}(K\boldsymbol{U}, \boldsymbol{U}), \\ -F(\boldsymbol{U}) = -\displaystyle\int_i F(\boldsymbol{U}^{(i)}) = -\int_i (\boldsymbol{F}^{(i)}, U^{(i)}) = -(\boldsymbol{F}, \boldsymbol{U}). \end{cases} \tag{2.36}$$

The total potential energy is

$$J(U) = P(U) - F(U) = \frac{1}{2}(KU, U) - (F, U), \qquad (2.37)$$

where the matrix K is the total stiffness matrix of the system of rods and F is the total load column matrix. Note that K is still symmetric positive semi-definite.

According to the equivalence mentioned several times above, the following three problems:

1. $J(U) = \frac{1}{2}(KU, U) - (F, U) = \text{Min},$ \hfill (2.38)
2. $(KU, V) - (F, V) = 0, \quad$ for any $V,$ \hfill (2.39)
3. $KU = F.$ \hfill (2.40)

Eigher all have the same solution, or none of them has a solution. Necessary and sufficient conditions for the existence of solutions are

$$(F, V^i) = 0, \qquad i = 1, 2, \cdots, l \qquad (2.41)$$

where $V^i, i = 1, \cdots, l$, constitute the linearly independent solution of the homogeneous equations $KV = 0$.

Since

$$KV = 0 \Longleftrightarrow (KV, V) = 0^{1)} \qquad (2.42)$$

i.e., the strain energy vanishes, so the nontrivial solutions V^i of the homogeneous equations $KV = 0$ are also called the strainless state.

The coefficient matrix of the equilibrium equations (2.40) is also called the coefficient matrix of the quadratic form of the total strain energy, i.e., the global stiffness matrix. The stiffness matrices of the individual rods

---

[1] The proof of the equivalence is as follows: If $KV = 0$, then obviously $(KV, V) = 0$. Conversely, if $(KV, V) = 0$, then the quadratic form for arbitrarily given vector U and real number t,

$$(K(V + tU), V + tU) = (KV, V) + 2t(KV, U) + t^2(KU, U) \geq 0,$$

because the stiffness matrix K is symmetric positive semi-definite. Hence the discriminate of two quadratic equation in t is nonpositive, so

$$(KV, U)^2 \leq (KV, V)(KU, U) = 0.$$

Consequently,
$$(KV, U) = 0 \quad \text{for any } U.$$

Hence $KV = 0$. Q. E. D.

that are components of the system of the rods are generally called the element stiffness matrices. The global stiffness matrix is assembled by superimposing the element stiffness matrices according to the connections of the end points of the rods. Similarly, the load column matrix F is also assembled by superimposing the load column matrices of each rod element according to the same rule.

In the foregoing, we have not considered the case of displacement constraints at the fixed end. Three methods often used to deal with such geometrical constraints will be introduced in the following. Suppose the displacement component $V_i^{(i)}$ at the first nodal point of the i-th rod is fixed: $V_1^{(i)} = \overline{V}_1^{(i)}$ (a given value), and let the displacement component $V_1^{(i)}$ be the k-th component of vector U.

The first method: Delete the k-th row including the right-hand term in the equilibrium equations $KU = F$, set the unknown $V_1^{(i)}$ in the k-th column on the left-hand side equal to $\overline{V}_1^{(i)}$ and transpose this column to the right-hand side. When $\bar{V}_1^{(i)} = 0$, this is just a matter of deleting the k-th column. The rank of the system of equations is thereby reduced by one. This operation must be repeated several times if there are several fixed displacement components, and the rank of the coefficient matrix is reduced by one each time.

This approach has the advantage of saving computer memory. However, programming it is somewhat complicated because it needs to rearrange the coefficient matrix of equations.

The second method: Set all the off-diagonal entries in the k-th row and in the k-th column of the global stiffness matrix K to zero, the corresponding diagonal entry equal to 1, and set the k-th component of the vector F on the right side equal to $\overline{V}_1^{(i)}$. The rank of the system of equations remains unchanged. The advantage of this method is that, it is comparatively simple to program. But on the other hard, some computer memory as well as computational effort is wasted.

The third method: Set the diagonal entry at the crossing of the k-th row and column of K equal to a large number L and the k-th component of the vector F on the right-hand side equal to $L\overline{V}_1^{(i)}$, while all other entries are kept unchanged. Thus, all other terms except the diagonal term of the k-th equation are relatively small quantities and play an insignificant role. Hence $LV_1^{(i)} \approx L\overline{V}_1^{(i)}$, i.e., $V_1^{(i)} \approx \overline{V}_1^{(i)}$. The advantage of this method is that the programming becomes even simpler. But memory as well as computational effort is wasted as in the second method, and some error will be induced as well.

In addition, there is a method to deal directly with the displacement constraint at the fixed end, i.e., to distinguish whether the rod has a fixed end or not before constructing each element stiffness matrix. If the rod has a fixed end, the stiffness matrix for an element with fixed end should be used, e.g., if the left end is fixed, then the strain energy of the rod is

$$P(\boldsymbol{u}) = P(u_2) = \frac{1}{2}cu_2^2,$$

where c is the tensile rigidity, the element stiffness matrix is $C = c$, the displacement vector is $\boldsymbol{u} = u_2$, and the load vector is $\boldsymbol{f} = f_2$. The element stiffness matrix in global coordinates is

$$K = c \begin{bmatrix} \cos^2\theta & \sin\theta\cos\theta \\ \sin\theta\cos\theta & \sin^2\theta \end{bmatrix},$$

the load vector is

$$\boldsymbol{F} = \begin{bmatrix} f_2\cos\theta \\ f_2\sin\theta \end{bmatrix},$$

and the displacement vector is

$$\boldsymbol{U} = \begin{bmatrix} U_2 \\ V_2 \end{bmatrix}, \quad \boldsymbol{u} = \begin{bmatrix} \cos\theta \\ \sin\theta \end{bmatrix}^T \boldsymbol{U}.$$

The advantages of the method are the compactness and the saving of computer memory. The shortcoming is that, the kind of stiffness matrix to be used must be determined for each rod in advance. Consequently the process is not as uniform as before.

By treating the displacement constraint at the fixed ends may make the global stiffness matrix positive definite. Then the equilibrium equations have a unique displacement solution. However, it still may be positive semi-definite, then the displacement solution is not unique and may differ by an arbitrary strainless state.

The main steps for analyzing the system of rods in plane tension can be summarized as follows:

1. Select the global coordinates and the local coordinates for each rod.

2. Construct the element stiffness matrices and load column matrices of the individual rods, and transform them into global coordinates.

3. According to the connections of the rods, assemble all the element stiffness matrices and load column matrices in the global coordinates to form the global stiffness matrix and load column matrices.

4. Treat the displacement constraints at fixed ends. This step may be omitted if the displacement constraint has already been considered during the construction of the element stiffness matrix.

5. Solve the equilibrium equations – the linear system of algebraic equations-and determine all the nodal displacements in the global coordinates. If one desires to evaluate the displacement components in the local coordinates, then use transformation (2.32).

We illustrate the details of how to assemble to element stiffness matrices and load column matrices to form the global stiffness matrix and load column matrices with the following examples.

Example 1. An equilaterally triangular system of rods (Fig. 7).

Suppose an equilateral triangle ABC is composed of three uniform rods numbered 1, 2, 3, respectively. The local x-directional coordinate of each rod is shown by an arrow. The global coordinates and the local coordinates of rod 1 are coincident. Let the external force applied of each nodal point be decomposed into the axial forces of the two rods which

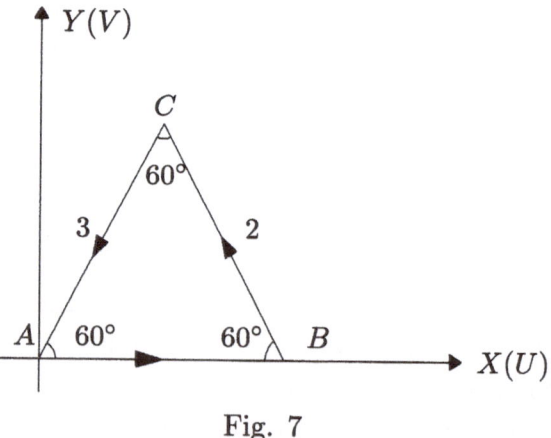

Fig. 7

are intersecting at this nodal point. Denote the axial forces applied to the ends of rods 1, 2, 3 by $f_A^{(1)}, f_B^{(1)}; f_B^{(2)}, f_C^{(2)}; f_C^{(3)}, f_A^{(3)}$; respectively, where the superscripts denote the rod, while the subscripts denote the number of the nodal point at which the force is applied. The tensile rigidities are c_1, c_2 and c_3, respectively.

At first, we will not consider displacement constraints at fixed ends or the elastic supports not considered first. According to the steps summarized above, the analysis is as follows:

Steps 1, 2. Construct the element stiffness matrices and load column matrices of the individual rods.

Denote the intersection angle between the local x-coordinate and the global X-coordinate by θ. Then we have

Rod 1: $\theta_1 = 0, \cos\theta_1 = 1, \sin\theta_1 = 0$.

The transformation matrix is

$$R_1 = \begin{bmatrix} \cos\theta_1 & \sin\theta_1 & 0 & 0 \\ 0 & 0 & \cos\theta_1 & \sin\theta_1 \end{bmatrix} = \begin{bmatrix} 1 & 0 & 0 & 0 \\ 0 & 0 & 1 & 0 \end{bmatrix};$$

the local element stiffness matrix is

$$C_1 = c_1 \begin{bmatrix} 1 & -1 \\ -1 & 1 \end{bmatrix};$$

the load column matrix is

$$\boldsymbol{f}^{(1)} = \begin{bmatrix} f_A^{(1)} \\ f_B^{(1)} \end{bmatrix};$$

the displacement vector is

$$\boldsymbol{u}^{(1)} = \begin{bmatrix} u_A^{(1)} \\ u_B^{(1)} \end{bmatrix};$$

the global element stiffness matrix is

$$K^{(1)} = R_1^T C_1 R_1 = c_1 \left[\begin{array}{cc|cc} 1 & 0 & -1 & 0 \\ 0 & 0 & 0 & 0 \\ \hline -1 & 0 & 1 & 0 \\ 0 & 0 & 0 & 0 \end{array} \right];$$

the global load column matrix is

$$\boldsymbol{F}^{(1)} = R_1^T \boldsymbol{f}^{(1)} = (f_A^{(1)}, 0, f_B^{(1)}, 0)^T;$$

and the displacement vector is

$$\boldsymbol{U}^{(1)} = (U_A, V_A, U_B, V_B)^T, \quad \boldsymbol{u}^{(1)} = R_1 \boldsymbol{U}^{(1)}.$$

Rod 2: $\theta_2 = 120°$, $\cos\theta_2 = -\dfrac{1}{2}$, $\sin\theta_2 = \dfrac{\sqrt{3}}{2}$.

$$
R_2 = \begin{bmatrix} -\dfrac{1}{2} & \dfrac{\sqrt{3}}{2} & 0 & 0 \\[2mm] 0 & 0 & -\dfrac{1}{2} & \dfrac{\sqrt{3}}{2} \end{bmatrix}, \quad C_2 = c_2 \begin{bmatrix} 1 & -1 \\ -1 & 1 \end{bmatrix},
$$

$$
\boldsymbol{f}^{(2)} = \begin{bmatrix} f_B^{(2)} \\[1mm] f_C^{(2)} \end{bmatrix}, \quad \boldsymbol{u}^{(2)} = \begin{bmatrix} u_B^{(2)} \\[1mm] u_C^{(2)} \end{bmatrix},
$$

$$
K^{(2)} = R_2^T C_2 R_2 = c_2 \left[\begin{array}{cc|cc} \dfrac{1}{4} & -\dfrac{\sqrt{3}}{4} & -\dfrac{1}{4} & \dfrac{\sqrt{3}}{4} \\[2mm] -\dfrac{\sqrt{3}}{4} & \dfrac{3}{4} & \dfrac{\sqrt{3}}{4} & -\dfrac{3}{4} \\[1mm] \hline -\dfrac{1}{4} & \dfrac{\sqrt{3}}{4} & \dfrac{1}{4} & -\dfrac{\sqrt{3}}{4} \\[2mm] \dfrac{\sqrt{3}}{4} & -\dfrac{3}{4} & -\dfrac{\sqrt{3}}{4} & \dfrac{3}{4} \end{array} \right],
$$

$$
\boldsymbol{F}^{(2)} = R_2^T \boldsymbol{f}^{(2)} = \left(-\frac{1}{2}f_B^{(2)}, \frac{\sqrt{3}}{2}f_B^{(2)}, -\frac{1}{2}f_C^{(2)}, \frac{\sqrt{3}}{2}f_C^{(2)} \right)^T,
$$

$$
\boldsymbol{U}^{(2)} = (U_B, V_B, U_C, V_C)^T, \quad \boldsymbol{u}^{(2)} = R_2 U^{(2)}.
$$

Rod 3: $\theta_3 = 240°$, $\cos\theta_3 = -\dfrac{1}{2}$, $\sin\theta_3 = -\dfrac{\sqrt{3}}{2}$.

$$
R_3 = \begin{bmatrix} -\dfrac{1}{2} & -\dfrac{\sqrt{3}}{2} & 0 & 0 \\[2mm] 0 & 0 & -\dfrac{1}{2} & -\dfrac{\sqrt{3}}{2} \end{bmatrix},
$$

$$
C_3 = c_3 \begin{bmatrix} 1 & -1 \\ -1 & 1 \end{bmatrix}, \quad \boldsymbol{f}^{(3)} = \begin{bmatrix} f_C^{(3)} \\[1mm] f_A^{(3)} \end{bmatrix}, \quad \boldsymbol{u}^{(3)} = \begin{bmatrix} u_C^{(3)} \\[1mm] u_A^{(3)} \end{bmatrix},
$$

$$
K^{(3)} = R_3^T C_3 R_3 = c_3 \left[\begin{array}{cc|cc} \dfrac{1}{4} & \dfrac{\sqrt{3}}{4} & -\dfrac{1}{4} & -\dfrac{\sqrt{3}}{4} \\[2mm] \dfrac{\sqrt{3}}{4} & \dfrac{3}{4} & -\dfrac{\sqrt{3}}{4} & -\dfrac{3}{4} \\[1mm] \hline -\dfrac{1}{4} & -\dfrac{\sqrt{3}}{4} & \dfrac{1}{4} & \dfrac{\sqrt{3}}{4} \\[2mm] -\dfrac{\sqrt{3}}{4} & -\dfrac{3}{4} & \dfrac{\sqrt{3}}{4} & \dfrac{3}{4} \end{array} \right],
$$

$$F^{(3)} = R_3^T f^{(3)} = \left(-\frac{1}{2} f_C^{(3)}, -\frac{\sqrt{3}}{2} f_C^{(3)}, -\frac{1}{2} f_A^{(3)}, -\frac{\sqrt{3}}{2} f_A^{(3)} \right)^T,$$

$$U^{(3)} = (U_C, V_C, U_A, V_A)^T, \quad u^{(3)} = R_3 U^{(3)}.$$

Step 3. Construct the global stiffness matrix and the load column matrices.

We shall explain through diagrams the process of assembling the individual element stiffness matrices through diagrams. Suppose the global stiffness matrix is K, the load vector is F, and the displacement vector is $U = (U_A, V_A, U_B, V_B, U_C, V_C)^T$. Arrange the quadratic form of the total strain energy (KU, U) and the negative potential energy of the external work (F, U) into a tabular form, and start by setting all the entries to zero.

	U_A	V_A	U_B	V_B	U_C	V_C	F
U_A	0	0	0	0	0	0	0
V_A	0	0	0	0	0	0	0
U_B	0	0	0	0	0	0	0
V_B	0	0	0	0	0	0	0
U_C	0	0	0	0	0	0	0
V_C	0	0	0	0	0	0	0

	U_A	V_A	U_B	V_B	$F^{(1)}$
U_A	c_1	0	$-c_1$	0	$f_A^{(1)}$
V_A	0	0	0	0	0
U_B	$-c_1$	0	c_1	0	$f_B^{(1)}$
V_B	0	0	0	0	0

Now arrange the quadratic form of the strain energy $(K^{(1)}U^{(1)}, U^{(1)})$ and the negative potential energy of the external work $(F^{(1)}, U^{(1)})$ of rod 1 into a similar tabular form, in which the part of the square block is actually the element stiffness matrix $K^{(1)}$ and the last column is the load column matrix $F^{(1)}$. Superimpose the individual entries in this table on the corresponding locations in the large table for (KU, U) and (F, U). For example, the entry at the crossing of row U_A and column U_B is $-c_1$. Superimpose this entry on the entry (being zero at present) at the crossing of row U_A and column U_B in the large table, and assemble the result into the large table. Similarly, the entry at the crossing of row U_B and column $F^{(1)}$ is $f_B^{(1)}$. Superimpose this entry on the entry (being zero at present) at the crossing of row U_B and column F in the large table, and assemble the result into the large table, etc.

Continue by applying the same treatment to rods 2 and 3. After finishing the superimposition of these three rods, we obtain a large table as follows:

	U_A	V_A	U_B	V_B	U_C	V_C	F
U_A	$c_1 + \frac{1}{4}c_1$	$\frac{\sqrt{3}}{4}c_3$	$-c_3$	0	$-\frac{1}{4}c_3$	$-\frac{\sqrt{3}}{4}c_3$	$f_A^{(1)} - \frac{1}{2}f_A^{(3)}$
V_A		$\frac{3}{4}c_3$	0	0	$-\frac{\sqrt{3}}{4}c_3$	$-\frac{3}{4}c_3$	$-\frac{\sqrt{3}}{2}f_A^{(3)}$
U_B			$c_1 + \frac{1}{4}c_2$	$-\frac{\sqrt{3}}{4}c_2$	$-\frac{1}{4}c_2$	$\frac{\sqrt{3}}{4}c_2$	$f_B^{(1)} - \frac{1}{2}f_B^{(2)}$
V_B				$\frac{3}{4}c_2$	$\frac{\sqrt{3}}{4}c_2$	$-\frac{3}{4}c_2$	$\frac{\sqrt{3}}{2}f_B^{(2)}$
U_C		symmetric			$\frac{1}{4}c_2 + \frac{1}{4}c_3$	$-\frac{\sqrt{3}}{4}c_2 + \frac{\sqrt{3}}{4}c_3$	$-\frac{1}{2}f_C^{(2)} - \frac{1}{2}f_C^{(3)}$
V_C						$\frac{3}{4}c_2 + \frac{3}{4}c_3$	$\frac{\sqrt{3}}{2}f_C^{(2)} - \frac{\sqrt{3}}{2}f_C^{(3)}$

The square block in the left-hand part of the table is just the global stiffness matrix K which is to be determined. The entries in the lower triangular part are omitted due to symmetry. The last column is the global load column matrix F.

Such a mechanical and monotonous operation is well suited for implementation on computers.

Step 4. Consider the strainless state.

Because

$$(KV, V) = 0 \Longleftrightarrow (K^{(1)}V^{(1)}, V^{(1)})$$

$$= (K^{(2)}V^{(2)}, V^{(2)}) = (K^{(3)}V^{(3)}, V^{(3)})$$

$$= 0 \Longleftrightarrow \begin{cases} U_A = U_B, \\ \dfrac{1}{4}U_B - \dfrac{\sqrt{3}}{4}V_B - \dfrac{1}{4}U_C + \dfrac{\sqrt{3}}{4}V_C = 0, \\ \dfrac{1}{4}U_C + \dfrac{\sqrt{3}}{4}V_C - \dfrac{1}{4}U_A - \dfrac{\sqrt{3}}{4}V_A = 0, \end{cases}$$

this system of equations has six unknowns $U_A, V_A, U_B, V_B, U_C, V_C$, but only three independent equations, so there must be three linearly independent solutions:

(I) $U_A = U_B = U_C = 1$, $V_A = V_B = V_C = 0$.

In terms of mechanics, this solution represents a translation in the X direction.

(II) $U_A = U_B = U_C = 0$, $V_A = V_B = V_C = 1$.

This is solution represents a translation in the Y direction.

(III) $U_A = 0$, $V_A = 0$; $U_B = 0$, $V_B = 1$; $U_C = -\dfrac{\sqrt{3}}{2}$, $V_C = \dfrac{1}{2}$.

The third solution represents an infinitesimal rotation about the point A in the plane of the system of rods. That point can be verified directly as follows: suppose a point (X, Y) in the plane coordinate system XOY makes an infinitesimal rotation about the origin O and the angle of rotation is ω. Point (X, Y) becomes point (X', Y') owing to the rotation. According to the transformation relations of coordinates, we have

$$\begin{cases} X' = X \cos\omega - Y \sin\omega, \\ Y' = X \sin\omega + Y \cos\omega. \end{cases}$$

Hence, point (X, Y) sufters a displacement

$$U(X, Y) = X' - X = X(\cos\omega - 1) - Y \sin\omega,$$

$$V(X, Y) = Y' - Y = X \sin\omega + Y(\cos\omega - 1).$$

When the angle of rotation ω is very small, $\cos\omega \approx 1$, and $\sin\omega \approx \omega$. Hence,

$$U(X, Y) = -Y\omega, \quad V(X, Y) = X\omega. \tag{2.43}$$

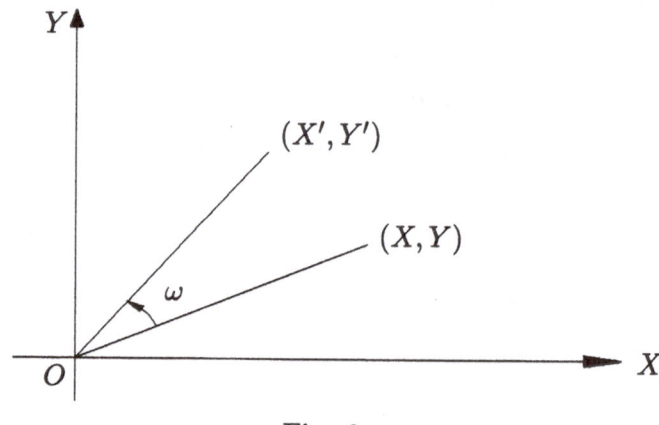

Fig. 8

Now the coordinates of point A are $X_A = Y_A = 0$, so $U_A = V_A = 0$. The coordinates of point B are $X_B = L$ (the length of the rod), $Y_B = 0$, so

$$U_B = 0, \quad V_B = L\omega.$$

The coordinates of point C are $X_C = \dfrac{1}{2}L$, $Y_C = \dfrac{\sqrt{3}}{2}L$, so

$$U_C = \frac{\sqrt{3}}{2}L\omega, \quad V_C = \frac{1}{2}L\omega.$$

Thus, when the triangle ABC makes an infinitesimal rotation ω about point A, the displacements yielded at the three vertices differ from the solution (III) only by a constant factor $L\omega$.

The above three linearly independent solutions are just three independent strainless states of the triangular system of rods: two translations and one rotation. Let them be abbreviated by V^1, V^2, V^3, respectively. Then the necessary and sufficient conditions for the existence of solutions of the equilibrium equations $KU = F$ are

$$(F, V^i) = 0, \qquad i = 1, 2, 3,$$

written concretely as follows:

$$(F, V^1) = f_A^{(1)} - \frac{1}{2}f_A^{(3)} + f_B^{(1)} - \frac{1}{2}f_B^{(2)} - \frac{1}{2}f_C^{(2)} - \frac{1}{2}f_C^{(3)} = 0.$$

This denotes the X-directional equilibrium of the external forces;

$$(F, V^2) = -\frac{\sqrt{3}}{2}f_A^{(3)} + \frac{\sqrt{3}}{2}f_B^{(2)} + \frac{\sqrt{3}}{2}f_C^{(2)} - \frac{\sqrt{3}}{2}f_C^{(3)} = 0.$$

After simplification, this becomes

$$-(f_A^{(3)} + f_C^{(3)}) + (f_B^{(2)} + f_C^{(2)}) = 0,$$

which denotes the Y-directional equilibrium of the external forces.

$$(\boldsymbol{F}, \boldsymbol{V}^3) = \frac{\sqrt{3}}{2} f_B^{(2)} - \frac{\sqrt{3}}{2}\left(-\frac{1}{2} f_C^{(2)} - \frac{1}{2} f_C^{(3)}\right)$$
$$+ \frac{1}{2}\left(\frac{\sqrt{3}}{2} f_C^{(2)} - \frac{\sqrt{3}}{2} f_C^{(3)}\right)$$
$$= 0.$$

After simplification, this becomes $f_B^{(2)} + f_C^{(2)} = 0$, which denotes the moment equilibrium of the external forces about point A.

These three self-equilibrating conditions of the loads can be written in another equivalent form. Substituting $f_B^{(2)} + f_C^{(2)} = 0$ obtained from $(\boldsymbol{F}, \boldsymbol{V}^3) = 0$ into expression $(\boldsymbol{F}, \boldsymbol{V}^2) = 0$, we obtain $f_A^{(3)} + f_C^{(3)} = 0$. Furthermore, substituting the two expressions of $f_B^{(2)} + f_C^{(2)} = 0$ and $f_A^{(3)} + f_C^{(3)} = 0$ into expression $(\boldsymbol{F}, \boldsymbol{V}^1) = 0$, we obtain $f_A^{(1)} + f_B^{(1)} = 0$. The latter three conditions mean that, the axial loads applying on each rod are self-equilibrating. These two sets of self-equilibrating conditions of the loads are equivalent.

When the self-equilibrating condition on the loads is satisfied, there exists a solution of the equilibrium equations of the unrestrained system of rods, but the displacement solution is not unique and may differ by an arbitrary strainless state, which in the example is just a plane rigid displacement. We can eliminate such arbitrariness by simply fixing three of the the six displacement components (or linear combination of them) of the displacement vector $\boldsymbol{U}$. In other words, by adding three geometrical constraint conditions, the global stiffness matrix can be made positive definite instead of positive semi-definite. In terms of mechanics, to restrain the plane rigid motion of the triangular system of rods, we have imposed three displacement constraints.

To treat fixed displacement, any one of the methods introduced above can be used.

Step 5. The last step is to solve the symmetric positive definite system of the algebraic equations. We will not discussed this here. Readers are referred to the relevant literature on computational linear algebraic.

Example 2. A rectangular system of rods (Fig. 9).

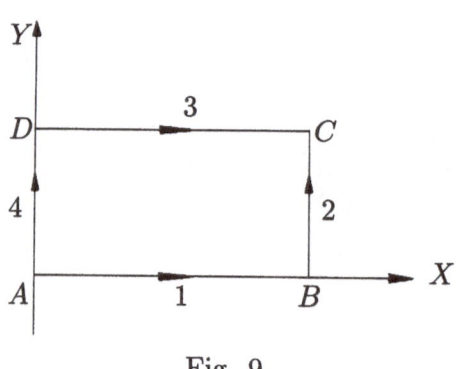

Fig. 9

For simplicity, assume that the tensile rigidities of the individual rods are all the same. In the present example, these four rods are orthogonal to each other, hence the local coordinates are no longer necessary. The element stiffness matrices and load vectors can be constructed directly in the global coordinates. In the following, the notation is the same as in Example 1.

Rod 1 : $K^{(1)} = c \begin{bmatrix} 1 & -1 \\ -1 & 1 \end{bmatrix}$, $F^{(1)} = \begin{bmatrix} f_A^{(1)} \\ f_B^{(1)} \end{bmatrix}$, $U^{(2)} = \begin{bmatrix} U_A \\ U_B \end{bmatrix}$.

Rod 2 : $K^{(2)} = c \begin{bmatrix} 1 & -1 \\ -1 & 1 \end{bmatrix}$, $F^{(2)} = \begin{bmatrix} f_B^{(2)} \\ f_C^{(2)} \end{bmatrix}$, $U^{(2)} = \begin{bmatrix} V_B \\ V_C \end{bmatrix}$.

Rod 3 : $K^{(3)} = c \begin{bmatrix} 1 & -1 \\ -1 & 1 \end{bmatrix}$, $F^{(3)} = \begin{bmatrix} f_D^{(3)} \\ f_C^{(3)} \end{bmatrix}$, $U^{(3)} = \begin{bmatrix} U_D \\ U_C \end{bmatrix}$.

Rod 4 : $K^{(4)} = c \begin{bmatrix} 1 & -1 \\ -1 & 1 \end{bmatrix}$, $F^{(4)} = \begin{bmatrix} f_A^{(4)} \\ f_D^{(4)} \end{bmatrix}$, $U^{(4)} = \begin{bmatrix} V_A \\ V_D \end{bmatrix}$.

The global stiffness matrix is

$$K = c \begin{bmatrix} 1 & 0 & -1 & 0 & 0 & 0 & 0 & 0 \\ & 1 & 0 & 0 & 0 & 0 & 0 & -1 \\ & & 1 & 0 & 0 & 0 & 0 & 0 \\ & & & 1 & 0 & -1 & 0 & 0 \\ & & & & 1 & 0 & -1 & 0 \\ & \text{sym.} & & & & 1 & 0 & 0 \\ & & & & & & 1 & 0 \\ & & & & & & & 1 \end{bmatrix},$$

$$F = (f_A^{(1)}, f_D^{(4)}, f_B^{(1)}, f_B^{(2)}, f_C^{(3)}, f_C^{(2)}, f_D^{(3)}, f_A^{(4)})^T,$$

$$U = (U_A, V_A, U_B, V_B, U_C, V_C, U_D, V_D)^T.$$

The strainless states are given by:

$$(KV, V) = 0 \iff (K^{(1)}V^{(1)}, V^{(1)}) = (K^{(2)}V^{(2)}, V^{(2)})$$
$$= (K^{(3)}V^{(3)}, V^{(3)}) = (K^{(4)}V^{(4)}, V^{(4)})$$
$$= 0$$
$$\iff U_A = U_B, V_B = V_C, U_D = U_C, V_D = V_A.$$

Hence there are four linearly independent strainless states:

(1) an X-directional translation,
$U_A = U_B = U_C = U_D = 1, \ V_A = V_B = V_C = V_D = 0.$
(2) a Y-directional translation,
$U_A = U_B = U_C = U_D = 0, \ V_A = V_B = V_C = V_D = 1.$
(3) an infinitesimal plane rotation about point A,
$U_A = 0, \ V_A = 0, \ U_B = 0, \ V_B = 1; \ U_C = -1,$
$V_C = 1; \ U_D = -1, \ V_D = 0.$
(4) $U_A = 0, \ V_A = 0; \ U_B = 0, \ V_B = 0; \ U_C = 1, \ V_C = 0;$
  $U_D = 1, \ V_D = 0.$

The points A and B are fixed, while the points C and D make an equidistant horizontal slide to the right.

Denote these four independent strainless states by V^1, V^2, V^3, V^4 respectively. Then the necessary and sufficient conditions for making the equilibrium equations of the unrestrained rectangular system of rods solvable are

$$(F, V^i) = 0, \qquad i = 1, 2, 3, 4.$$

or the X-directional equilibrium of the external forces,

$$(F, V^1) = f_A^{(1)} + f_B^{(1)} + f_C^{(3)} + f_D^{(3)} = 0;$$

the Y-directional equilibrium of the external forces,

$$(F, V^2) = f_D^{(4)} + f_A^{(4)} + f_B^{(2)} + f_C^{(2)} = 0;$$

the equilibrium of moments about point A;

$$(F, V^3) = f_B^{(2)} - f_C^{(3)} + f_C^{(2)} - f_D^{(3)} = 0;$$

and the equilibrium of the axial loads exerted on rod 3,

$$(F, V^4) = f_C^{(3)} + f_D^{(3)} = 0.$$

Substituting $f_C^{(3)} + f_D^{(3)} = 0$ into $(\boldsymbol{F}, \boldsymbol{V}^3) = 0$ and $(\boldsymbol{F}, \boldsymbol{V}^1) = 0$, respectively, we obtain

$$f_B^{(2)} + f_C^{(2)} = 0,$$
$$f_A^{(1)} + f_B^{(1)} = 0.$$

These two expressions denote the equilibrium of the axial loads on rods 1 and 2. Furthermore, substituting $f_B^{(2)} + f_C^{(2)} = 0$ into $(\boldsymbol{F}, \boldsymbol{V}^2) = 0$, we obtain $f_D^{(4)} + f_A^{(4)} = 0$, which denotes the equilibrium of the axial loads on the rod 4.

Hence, only if the axial loads on each rod are self-equilibrating does there exist a solution of the equilibrium equations of the rectangular system of rods. This conclusion is exactly the same as that of the triangular system of rods in Example 1. In Example 1, the displacement solution is not unique and may differ by an arbitrary strainless state. In Example 2, there are four independent strainless states, i.e., two translationstates, one rotation state, and one slide state in which one of the edges is fixed, and the opposite edge makes an equidistant parallel slide.

In order to make the equilibrium equations have a unique displacement solution, it is necessary to impose four constraint conditions so as to eliminate the four strainless states. However, only to fix, e.g., four displacement components of the two points A and B is not enough to eliminate the strainless states, because points C and D can still slide horizontally. Therefore, these four constraint conditions are actually not independent: only three of them are independent. However, if two displacement components of point A, the Y-directional displacement component of point B or C, and the X-directional displacement component of point C or D are fixed, then the global stiffness matrix becomes positive definite, and there exists a unique displacement solution of the equilibrium equation.

§3   Stretching and Compression of Nonuniform Rods

3.1   Deformation modes

The elastic deformation problem for uniform or piecewise uniform rods in tension has been discussed in §2. When the rod is subjected to a longitudinal body force in addition to the end loads, or the area $A = A(x)$ of the cross section S_x is varying, the stress distribution is longitudinally nonuniform and may be a function of the coordinate x:

$$\sigma = \sigma(x), \quad \varepsilon = \varepsilon(x) = u'(x), \tag{3.1}$$

although the stress distribution within the rod may be considered as uniform over each cross section (or is only concerned with its mean value over the cross section). The displacement $u(x)$ is no longer linear. Hence the whole system can no longer be described by a finite number of degrees of freedom. This is a problem with infinite degrees of freedom.

The internal force on a cross section S_x is

$$Q(x) = A(x)\sigma(x). \tag{3.2}$$

According to the Hooke's law $\sigma = E\varepsilon$, we obtain

$$Q = A\sigma = EA\varepsilon = EAu'. \tag{3.3}$$

This can be thought of as Hooke's law for the deformation mode of stretching and compression of rods. The internal force Q is proportional to the strain ε, and the coefficient of ratio EA is called the tensile rigidity of the cross section, which may vary with x.

From §2 we see that the volume density of the strain energy of the stretching and compression of rods is

$$W = \frac{1}{2}\sigma\varepsilon = \frac{1}{2}E\varepsilon^2.$$

Intergrating W over the cross section s, we obtain the line density of the strain energy

$$\overline{W} = \overline{W}(x) = \iint_s W ds = \frac{1}{2}EA\varepsilon^2,$$

and

$$Q = \frac{\partial \overline{W}}{\partial \varepsilon}, \quad \overline{W} = \frac{1}{2}Q\varepsilon. \tag{3.4}$$

Intergrating W over the length of the rod, we obtain the total strain energy of the rod

$$P(u) = \int_a^b \overline{W} dx = \frac{1}{2}\int_a^b Q(u)\varepsilon(u)dx$$

$$= \frac{1}{2}\int_a^b EA\varepsilon^2(u)dx = \frac{1}{2}\int_a^b EAu'^2 dx.$$

3.2 Variatoinal principles

For a given displacement function $u = u(x)$, no matter whether it is an equilibrium configuration or not, there always exists a corresponding strain energy

$$P(u) = \frac{1}{2} \int_a^b EAu'^2 dx. \tag{3.5}$$

Hence $P(u)$ is "a function of the function $u(x)$" called a functional.

Introduce the notation

$$D(u,v) = \int_a^b EAu'v' dx. \tag{3.6}$$

Obviously, D is symmetric

$$D(u,v) = D(v,u), \tag{3.7}$$

and is bilinear, i.e., linear to both "variables" u and v, respectively:

$$\begin{cases} D(tu,v) = tD(u,v), \text{ for any real number } t, \\ D(u+w,v) = D(u,v) + D(w,v). \end{cases} \tag{3.8}$$

$D(u,v)$ is called the bilinear symmetric functional of u and v. Hence the corresponding strain energy is

$$P(u) = \frac{1}{2}D(u,u),$$

and

$$D(u,u) \geq 0, \text{ for any displacement function } u.$$

In the case of piecewise uniform rods in tension, the strain energy

$$P = P(u_1, \cdots, u_N)$$

is a homogeneous quadratic form of the variables $\boldsymbol{u} = (u_1, \cdots, u_N)^T$, while in the case of nonuniform rods the strain energy $P = P(u)$ is a homogeneous quadratic functional of the variable $u = u(x)$ satisfying

$$\begin{cases} D(tu,tu) = t^2 D(u,u), \\ D(u+v,u+v) = D(u,u) + 2D(u,v) + D(v,v). \end{cases} \tag{3.9}$$

On the other hand, suppose the rod is subjected to body loads, with force per unit length $f = f(x)$. Then the potential energy of the external work is

$$-F(u) = -\int_a^b fu dx, \tag{3.10}$$

$F(u)$ is also a functional of the variable $u = u(x)$, but it is a first order homogeneous, i.e., linear functional satisfying

$$\begin{cases} F(tu) = tF(u), \\ F(u + v) = F(u) + F(v). \end{cases} \tag{3.11}$$

Hence the total potential energy

$$J(u) = \frac{1}{2}D(u, u) - F(u)$$

is a quadratic nonhomogeneous functional of u. The strain energy $\frac{1}{2}D(u, u)$ is the quadratic homogeneous part and the potential energy of the external work $-F(u)$ is the linear homogeneous part.

In §2, we described in matrix form the variational principles for equilibrium problems with finite degrees of freedom. In order to make this theory suitable for all kinds of equilibrium problems nomatter what the concrete form of the strain energy and the potential energy of external work, we formulate the variational principles in more abstract form. The general characteristics of the strain energy and the potential energy of the external work are

1. $D(u, u) \geq 0$, and $D(u, v)$ is a bilinear symmetric functional satisfying the conditions (3.8).
2. $F(u)$ is a linear functional satisfying the conditions (3.11).

In linear elasticity, these two requirements are satisfied generally.

In some equilibrium problems, geometrical constraints are often prescribed for the displacement u, e.g., the value of the displacement at the boundary point $x = a$ is prescribed to be $u(a) = \bar{u}_a$ (a given value), and so on. The set of all displacements u satisfying the prescribed constraint is denoted by K.

If the displacements $u, w \in K$, then the increment $v = u - w$ satisfies the annihilating constraint corresponding to the prescribed constraint. For example, if the prescribed constraint is $u(a) = \bar{u}_a$, then the corresponding annihilating constraint is

$$v(a) = u(a) - w(a) = 0.$$

This kind of displacement increment is called the virtual displacement in elasticity. In fact it is the increment within the admissible domain of the prescribed constraint. The set of all virtual displacements v is denoted by K_0. Hence, if $u \in K, v \in K_0$, then $u + v \in K$.

If no constraint has been imposed on the displacement in the problem, then K can be understood as the set of all displacements, and then $K_0 = K$.

For the sake of convenience, we introduce the following definitions are introduced.

Definitions:

1. The strain energy (neglecting the factor of $\frac{1}{2}$, similarly below) $D(v, v)$ is called non-negative under the prescribed constraint if $D(v, v) \geq 0$ for any $v \in K_0$.

2. The strain energy $D(v, v)$ is called positive definite under the prescribed constraint if $D(v, v) > 0$ for any $v \in K_0, v \not\equiv 0$.

3. The strain energy $D(v, v)$ is called degenerative under the prescribed constraint, if there exists $v \in K_0, v \not\equiv 0$ such that $D(v, v) = 0$, v is called the strainless virtual displacement or strainless state.

The principle of minimum potential energy states that, under the prescribed constraint, the displacement making the total potential energy a minimum is just the equilibrium configuration:

$$J(u) = \frac{1}{2}D(u, u) - F(u) = \text{Min}.$$

Here and below we always suppose $u \in K$.

Proposition 1.1   Suppose $D(v, v)$ is non-negative. Then the solutions of the following two variational problems are identical:

1. $J(u) = \frac{1}{2}D(u, u) - F(u) = \text{Min}$ (the principle of minimum potential energy),

2. $D(u, v) - F(v) = 0$, for any $v \in K_0$ (the principle of virtual work).

Proof.   Suppose u_0 is a solution of problem 1, i.e., u_0 is a minimal point of $J(u)$. Choose an arbitrary take $v \in K_0$. Then $u_0 + tv \in K$ for any real number t. Let

$$\varphi(t) = J(u_0 + tv).$$

Expanding $\varphi(t)$, we obtain

$$\varphi(t) = J(u_0) + t[D(u_0, v) - F(v)] + \frac{t^2}{2}D(v, v), \quad -\infty < t < \infty.$$

Because u_0 is the minimal point of the functional $J(u), t = 0$ is the minimum of the function $\varphi(t)$ of single variable. Hence

$$\varphi'(0) = D(u_0, v) - F(v) = 0, \text{ for any } v \in K_0,$$

i.e., u_0 is a solution of the problem 2 as well.

Conversely, suppose u_0 is a solution of the problem $2°$. Taking $w \in K$, let

$$v = w - u_0 \in K_0,$$

then

$$J(w) = J(u_0 + v) = J(u_0) + [D(u_0, v) - F(v)] + \frac{1}{2}D(v, v)$$

$$= J(u_0) + \frac{1}{2}D(v, v).$$

Since D is non-negative from the assumption, i.e.,

$$D(v, v) \geq 0, \text{ for any } v \in K_0,$$

so

$$J(w) \geq J(u_0).$$

Hence U_0 is a solution of the problem $1°$ as well. Q.E.D.

This is similar to the case of finitely many degrees of freedom, $D(u, v)$ is just the work done by the internal force of the displacement "applied over the virtual displacement v, hence it is called the functional of virtual work. $F(v)$ is the work done by the external force applied over the virtual displacement v, so that

$$D(u_0, v) - F(v) = 0, \text{ for any } v \in K.$$

This means that the virtual work done by the internal force of the equilibrium configuration u_0 is identically equal to the virtual work done by the external load. This is exactly the principle of virtual work. Hence the principle of virtual work is an equivalent form of the principle of minimum potential energy.

Proposition 1.1 proves that, when $D(v, v)$ is nonnegative, the solutions of these two variational problems are identical. We still have not answered the question of whether or not exists a solution for a given variational problem. The following two propositions will answer this question from both positive and negative sides.

Proposition 1.2   Suppose $D(v, v)$ is positive definite. Then there exists a solution of the variational problems in Proposition 1.1, and the solution is unique.

Proof. The proof of the existence of a solution is rather difficult. It needs considerable mathematical preliminaries, which are out of the scope of the book. Hence we omit it here. However, the proof of the uniqueness of the solution is quite simple. In fact, suppose u_1 and u_2 are two solutions of the variational problem $2°$, i.e.,

$$D(u_1, v) - F(v) = 0, \text{ for any } v \in K_0,$$

$$D(u_2, v) - F(v) = 0, \text{ for any } v \in K_0.$$

Substracting of one from another of the above two expressions, we obtain

$$D(u_1 - u_2, v) = 0, \text{ for any } v \in K_0.$$

$u_1 - u_2 \in K_0$ because $u_1, u_2 \in K$. Taking

$$v = u_1 - u_2,$$

in the above expression,

$$D(u_1 - u_2, u_1 - u_2) = 0.$$

Since D is positive definite,

$$u_1 - u_2 \equiv 0,$$

hence the solution is unique. Q.E.D.

Proposition 1.3   Suppose $D(v, v)$ is degenerative and that D has p linearly independent strainless virtual displacements: $v^{(i)}, i = 1, \cdots, p$. That is, there exist p linearly independent $v^{(i)} \in K_0$ such that $D(v^{(i)}, v^{(i)}) = 0, i = 1, \cdots, p$. Then the necessary and sufficient condition for the existence of solutions of the variational problems is

$$F(v^{(i)}) = 0, \qquad i = 1, \cdots, p. \tag{3.12}$$

The displacement solution is not unique and may differ by a strainless virtual displacement when the solution exists.

Proof. Let us first prove a relation, which we will use below. For an arbitrary $u \in K$ and strainless virtual displacement $v^{(i)}$, we always have

$$D(u, v^{(i)}) = 0. \qquad (3.13)$$

By the nonegative definiteness of D and the defintion of the strainless virtual displacement, for any real number t we have

$$D(u + tv^{(i)}, u + tv^{(i)}) = D(u, u) + 2tD(u, v^{(i)}) \geq 0.$$

Hence

$$D(u, v^{(i)}) = 0.$$

Necessity. Suppose u_0 is a solution of the variational problems, i.e.,

$$D(u_0, v) - F(v) = 0, \text{ for any } v \in K_0.$$

Taking $v = v^{(i)}$, from (3.13) $D(u_0, v^{(i)}) = 0$, we have

$$F(v^{(i)}) = 0, \qquad i = 1, \cdots, p.$$

Sufficiency. Suppose the condition (3.12) holds. Impose p additional constraints

$$(u, v^{(i)}) = 0, \qquad i = 1, \cdots, p \qquad (3.14)$$

to the variational problem 2 so as to make it a new variational problem:

$$\begin{cases} D(u, v) - F(v) = 0, \text{ for any } v \in K_0, \\ (v, v^{(i)}) = 0, \qquad i = 1, \cdots, p, \\ (u, v^{(i)}) = 0. \end{cases} \qquad (3.15)$$

Here the notation (u, v) denotes the integral $\int_Q uvdx$, and $\int_Q \cdots dx$ denotes the integral over the two or three dimensional domain Q of the two or three dimensional problem, respectively.

Now we prove that, by imposing p new constraints in addition to the original ones, $D(v, v)$ becomes positive definite. This is because

$$D(v, v) = 0 \Longleftrightarrow v = \sum_{j=1}^{p} c_j v^j,$$

and the constraints on the virtual displacement v are $(v, v^{(i)}) = 0, i = 1, \cdots, p$, in the new variational problem (3.15). Hence

$$(v, v^{(i)}) = \Big(\sum_{j=1}^{p} c_j v^{(j)}, v^{(i)} \Big) = \sum_{i=1}^{p} c_j (v^{(i)}, v^{(j)}) = 0,$$

$$i = 1, \cdots, p . \qquad (3.16)$$

One can prove (as we will give later on) that the determinant

$$|(v^{(i)}, v^{(j)})|_{i,j=1,\cdots,p} \neq 0. \qquad (3.17)$$

Hence the homogeneous equations (3.16) can have only the zero solution

$$c_1 = c_2 = \cdots = c_p = 0.$$

Thus $v \equiv 0$, hence D is positive definite.

From Proposition 2, we see that there exists a unique solution u_0 of the variational problem (3.15). It remains to prove that u_0 is also a solution of the original variational problem.

An arbitrary displacement $u \in K$ can be decomposed into

$$u = w + v^*, \ w \in K, \ v^* \in K_0,$$

where w satisfies the constraint conditions

$$(w, v^{(i)}) = 0, \qquad i = 1, \cdots, p, \qquad (3.18)$$

while v^* is a strainless virtual displacement.

In fact, let

$$w = u - \sum_{j=1}^{p} c_i v^{(j)},$$

where the coefficients c_j are to be determined. From conditions (3.18), we obtain

$$\sum_{j=1}^{p} c_j (v^{(i)}, v^{(j)}) = (u, v^{(i)}), \qquad i = 1, \cdots, p.$$

Since the coefficient determinant

$$|(v^{(i)}, v^{(j)})| \neq 0,$$

there exists a unique solution of the above equations $c_j = c_j^*$. Let

$$v^* = \sum_{j=1}^{p} c_j^* v^{(j)}, \quad w = u - v^*.$$

This is the required decomposition.

Decomposing u into $w + v^*$, we have

$$J(u) = J(w + v^*) = J(w) + [D(w, v^*) - F(v^*)] + \frac{1}{2}D(v^*, v^*).$$

Since v^* is a strainless virtual displacement, $D(v^*, v^*) = 0$. And since $w \in K, D(w, v^*) = 0$, by (3.13). Furthermore, it follows from (3.12) that,

$$F(v^*) = F\left(\sum_{j=1}^{p} c_j^* v^{(j)}\right) = \sum_{j=1}^{p} c_j^* F(v^{(j)}) = 0.$$

Thus we have

$$J(u) = J(w).$$

w is a displacement function of the variational problem (3.15) because it satisfies the constraint condition (3.18), and u_0 is just the solution of that problem. Hence, from the principle of minimum potential energy

$$J(w) \geq J(u_0).$$

Hence, for an arbitrary $u \in K$, we always have

$$J(u) = J(w) \geq J(u_0).$$

That is to say, u_0 is also a solution of the original variational problem.

One can also see that, if u_0 is a solution, then $u_0 + v^*$ is also a solution, where v^* is an arbitrary strainless virtual displacement. In fact

$$J(u_0 + v^*) = J(u_0) + [D(u_0, v^*) - F(v^*)] + \frac{1}{2}D(v^*, v^*)$$

$$= J(u_0).$$

Hence the displacement solution is not unique and may differ by an arbitrary strainless virtual displacement.

What remains to be proved is that, if $v^{(1)}, \cdots, v^{(p)}$ are linearly independent, then the determinant

$$|(v^{(i)}, v^{(j)})| \neq 0.$$

Arbitrarily take a set of real numbers $t_1, \cdots, t_p$ not all zero. Since $v^{(1)}, \cdots, v^{(p)}$ are linearly independent, the linear combination

$$f = \sum_{i=1}^{p} t_i v^{(i)} \neq 0.$$

Consequently,

$$(f, f) = \left(\sum_{i=1}^{p} t_i v^{(i)}, \sum_{j=1}^{p} t_j v^{(j)} \right) = \sum_{i,j=1}^{p} (v^{(i)}, v^{(j)}) t_i t_j > 0.$$

This is a positive definite quadratic form with respect to t_i and t_j. The refore the determinant of the coefficient matrix

$$|(v^{(i)}, v^{(j)})| > 0.$$

Q.E.D..

3.3   Boundary value problems

Now let us apply the above-mentioned variational principles to the concrete problem of elastic equilibrium of nonuniform rods in tension. We will discuss three kinds of problems according to their different respective boundary conditions.

1. First kind of boundary condition: The prescribed displacement. Suppose $u(a) = \bar{u}_a, u(b) = \bar{u}_b$, where $\bar{u}_a$ and $\bar{u}_b$ are given values. The strain energy:

$$\frac{1}{2} D(u, u) = \frac{1}{2} \int_a^b E A u'^2 dx,$$

the functional of virtual work:

$$D(u, v) = \int_a^b E A u' v' dx,$$

and the potential energy of the external work:

$$-F(u) = -\int_a^b f u \, dx,$$

$$K = (u(x) | u(a) = \bar{u}_a, u(b) = \bar{u}_b),$$

$$K_0 = (v(x) | v(a) = v(b) = 0).$$

According to the variational principle, the displacement of the equilibrium configuration satisfies

$$D(u, v) - F(v) = 0, \text{ for any } v \in K_0. \tag{3.19}$$

Integrating by parts and noting the boundary conditions, we have

$$D(u, v) = \int_a^b EAu'v'dx = [EAu'v]_a^b - \int_a^b (EAu')'vdx$$

$$= -\int_a^b (EAu')'vdx,$$

consequently,

$$D(u, v) - F(v) = \int_a^b [-(EAu')' - f]vdx = 0, \text{ for any } v \in K_0.$$

Since $v(x)$ is arbitrary in the open-interval (a, b), the value included in the brackets $[\cdots]$ under the above integral sign must vanish according to the fundamental lemma of the variational method. Hence, the displacement u of the equilibrium configuration satisfies the following equation

$$a < x < b : -(EAu')' = f. \tag{3.20}$$

This is a second order ordinary differential equation. In order to determine the solution, in general the boundary conditions at the two ends are needed. This is a two-point boundary value problem. The geometrical constraints prescribed in the variational problem provide the following two conditions

$$x = a : \quad u(a) = \bar{u}_a, \tag{3.21}$$

$$x = b : \quad u(b) = \bar{u}_b. \tag{3.22}$$

Conditions (3.21) and (3.22) are called the geometrical boundary conditions or the essential boundary conditions. By saying "essential," we mean that these boundary conditions must be imposed on the variational problem.

Conversely, suppose u is a solution of the above boundary value problem. Multiplying both sides of Equation (3.20) by an arbitrary $v \in K_0$ and then integrating over x, we obtain

$$\int_0^b (-EAu')'vdx = \int_a^b fvdx.$$

Using integration by parts and the boundary conditions, we have

$$\int_a^b -(EAu')'v\,dx = \left| - (EAu')v \right|_a^b + \int_a^b EAu'v'\,dx$$

$$= \int_a^b EAu'v'\,dx = \int_a^b fv\,dx,$$

i.e.,

$$D(u, v) - F(v) = 0, \text{ for any } v \in K_0.$$

Hence u is also a solution of the variational problem. This shows that variational problem (3.19) is equivalent to the boundary value problem (3.20)–(3.22).

2. The second kind of boundary conditions: The loading support.

Suppose the load applying at the end point $x = a$ is g_a, the load at the end point $x = b$ is g_b. Their contributions to the potential energy of external work are $-g_a u(a)$ and $-g_a u(b)$, respectively. Then the total potential energy of external work is

$$-F(u) = -\left(\int_a^b fu\,dx + g_a u(a) + g_b u(b) \right).$$

Since there are no constraints on the positions of the two ends, K and K_0 are both equal to the set of all displacement functions. Starting from the variational principle

$$D(u, v) - F(v) = 0, \text{ for any } v, \tag{3.23}$$

and applying integration by parts,

$$D(u, v) = \int_a^b EAu'v'\,dx = [EAu'v]_a^b - \int_a^b (EAu')'v\,dx,$$

we obtain

$$D(u, v) - F(v) = [(EAu')_b - g_b]v(b) + [-(EAu') - g_a]v(a)$$

$$+ \int_a^b [-(EAu')' - f]v\,dx = 0, \text{ for any } v.$$

Since $v(x)$ within (a, b) and the values $v(a)$ and $v(b)$ at the two end points are arbitrary, the terms included in the above three brackets $[\cdots]$ must vanish for the same reason as in 1. Then the equations satisfied by the displacement u of the equilibrium configuration and the boundary conditions are,

$$
\begin{cases}
a < x < b: & -(EAu')' = f, \\
x = a: & -(EAu')_a = g_a, \\
x = b: & (EAu')_b = g_b.
\end{cases}
\tag{3.24}
$$

$$\tag{3.25}$$

Conversely, one can prove using a simular method that the solution of the above boundary value problem is exactly the solution of the variational problem (3.23). Thus these two problems are equivalent. Note that here the boundary conditions (3.24) and (3.25) are derived automatically from the variational principle, so they are also called the natural boundary conditions, so as to be distinguished from the essential boundary conditions of the first kind. Thus, the variational principle automatically provides the mechanical boundary conditions at the end points when no constraints exist.

3. The third kind of boundary conditions: Elastic support.

Suppose the rod is elastically coupled with the external region at its end points. The rod is subjected to elastic reaction forces which are proportional to the deviations of the displacements. For example, the elastic reaction force at $x = a$ is

$$-c_a(u(a) - \bar{u}_a) = -c_a u(a) + g_a, \quad g_a = c_a \bar{u}_a,$$

and the elastic reaction force at $x = b$ is

$$-c_b(u(b) - \bar{u}_b) = -c_b u(b) + g_b, \quad g_b = c_b \bar{u}_b,$$

where $c_a, c_b > 0$. c_a and c_b are constants of the elastic coupling. $\bar{u}_a$ and $\bar{u}_b$ are reference displacements. g_a and g_b are given values. At that time, the strain energy will be increased by an amount $\frac{1}{2}c_a u^2(a) + \frac{1}{2}c_b u^2(b)$. Correspondingly, the functional of virtual work $D(u, v)$ is increased by an amount of $c_a u(a) + c_b u(b)$. The potential energy of external work is

$$-F(u) = -\left(\int_a^b fv dx + g_a u(a) + g_b u(b) \right).$$

The two end points are still unrestrained: K and K_0 are equal the set of all displacement functions. Similarly, the variational principle implies

$$D(u, v) - F(v) = 0, \quad \text{for any } v.$$

The equivalent boundary value problem can be derived through integrations by parts as

$$\begin{cases} a < x < b: & -(EAu')' = f, \\ x = a: & -(EAu')_a + c_a u(a) = g_a, \\ x = b: & (EAu')_b + c_b u(b) = g_b. \end{cases} \tag{3.26}$$

Here, the boundary conditions (3.26) and (3.27) are also automatically derived from the variational principle and need not be listed explicitly as prescribed constraints in the variational principle. Hence they too can be considered to be natural boundary conditions.

Sometimes the rod is coupled elastically with the external region longitudinally elastically along its length. The elastic reaction force per unit length of the rod, which is a applied to the whole length or a segment $a' \leq x \leq b'(a \leq a' < b' \leq b)$, is

$$-c(u - \bar{u}) = -cu + \bar{f}, \quad \bar{f} = c\bar{u},$$

where $c > 0$ is the constant of elastic coupling. $\bar{f}$ may be merged into the line load and then the elastic reaction force per unit length may be regarded as $-cu$. Then, the strain energy should be increased by adding a term $\dfrac{1}{2} \displaystyle\int_a^b cu^2 dx$. This corresponds to adding to $D(u, v)$ a term $\displaystyle\int_a^b cuvdx$, i.e.,

$$D(u, v) = \int_a^b EAu'v'dx + \int_a^b cuvdx.$$

We can set $c = 0$ in the region without elastic support. The potential energy of external work is

$$-F(u) = -\left(\int_a^b fudx + g_a u(a) + g_b u(b) \right),$$

and both end points are still unrestrained. Then the equivalent boundary value problem can be derived from the variational principle as

$$\begin{cases} a < x < b: & -(EAu')' + cu = f, \\ x = a: & -(EAu')_a = g_a, \\ x = b: & (EAu')_b = g_b. \end{cases} \tag{3.27}$$

The above three kinds of boundary conditions can also be mixed together, i.e., any one of these three kinds of boundary conditions can be
taken at each end point, and different modes can be prescribed at the two
end points as well.

Interface conditions. Suppose that the tensile rigidity EA is discontinuous at some cross section $x = p$ (Fig. 10a), or a concentrated load
g_p acts at a cross section $x = p$ (Fig. 10b), or these two cases arise simultaneously. Then the form of the strain energy remains unchanged, while
the potential energy of the external work should be modified to

$$-F(v) = -\left(\int_a^b fvdx + g_a v(a) + g_b v(b) + g_p v(p) \right).$$

At the same time, owing to the discontinuity of EA at $x = p$, the integration of $\int_a^b EAu'v'dx$ by parts must be performed on the two segments
individually. Hence we have

$$D(u,v) = \int_a^b EAu'v'dx = \int_a^{p-0} EAu'v'dx + \int_{p+0}^b EAu'v'dx$$

$$= \int_0^{p-0} -(EAu')'vdx + \int_{p+0}^b -(EAu')'vdx$$

$$-(EAu')_a v(a) + (EAu')_b v(b) - [EAu']_{p-0}^{p+0}(p).$$

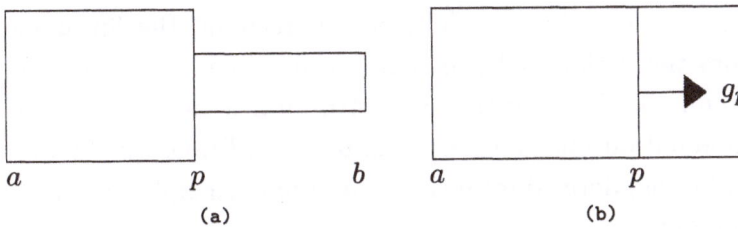

(a)   (b)

Fig. 10

Then from the variational principle,

$$D(u,v) - F(v) = \int_a^{p-0}[-(EAu')' - f]vdx + \int_{p+0}^b[-(EAu')' - f]vdx$$

$$+[-(EAu')_a - g_a]v(a) + [(EAu')_b - g_b]v(b)$$

$$+[-[EAu']_{p-0}^{p+0} - g_p]v(p) = 0, \text{ for any } v.$$

we obtain

$$\begin{cases} a < x < p, \ p < x < b : & -(EAu')' = f, \\ x = a : & -(EAu')_a = g_a, \\ x = b : & (EAu')_b = g_b, \\ x = p : & -[EAu']_{p-0}^{p+0} = g_p. \end{cases} \tag{3.28}$$

Besides this, it is still necessary to impose the constraint condition for the continuity of displacement u at $x = p$

$$x = p : \quad [u]_{p-0}^{p+0} = 0. \tag{3.29}$$

(3.28) and (3.29) are two interface conditions. (3.28) is automatically derived from the variational principle, so it too is also a kind of natural boundary condition. However, in the boundary value problems of the differential equation, (3.28) is constituent part which is the necessary condition to get a well-defined solution.

The continuity condition (3.29) need not be set out as a essential condition if the convention has been made such that the displacement is single-valued at the discontinuity.

The case of serveral discontinuity points is handlied like the case of a single discontinuity point.

3.4   Equilibrium equations

In the case with finitely many degrees of freedom, the linear equations derived from the variational principle are just the equilibrium equations of the system. Similarly, in the case with infinite degree of freedom, the differential equation and natural boundary conditions (including the interface conditions) derived from the variational principle are just the equilibrium equations.

Arbitrarily taking a segment (x_1, x_2) of the rod, we consider the equilibrium relation of the longitudinal forces (Fig. 11).

$$0 = Q(x_2) - Q(x_1) + \int_{x_1}^{x_2} f dx = \int_{x_1}^{x_2} (Q' + f) dx.$$

The equation holds for any x_1 and x_2 within the interval (a, b), hence the equilibrium equation can be obtained

$$a < x < b : \quad -Q' = f.$$

From (3.3), $Q = EAu'$. In terms of the displacement u, the equilibrium equation becomes

$$a < x < b: \quad -(EAu')' = f.$$

This is equation (3.20), satisfied by the displacement u of the equilibrium configuration derived from the variational principle.

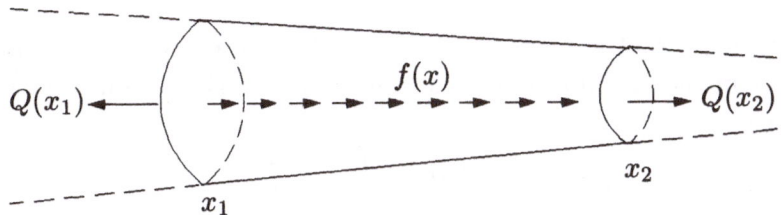

Fig. 11

For uniform cross sections, the equilibrium equation becomes

$$-EAu'' = f.$$

If there is no body load, $f = 0$ and the displacement

$$u = a_0 + a_1 x$$

is a linear function of x. The strain $\varepsilon = u' = a_1$ and the stress $\sigma = E\varepsilon = Ea_1$ The problem is then reduced to a problem with one degree of freedom, which is just the uniformly deformed rod discussed in §2. When $f =$ is constant., the displacement

$$u = a_0 + a_1 x + a_2 x^2$$

is a quadratic function of x, while ε and σ are linear functions of x. This is a case of two degrees of freedom. For general f or EA, the system has infinitely many degree of freedom, which can not described by a finite number of parameters.

Now we consider the equilibrium relations of the forces acting at the end points as well as at the interface point. Suppose there exists a point load g_b at $x = b$. Take a small interval of length Δx in the neighborhood of $x = b$, on which the equilibrium relation of the longitudinal forces is

$$-Q(b - \Delta x) + g_b + \int_{b-\Delta x}^{b} f\,dx = 0.$$

Letting $\Delta x \to 0$, we obtain

$$-Q(b) + g_b = 0.$$

This is the natural boundary condition (3.25):

$$(EAu')_b = g_b.$$

Similarly, suppose there exists a point load g_a at $x = a$. Take a small interval Δx in the neighborhood of $x = a$, on which the equilibrium relation of forces is

$$Q(a + \Delta x) + g_a + \int_a^{a+\Delta x} f\,dx = 0.$$

Letting $\Delta x \to 0$, we obtain

$$Q(a) + g_a = 0.$$

This is also the natural boundary condition (3.24):

$$-(EAu')_a = g_a.$$

If there exist elastic supports at the end points, then terms belonging to elastic reaction forces should be added to the above equilibrium relations as well. This yields the natural boundary conditions (3.26) and (3.27).

We continue by discussing the equilibrium of forces at an interface point within the rod. Suppose there exists a concentrated load g_p at $x = p$. Taking a small interval $(p - \Delta x, p + \Delta x)$ including $x = p$, we then have the equilibrium relation

$$Q(p + \Delta x) - Q(p - \Delta x) + g_p + \int_{p-\Delta x}^{p+\Delta x} f\,dx = 0.$$

Letting $\Delta x \to 0$, we obtain

$$Q(p + 0) - Q(p - 0) + g_p = 0.$$

This is the interface condition (3.28). Hence we obtain the following two corollaries:

1. Suppose there exists the concentrated load $g_p = 0$ at the cross section $x = p$ and the tensile rigidity EA has a jump across the section. Then

$$[Q]_{p-0}^{p+0} = [EAu']_{p-0}^{p+0} = 0.$$

This shows that at $x = p$ the internal force Q is continuous, but the strain $\varepsilon = u'$ is discontinuous.

2. Suppose that EA is continuous at the cross section and that $g_p \neq 0$. Then

$$-[Q]_{p-0}^{p+0} = -EA[u']_{p-0}^{p+0} = g_p.$$

This shows that a concentrated load g_p leads to the discontinuity of the internal force Q and the strain ε.

3.5   Strainless states

According to the variational principle in Section 3.2, whether or not there exists a solution of the variational problem and its equivalent equilibrium equations and whether or not the solution is unique depends on whether the quadratic functional of the strain energy $D(v, v)$ is positive definite or degenerate, and the latter is closely related to the statement of the boundary conditions. Now let us discuss this problem according to the three kinds of boundary conditions described in Section 3.3, respectively.

1. Prescribed displacement.

Prescribed constraints at both ends:

$$K_0 = (v(x)|_{v(a)=v(b)=0}),$$

or a prescribed constraint at one end:

$$K_0 = (v(x)|_{v(a)=0 \text{ or } v(b)=0}).$$

The strainless virtual displacement

$$v \in K_0: \ D(v, v) = \int_a^b EAv'^2 dx = 0 \Longleftrightarrow v' \equiv 0 \Longleftrightarrow v \equiv 0.$$

Hence D is always positive definite no matter whether the constraint is prescribed at both ends or at one end. There exists a unique solution of the equilibrium problem.

2. Loading support.

The displacement is uncanstrained. The strainless state

$$v: \ D(v, v) = \int_a^b EAv'^2 dx = 0 \Longleftrightarrow v' \equiv 0 \Longleftrightarrow v = k(\text{const.}),$$

hence D is degenerate. The strainless state has one degree of freedom: $v^{(1)} = 1$, which corresponds to a rigid translation. Hence the displacement solution u of the variational problem is not unique and may differ by a rigid translation according to the theorm 3 in Section 3.2. However the rigid translation contributes nothing to the internal force or the strain ε. The stress solution of the equilibrium state is still unique even in the degenerative case. The necessary and sufficient condition for the existence of solutions is

$$F(v^{(1)}) = F(1) = 0,$$

i.e.,

$$\int_a^b f \, dx + g_a + g_b = 0. \tag{3.30}$$

(3.30) indicates that the external loads are self-equilibrating. This is the prerequisite for achieving equilibrium in the unrestrained case. There are no solutions if the prerequisite is not satisfied.

3. Elastic support.

The displacement is still unrestrained. The strainless state

$$v : \ D(v, v) = \int_a^b EA v'^2 \, dx + c_a v^2(a) + c_b v^2(b) = 0$$

$$\Longleftrightarrow v' \equiv 0, \ v(a) = v(b) = 0$$

$$\Longleftrightarrow v \equiv 0.$$

Hence D is positive definite and there exists a unique solution of the equilibrium problem. It is the same as the case of the prescribed displacement, D is actually positive definite so long as one of its ends is elastically supported.

4. Rods on an elastic foundation.

Since

$$D(v, v) = \int_a^b EA v'^2 \, dx + \int_a^b c v^2 \, dx,$$

so long as the constant of elastic coupling $c > 0$ unclear the strainless state

$$v : D(v, v) = 0 \Longleftrightarrow v' \equiv 0 \quad (a \leq x \leq b),$$

$$v \equiv 0 \quad (a' \leq x \leq b') \Longleftrightarrow v \equiv 0 \quad (a \leq x \leq b),$$

i.e., D is positive definite, and there exists a unique solution of the equilibrium problem.

Two conclusions follow from the above discussions:

1. The equilibrium problem in which the boundary is supported by loads (i.e., the pure mechanical boundary condition) is degenerate. In that case, the external loads must be self-equilibrating for achieving equilibrium.

2. Adding geometrical constraints or elastic supports always reduces the number of degrees of freedom of the strainless states, i.e., brings the strain energy closer to positive definiteness.

These are two general rules of elasticity.

For rods in tension, since the strainless state of the degenerate problem has only one degree of freedom, the strain energy is positive definite provided that one of the ends has been geometrically restrained or elastically supported.

To sum up, starting from the concept of energy and establishing the energy equation for a given deformation mode, we can derive all equations about elastic equilibrium of rods in tension from the variational principle, including the equilibrium equation over the rod, the mechanical boundary conditions for all types of loading and elastic support, the interface conditions at discontinuity, and so on. In contrast with the statement of equilibrium in terms of a differential equation boundary value problem, the variational problem only needs the geometrical boundary conditions as constraint conditions. In the variational problem, if we put the all relevant contributions into the energy functional then all other mechanical boundary conditions, as well as the equilibrium equation, are satisfied automatically when the potential energy achieves a minimum. The mechanical boundary conditions conditions involve the derivatives of u and and the from are comparatively complex. In particular, the interface conditions are the natural boundary conditions. Note that the retained essential boundary conditions in the variational problem only involve the function u itself. Their form is comparatively simple, while the natural boundary conditions caused by the discontinuity of coefficients at an interface point are more tedious. These conditions can not be neglected in the statement in terms of the boundary value problem of differential equations, but these complications be simplified or neglected altogether in the statement in terms of the variational problem. Therefore, the variational principle leads to great simplification in practice. It puts complete information about equilibrium into an extremely compact energy expression. Here we state the variational principle in a quite abstract form, which makes it suitable for various kinds of elastic equilibrium problems. The conclusions derived from the equilibrium problems of rods in tension are generally valid in elasticity. Though the idea of the variational principle can clearly be explained through simple cases, its advantages become clear only when the problem is more complex. The following discussions, will gradually make these advantages clear.

§4   Stretching and Compression in Various Directions

4.1   Hook's law and strain energy

We can now proceed to discuss three dimensional stretching and compression under three dimensional loads in the light of our knowledge of the simple stretching and compression under one dimensional loads. Consider a rectangular parallelopiped whose the lengths in the x, y and z directions are L_x, L_y, L_z respectively. The loads F_x, F_y, F_z are applied uniformly in the x, y and z directions. We discuss the stress and strain states within its body. This problem can be considered as the superimposition of three independent problems with unidirectional loading.

The stresses and the strains caused by the x-directional load F_x alone are

$$\sigma_x^{(1)} = F_x/A_x = F_x/L_y L_z, \ \ \sigma_y^{(1)} = 0, \ \ \sigma_z^{(1)} = 0,$$
$$\varepsilon_x^{(1)} = \sigma_x^{(1)}/E, \ \ \varepsilon_y^{(1)} = -\nu\varepsilon_x^{(1)}, \ \ \varepsilon_z^{(1)} = -\nu\varepsilon_x^{(1)}.$$

Similarly,

$$\sigma_x^{(2)} = 0, \ \ \sigma_y^{(2)} = F_y/A_y = F_y/L_z L_x, \ \ \sigma_z^{(2)} = 0,$$
$$\varepsilon_x^{(2)} = -\nu\varepsilon_y^{(2)}, \ \ \varepsilon_y^{(2)} = \sigma_y^{(2)}/E, \ \ \varepsilon_z^{(2)} = -\nu\varepsilon_y^{(2)}.$$

And

$$\sigma_x^{(3)} = 0, \ \ \sigma_y^{(3)} = 0, \ \ \sigma_z^{(3)} = F_z/A_z = F_z/L_x L_y,$$
$$\varepsilon_x^{(3)} = -\nu\varepsilon_z^{(3)}, \ \ \varepsilon_y^{(3)} = -\nu\varepsilon_z^{(3)}, \ \ \varepsilon_z^{(3)} = \varepsilon_z^{(3)}/E.$$

The results of the linear superimposition are

$$\sigma_x = \sum_{i=1}^{3} \sigma_x^{(i)} = \sigma_x^{(1)}, \sigma_y = \sum_{i=1}^{3} \sigma_y^{(i)} = \sigma_y^{(2)},$$

$$\sigma_x = \sum_{i=1}^{3} \sigma_z^{(i)} = \sigma_x^{(3)},$$

$$\varepsilon_x = \sum_{i=1}^{3} \varepsilon_x^{(i)} = \frac{1}{E}\sigma_x^{(1)} - \frac{\nu}{E}(\sigma_y^{(2)} + \sigma_z^{(3)}),$$

$$\varepsilon_y = \sum_{i=1}^{3} \varepsilon_y^{(i)} = \frac{1}{E}\sigma_y^{(2)} - \frac{\nu}{E}(\sigma_z^{(3)} + \sigma_x^{(3)}),$$

$$\varepsilon_z = \sum_{i=1}^{i} \varepsilon_z^{(i)} = \frac{1}{E}\sigma_z^{(3)} - \frac{\nu}{E}(\sigma_x^{(1)} + \sigma_y^{(2)}).$$

Hence we obtain the Hooke's law in the case of three dimensional stretching and compression:

$$\varepsilon_x = \frac{1}{E}[\sigma_x - \nu(\sigma_y + \sigma_z)] = \frac{1}{E}[(1+\nu)\sigma_x - \nu(\sigma_x + \sigma_y + \sigma_z)],$$

$$\varepsilon_y = \frac{1}{E}[\sigma_y - \nu(\sigma_z + \sigma_x)] = \frac{1}{E}[(1+\nu)\sigma_y - \nu(\sigma_x + \sigma_y + \sigma_z)], \qquad (4.1)$$

$$\varepsilon_z = \frac{1}{E}[\sigma_z - \nu(\sigma_x + \sigma_y)] = \frac{1}{E}[(1+\nu)\sigma_z - \nu(\sigma_x + \sigma_y + \sigma_z)].$$

By further superimposing the above three expressions, a relation between the sum of the three dimensional strains and the sum of the three dimensional stresses can be obtained:

$$\varepsilon_x + \varepsilon_y + \varepsilon_z = \frac{1 - 2\nu}{E}(\sigma_x + \sigma_y + \sigma_x). \qquad (4.2)$$

Thus the inverse of (4.1) can be expressed as

$$\sigma_x = \frac{E}{(1+\nu)(1-2\nu)}[(1-\nu)\varepsilon_x + \nu(\varepsilon_y + \varepsilon_z)]$$

$$= \frac{E}{1+\nu}\varepsilon_x + \frac{E\nu}{(1+\nu)(1-2\nu)}(\varepsilon_x + \varepsilon_y + \varepsilon_z),$$

$$\sigma_y = \frac{E}{(1+\nu)(1-2\nu)}[(1-\nu)\varepsilon_y + \nu(\varepsilon_z + \varepsilon_x)]$$

$$= \frac{E}{1+\nu}\varepsilon_y + \frac{E\nu}{(1+\nu)(1-2\nu)}(\varepsilon_x + \varepsilon_y + \varepsilon_z),$$

$$\sigma_z = \frac{E}{(1+\nu)(1-2\nu)}[(1-\nu)\varepsilon_z + \nu(\varepsilon_x + \varepsilon_y)]$$

$$= \frac{E}{1+\nu}\varepsilon_z + \frac{E\nu}{(1+\nu)(1-2\nu)}(\varepsilon_x + \varepsilon_y + \varepsilon_z). \tag{4.3}$$

Since the direction of σ_x coincides with that of ε_x but is orthogonal to that of ε_y and ε_z, the contribution of σ_x to the strain energy per unit volume is $\frac{1}{2}\sigma_x\varepsilon_x$, according to (2.8). Similarly, there are contributions $\frac{1}{2}\sigma_y\varepsilon_y$ and $\frac{1}{2}\sigma_z\varepsilon_z$. Hence the volume density of the strain energy is

$$W = \frac{1}{2}(\sigma_x\varepsilon_x + \sigma_y\varepsilon_y + \sigma_z\varepsilon_z). \tag{4.4}$$

Substituting (4.3) into (4.4), we obtain

$$W = \frac{1}{2}\left\{\frac{E}{1+\nu}(\varepsilon_x^2 + \varepsilon_y^2 + \varepsilon_z^2) + \frac{E\nu}{(1+\nu)(1-2\nu)}(\varepsilon_x + \varepsilon_y + \varepsilon_z)^2\right\}. \tag{4.5}$$

W is a quadratic form (quadratic homogeneous expression) in $\varepsilon_x, \varepsilon_y, \varepsilon_z$, and we have

$$\sigma_x = \frac{\partial W}{\partial \varepsilon_x}, \quad \sigma_y = \frac{\partial W}{\partial \varepsilon_y}, \quad \sigma_z = \frac{\partial W}{\partial \varepsilon_z}.$$

Hence

$$W = \frac{1}{2}\left(\frac{\partial W}{\partial \varepsilon_x}\varepsilon_x + \frac{\partial W}{\partial \varepsilon_y}\varepsilon_y + \frac{\partial W}{\partial \varepsilon_z}\varepsilon_z\right). \tag{4.6}$$

4.2 Changes of volume

The relative change of volume can be expressed in terms of the three components of the strain. The original volume $V = L_xL_yL_z$. Suppose the elongations of the three edges after deformation are $\delta L_x, \delta L_y, \delta L_z$, respectively. Then the increment of volume is

$$\delta V = (\delta L_x)L_yL_z + (\delta L_y)L_zL_x + (\delta L_z)L_xL_y,$$

hence the relative increment of volume is

$$\frac{\delta V}{V} = \frac{\delta L_x}{L_x} + \frac{\delta L_y}{L_y} + \frac{\delta L_z}{L_z} = \varepsilon_x + \varepsilon_y + \varepsilon_z, \tag{4.7}$$

i.e., the sum of the strains in the x, y and z directions.

Consider the special case, of isotropic compression (or stretching). In this case $\sigma_x = \sigma_y = \sigma_z = \sigma$, hence $\varepsilon_x = \varepsilon_y = \varepsilon_z = \varepsilon$, and $\varepsilon = \dfrac{1-2\nu}{E}\sigma$. Thus

$$\frac{\delta V}{V} = 3\varepsilon = \frac{3(1-2\nu)}{E}\sigma,$$

or

$$\sigma = K\frac{\delta V}{V} = 3K\varepsilon, \tag{4.8}$$

where

$$K = \frac{E}{3(1-2\nu)} \tag{4.9}$$

is the bulk modulus, a constant describing the resistance of the solid material to compression of volume.

The volume density of the strain energy is then

$$W = \frac{1}{2}(\sigma_x\varepsilon_x + \sigma_y\varepsilon_y + \sigma_z\varepsilon_z) = \frac{3}{2}\sigma\varepsilon = \frac{9K\varepsilon^2}{2} = \frac{\sigma^2}{2K}. \tag{4.10}$$

The modulus K is always positive for any material. This means that the volume becomes smaller during compression and becomes larger during stretching. If $K < 0$, the elastic strain energy would be negative. Deformation would occur spontaneously and without limit because the material always tends towards the state of the minimum potential energy. But this is impossible.

So $K = \dfrac{E}{3(1-2\nu)} > 0$ and equivalently $1 - 2\nu > 0$, i.e., $\nu < \dfrac{1}{2}$. This is actually an upper bound of Poisson's ratio. For rubber, $\nu \sim \dfrac{1}{2}$, which is close upon this upper bound. On the other hand, for all known materials, $\nu > 0$. So far no materials with $\nu < 0$ have been discovered. (This would mean that when one direction is stretched, the other two directions also stretch.) Only for porous solids, such as cork, is Poisson's ratio close to the lower bound $\nu \sim 0$.

Another important special case of stretching is so-called "single-directional compression". A rectangular parallelopiped is compressed and shortened in the longitudinal direction, but the two transverse directions are restrained by rigid walls and cannot stretch or compress, i.e.,

$\varepsilon_y = \varepsilon_z = 0$. Then (4.3) implies that

$$\sigma_x = \frac{E(1-\nu)}{(1+\nu)(1-2\nu)}\varepsilon_x = E'\varepsilon_x,$$

$$E' = \frac{E(1-\nu)}{(1+\nu)(1-2\nu)},$$

$$\sigma_y = \sigma_z = \frac{E\nu}{(1+\nu)(1-2\nu)}\varepsilon_x,$$

$$W = \frac{1}{2}\sigma_x\varepsilon_x = \frac{1}{2}E'\varepsilon_x^2.$$

(4.11)

$E' > E$, since $0 < \nu < \frac{1}{2}$, $(1+\nu)(1-2\nu) = 1 - \nu - 2\nu^2 < 1 - \nu$. Thus we can see that the tensile rigidity of the transversely fixed rod, which is loaded longitudinally, is larger then that of the transversely free rod..

§5   Shear Deformations

5.1   Shearing stresses
In the above case of isotropic compression, the strains in all directions are identical. Hence the body's shape does not change—only its volume varies. The opposite occurs in another special case, i.e. the volume remains unchanged while the shape varieds. This is called a simple shear or shear deformation. In this case

$$\frac{\delta V}{V} = \varepsilon_x + \varepsilon_y + \varepsilon_z = 0.$$

(5.1)

By (4.2),

$$\sigma_x + \sigma_y + \sigma_z = 0.$$

(5.2)

In order to discuss the simple shear, it is necessary to discuss completely the stress defined in §4.1. Taking a cross section s normal to the $+x$-direction, denote by $\sigma_{xx}, \sigma_{yx}, \sigma_{zx}$ the components of force per unit area in the x, y, and z, directions, by which the positive side s^+ acts on the negative side s^- across s. Similarly, denote the three components of stress, respectively as $\sigma_{xy}, \sigma_{yy}, \sigma_{zy}$ when the normal direction of the cross section is $+y$. Similarly when the direction the cross section is $+z$ (Fig. 12).

Hence there are nine components of stress at each point. Among these, σ_{xx}, σ_{yy} and σ_{zz} are the components of normal stress and the other six $\sigma_{xy}, \sigma_{yz}, \cdots$ are the components of shear stress.

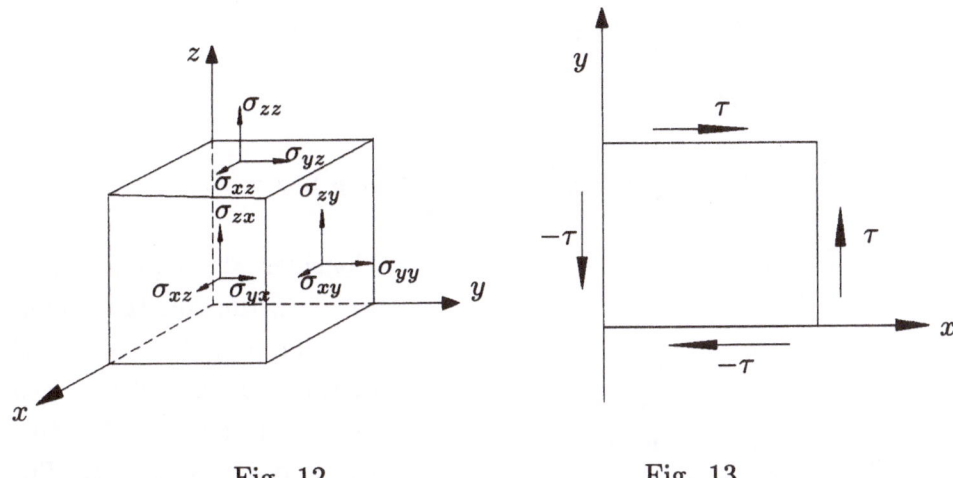

Fig. 12   Fig. 13

In the case of the isotropic stretching and compression in §4.2,

$$\sigma_{xx} = \sigma_x,\ \sigma_{yx} = \sigma_{zx} = 0,$$

$$\sigma_{yy} = \sigma_y,\ \sigma_{xy} = \sigma_{zy} = 0, \tag{5.3}$$

$$\sigma_{zz} = \sigma_z,\ \sigma_{xz} = \sigma_{yz} = 0,$$

i.e., there are only normal stresses but no shearing stress.

Now consider a cube of unit volume. A pair of uniformly distributed forces, which are tangential to the faces and have no component in the z-direction, acts on the left-right and front-back pairs of laterals. According to the equilibrium of forces and moments, the forces of each pair are equal in magnitude but reverse in direction (Fig. 13). We can assume that the stress distribution in the body is also uniform. Then we obtain

$$\sigma_{xx} = 0,\ \sigma_{yx} = \tau,\ \sigma_{zx} = 0,$$

$$\sigma_{xy} = \tau,\ \sigma_{yy} = 0,\ \sigma_{zy} = 0, \tag{5.4}$$

$$\sigma_{xz} = 0,\ \sigma_{yx} = 0,\ \sigma_{zz} = 0.$$

Hence the stresses acting on the cross sections with the normal directions x and y, respectively, are all shear stresses and there are no normal stresses.

5.2   Shear strains

If we rotate the coordinates axes in the xy plane through $45°$, the transformation relations between the new and the original coordinates are as follows:

$$x' = \frac{1}{\sqrt{2}}(x+y), \ y' = \frac{1}{\sqrt{2}}(-x+y), \ z' = z.$$

Inverting these relations yields

$$x = \frac{1}{\sqrt{2}}(x'-y'), \ y = \frac{1}{\sqrt{2}}(x'+y'), \ z = z'.$$

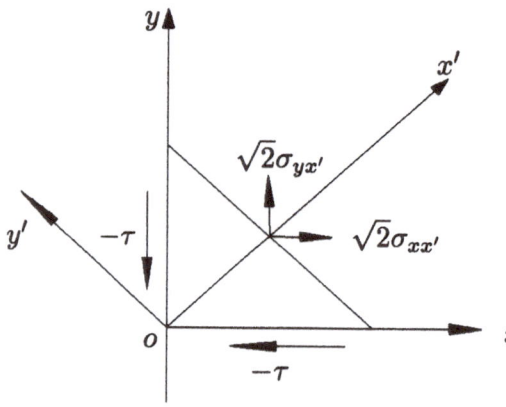

Consider the equilibrium of forces acting on the triangle shown in Fig. 14. The lengths of the two sides of the triangle are 1, and the length of the hypotenuse is $\sqrt{2}$. Then from the equilibria of the x-directional forces and of the y-directional forces, we get

$$\sqrt{2}\sigma_{xx'} - \sigma_{xy} = 0,$$
$$\sqrt{2}\sigma_{yx'} - \sigma_{yx} = 0,$$

Fig. 14

i.e.,

$$\sigma_{xx'} = \frac{\sigma_{xy}}{\sqrt{2}} = \frac{\tau}{\sqrt{2}},$$

$$\sigma_{yx'} = \frac{\sigma_{yx}}{\sqrt{2}} = \frac{\tau}{\sqrt{2}}.$$

Here, $\sigma_{xx'}$ and $\sigma_{yx'}$, respectively, stand for the x-and the y-directional components of the stress on a cross section normal to the x'-axis. Thus the x'-, y'-, and z'-directional components of the stress are

$$\sigma_{x'x'} = \frac{1}{\sqrt{2}}(\sigma_{xx'} + \sigma_{yx'}) = \tau,$$

$$\sigma_{y'x'} = \frac{1}{\sqrt{2}}(-\sigma_{xx'} + \sigma_{yx'}) = 0, \qquad (5.5)$$

$$\sigma_{x'x'} = 0.$$

For the same reason, the x'-, y'-, and z'-directional components of the stress on a cross section normal to the y'-axis are

$$\sigma_{x'y'} = 0, \ \sigma_{y'y'} = -\tau, \ \sigma_{x'y'} = 0. \qquad (5.6)$$

Moreover, since no forces act on the top and bottom,

$$\sigma_{x'z'} = 0, \ \sigma_{y'z'} = 0, \ \sigma_{z'z'} = 0. \tag{5.7}$$

We observed that in the new system of coordinates x', y', z' (x', y' correspond to the diagonal directions of the original square block), (5.5)–(5.7), which are just like to the stretching and compression in all directions as described in §4, stand for stretching in the x'-direction and compression in the y'-direction. Hooke's law takes the form

$$\frac{\partial u'}{\partial x'} = \varepsilon_{x'x'} = \frac{1}{E}(\sigma_{x'x'} - \nu\sigma_{y'y'}) = \frac{1+\nu}{E}\tau,$$

$$\frac{\partial v'}{\partial y'} = \varepsilon_{y'y'} = \frac{1}{E}(\sigma_{y'y'} - \nu\sigma_{x'x'}) = -\frac{1+\nu}{E}\tau, \tag{5.8}$$

$$\frac{\partial w'}{\partial z'} = \varepsilon_{z'z'} = \frac{1}{E}(\sigma_{z'z'} - \nu\sigma_{x'x'} - \nu\sigma_{y'y'}) = 0.$$

There is no change of volume because

$$\frac{\delta V}{V} = \varepsilon_{x'x'} + \varepsilon_{y'y'} + \varepsilon_{z'z'} = \frac{1+\nu}{E}\tau - \frac{1+\nu}{E}\tau + 0 = 0.$$

Return to the original system of coordinates x, y, z. In order to describe the shear deformation, in addition to the components of the normal strain

$$\varepsilon_{xx} = \frac{\partial u}{\partial x}, \ \varepsilon_{yy} = \frac{\partial v}{\partial y}, \ \varepsilon_{zz} = \frac{\partial w}{\partial z},$$

we also need to define the components of the shear strain as follows:

$$\varepsilon_{xy} = \varepsilon_{yx} = \frac{1}{2}\left(\frac{\partial v}{\partial x} + \frac{\partial u}{\partial y}\right),$$

$$\varepsilon_{yz} = \varepsilon_{zy} = \frac{1}{2}\left(\frac{\partial w}{\partial y} + \frac{\partial v}{\partial z}\right), \tag{5.9}$$

$$\varepsilon_{zx} = \varepsilon_{xz} = \frac{1}{2}\left(\frac{\partial u}{\partial z} + \frac{\partial w}{\partial x}\right).$$

In the case of small strains (See Fig. 15),

$$\frac{\partial v}{\partial x} = \text{tg}\alpha \approx \alpha,$$

$$\frac{\partial u}{\partial y} = \text{tg}\beta \approx \beta,$$

hence

$$\varepsilon_{xy} = \varepsilon_{yx} \approx \frac{1}{2}(\alpha + \beta)$$

$$= \frac{1}{2}\gamma. \tag{5.10}$$

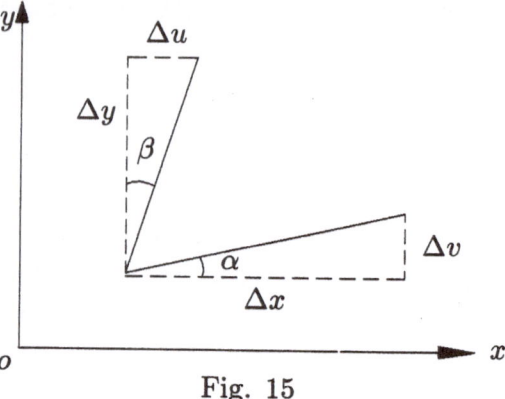

Fig. 15

After the deformation, the included angle of the local $x - y$ frame of axes, which is $\pi/2$ originally, is decreased by γ. Thus, the geometric meaning of the shear strain ε_{xy} is γ called the half angle of shear. $\varepsilon_{xy} > 0$ indicates that the right angle becomes acute, while $\varepsilon_{xy} < 0$ indicates that the right angle becomes obtuse. In Fig. 15, the angle of shear is $\gamma = \alpha + \beta$.

5.3   Hooke's law and the strain energy of shear deformations

Since

$$u' = \frac{1}{\sqrt{2}}(u + v), \ v' = \frac{1}{\sqrt{2}}(-u + v),$$

$$\frac{\partial}{\partial x'} = \frac{1}{\sqrt{2}}\Big(\frac{\partial}{\partial x} + \frac{\partial}{\partial y}\Big), \ \frac{\partial}{\partial y'} = \frac{1}{\sqrt{2}}\Big(-\frac{\partial}{\partial x} + \frac{\partial}{\partial y}\Big),$$

we get

$$\frac{\partial u'}{\partial x'} = \frac{1}{2}\Big(\frac{\partial u}{\partial x} + \frac{\partial v}{\partial y} + \frac{\partial v}{\partial x} + \frac{\partial u}{\partial y}\Big)$$

$$= \frac{1}{2}(\varepsilon_{xx} + \varepsilon_{yy} + 2\varepsilon_{xy}) = \varepsilon_{xy},$$

$$\frac{\partial v'}{\partial y'} = \frac{1}{2}\Big(\frac{\partial u}{\partial x} + \frac{\partial v}{\partial y} - \frac{\partial v}{\partial x} - \frac{\partial u}{\partial y}\Big)$$

$$= \frac{1}{2}(\varepsilon_{xx} + \varepsilon_{yy} - 2\varepsilon_{xy}) = -\varepsilon_{xy}. \tag{5.11}$$

Hooke's law for shear deformation can then be obtained from (5.8), as

$$\varepsilon_{xy} = \frac{1}{2G}\sigma_{xy}, \ G = \frac{E}{2(1 + v)}.$$

This indicates that the shear stress τ is proportional to the angle of shear γ

$$\tau = G_\gamma, \tag{5.12}$$

G is a material constant called the modulus of shear rigidity.

The volume density of strain energy of the pure shear deformation is

$$W = \frac{1}{2}(\sigma_{x'x'}\varepsilon_{x'x'} + \sigma_{y'y'}\varepsilon_{y'y'}) = \frac{1}{2}(\sigma_{xy}\varepsilon_{xy} + \sigma_{yx}\varepsilon_{yx})$$

$$= \sigma_{xy}\varepsilon_{xy} = \frac{1}{2}\tau\gamma = \frac{1}{2}G\gamma^2 = \frac{1}{2G}\tau^2. \tag{5.13}$$

§6   Torsion of Circular Rods

6.1   Deformation modes

Take a circular rod with radius a and length L. Suppose one of its ends $x = 0$ is fixed. A moment M is exerted at another end $x = L$ to twist it. The angle of rotation is Ω. For the sake of convenience, cylindrical coordinates (r, θ, x) are employed and the unit vectors e_r, e_θ, e_x constitute a right-hand system (Fig. 16).

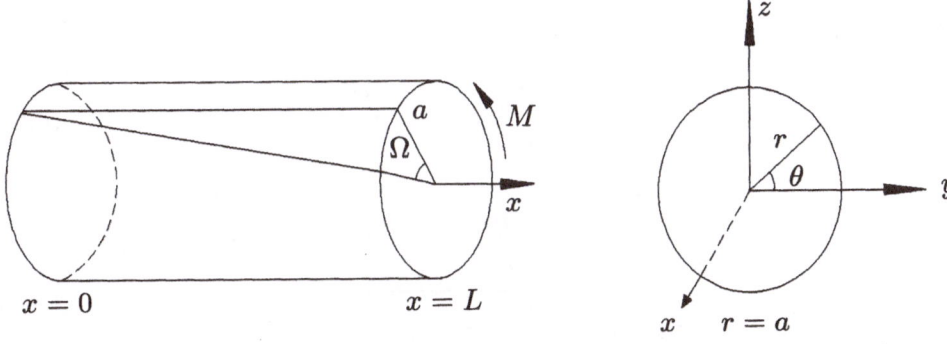

Fig. 16

After the torsional deformation, each cross section normal to the x-axis of the circular rod is twisted through an angle $\omega = \omega(x)$ relativize to the cross section at the fixed end. The angle of twist depends on the axial coordinate x of the cross section, and $\omega(0) = 0, \omega(L) = \Omega$. Since the cross section of the circular rod is uniform and no load has been applied at anywhere except at the two ends, the rate of twist will be uniform, i.e.

$$\xi = \frac{d\omega}{dx} = \frac{\Omega}{L}. \tag{6.1}$$

Inside the circular rod, the mass point located on every cylindrical surface with radius $r =$ constant will still be on that cylindrical surface after the twist. In fact, the mass points on each meridian circle still stay on that meridian circle and are displaced only in the θ-direction. The fibre parallel to the x-axis on every cylindrical surface is then deflected by an angle but still stays on that cylindrical surface. Hence a torsion deformation is a pure shear deformation without change of volume. The apparence of such a shear deformation is that the included angle of the local $(\boldsymbol{e}_\theta, \boldsymbol{e}_x)$ frame of axes is diminished by an angle γ, i.e., the component of strain is $\varepsilon_{\theta x} = \dfrac{1}{2}\gamma$.

Consider a cylindrical surface $r =$ constant between x and $x + dx$. After deformation, the direction of the line element on the meridian circle of that cylindrical surface, i.e. tangential to $\boldsymbol{e}_\theta$, is unchanged. The angle of deflection of the line element parallel to $\boldsymbol{e}_x$ is γ. The angle of rotation of the cross section at $x + dx$ relative to the cross section at x is $d\omega$. It can be seen from Fig. 17 that

$$l = \gamma dx = r d\omega.$$

Hence the angle of shear is

$$\gamma = r\frac{d\omega}{dx} = r\xi = r\frac{\Omega}{L}. \tag{6.2}$$

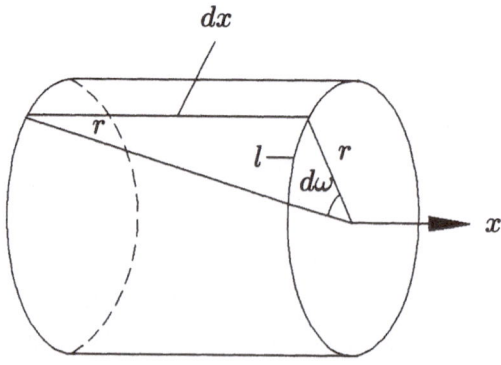

Fig. 17

Hence this is a nonuniform shear strain depending on the radial coordinate r.

The corresponding shearing stress $\sigma_{\theta x} = \tau$ is the θ-directional component of the force per unit area acting on the negative side from the positive side of the cross section. Here the normal direction is the $+x$-axis. Since no load has been applied in the r-and x-directions, the corresponding components are $\sigma_{rx} = \sigma_{xx} = 0$. According to Hooke's law for shear deformations,

$$\tau = G\gamma = G\xi r.$$

The moment of shearing stress per unit area is

$$\tau r = G\gamma r = G\xi r^2.$$

Hence the total moment of the shearing stress (of the positive side acting on the negative side) on the cross section S, i.e. the moment of torsion about the x-axis, is

$$M_x = \iint_S \tau r\, ds = \int_0^a G\xi r^2 \cdot 2\pi r\, dr = GJ\xi, \tag{6.3}$$

where

$$J = 2\pi \int_0^a r^3 dr = \frac{\pi}{2} a^4. \tag{6.4}$$

The relation (6.3) can be regarded as Hooke's law for torsion deformation of circular rods, where the "strain" is represented by the rate of twist $\xi = \dfrac{d\omega}{dx}$, and the "stress" is represented by the moment of torsion M_x. The coefficient GJ is called the torsional rigidity of the cross section, and J is called the geometric torsional rigidity. J depends only on the shape and measure of the geometry of the cross section, but is independent of the material of the circular rod.

In the present case, where the rate of twist is uniform the moment of torsion $M_x = GJ\xi$ is independent of the coordinate $x : M_x = M$. Hooke's law for uniform torsion can then be obtained from (6.2) as

$$\frac{M}{J} = G\frac{\Omega}{L},$$

which is similar in form to Hooke's law for uniform tension (2.1).

6.2   Variational principles and equilibrium equations

The density of strain energy for torsion of a circular rod is

$$W = \frac{1}{2}\tau\gamma = \frac{1}{2}G\xi^2 r^2.$$

Hence the line density of strain energy is

$$\overline{W} = \iint_s W ds = \int_0^a \frac{1}{2} G\xi^2 r^2 \cdot 2\pi r dr = \frac{1}{2} GJ\xi^2. \tag{6.5}$$

Clearly,

$$M = GJ\xi = \frac{\partial \overline{W}}{\partial \xi}, \quad \overline{W} = \frac{1}{2} M\xi, \tag{6.6}$$

and the total strain energy of the circular rod is

$$\int_0^L \overline{W} dx = \int_0^L \frac{1}{2} M\xi dx = \int_0^L \frac{1}{2} GJ\xi^2 dx$$

$$= \int_0^L \frac{1}{2} GJ\left(\frac{d\omega}{dx}\right)^2 dx$$

$$= \frac{1}{2} D(\omega, \omega). \tag{6.7}$$

The corresponding bilinear functional of virtual work is

$$D(\omega, \varphi) = \int_0^L GJ\frac{d\omega}{dx}\frac{d\varphi}{dx} dx. \tag{6.8}$$

Supposing a torsion load per unit length $\mu_x = \mu$ is exerted along the longitudinal length of a circular rod. The work done per unit length through the angle of rotation ω is $\mu\omega$. Hence the total work is $\int_0^L \mu\omega dx$. Furthermore, suppose two concentrated torsion moments m_0 and m_L act at the two ends $x = 0$, L and the work done by the two moments is $m_0\omega_0$ and $m_L\omega_L$, respectively. Hence the potential energy of external work is

$$-F(\omega) = -\left(\int_0^L \mu\omega dx + m_0\omega_0 + m_L\omega_L\right). \tag{6.9}$$

The total potential energy of the circular rod in torsion is then

$$J(\omega) = \frac{1}{2} D(\omega, \omega) - F(\omega)$$

$$= \frac{1}{2} \int_0^L GJ\left(\frac{d\omega}{dx}\right)^2 dx - \int_0^L \mu\omega dx - m_0\omega_0 - m_L\omega_L. \tag{6.10}$$

The mathematical form of (6.10) is identical with that of the rod in tension, and is only different in the mechanical meaning. Hence, by §3, the variational problem

$$J(\omega) = \frac{1}{2} D(\omega, \omega) - F(\omega) = \text{Min} \tag{6.11}$$

is equivalent to

$$D(\omega, \varphi) - F(\varphi) = 0, \text{ for all } \varphi, \tag{6.12}$$

and is also equivalent to

$$
\begin{cases}
-\dfrac{d}{dx}\left(GJ\dfrac{d\omega}{dx}\right) = \mu, & 0 < x < L, & (6.13) \\[2mm]
-GJ\dfrac{d\omega}{dx} = m_0, & x = 0, & (6.14) \\[2mm]
GJ\dfrac{d\omega}{dx} = m_L, & x = L. & (6.15)
\end{cases}
$$

(6.13) is just the equation of torsional equilibrium, while (6.14) and (6.15) are equilibrium conditions of the torsion moments at the two ends and are just the natural boundary conditions. The strainless state is $\omega = \varphi$(const.), i.e., the whole circular rod rotates through a constant angle φ.

The boundary conditions can also be changed into geometrically constrained, i.e. essential, conditions. For example, the natural boundary condition (6.14) can be replaced by the constraint $\omega_0 = \bar{\omega}_0$ and (6.15) can be replaced by the constraint $\omega_L = \bar{\omega}_L$.

6.3 Torsion of circular tubes

Although the above discussion concerns circular rods, all the arguments hold for circular tubes as well, so long as the cross sectional integral is modified into

$$\iint_S \cdots dS = \int_b^a \cdots 2\pi r dr,$$

where a is the external radius of the circular tube and b is the internal radius. Then except that the geometrical torsional rigidity (6.4) should be changed to

$$J = \int_b^a 2\pi r^3 dr = \frac{\pi}{2}(a^4 - b^4), \tag{6.16}$$

the other arguments and formulas are still valid. Let $A = \pi(a^2 - b^2)$ be the area of the cross section of the circular tube and $A_1 = \pi b^2$ be the area of the hole, then

$$J_{\text{tube}} = \frac{\pi}{2}(a^4 - b^4) = (A^2 + 2AA_1)/2\pi.$$

For the circular rod, because $b = 0$, $A_1 = 0$,

$$J_{\text{rod}} = \frac{\pi}{2}a^4 = A^2/2\pi.$$

Hence, under the condition that cross sectional area A is the same, the torsional rigidity

$$J_{\text{rod}} \leq J_{\text{tube}}.$$

It follows that, the circular tube has a greater torsional rigidity than that of the circular rod under the condition that the same amount of material has been used for both cases, and consequently, the circular tube undergoes a smaller deformation under the condition that the external forces are the same.

§7   Bending of Beams

7.1   Deformation modes

A slender rod (beam) yields by bending, which is a special kind of deformation, under the action of a bending moment or transverse load. Suppose the longitudinal direction of the beam is the x-axis. Then bending occurs in the xy-plane, and the coordinate axes x, y, z consist of a right-hand system. The longitudinal fibre element of the beam is a straight line segment before the deformation, and bends into an arc after deformation. Suppose its curvature is $K_z = \pm\dfrac{1}{R}$, where R is the radius of curvature. To be definite, suppose it is downwards convex (Fig. 18). Then $K_z = \dfrac{1}{R}$. The total length of the line element is fundamentally unchanged during the bending deformation. Speaking more exactly, there is a neutral surface within the body of the beam, and on that surface the length of the longitudinal fibre is unchanged. The longitudinal fibres elongate below the neutral surface and shorten above the neutral surface. This leads to a downwards convex condition.

Choose coordinate axes that make the neutral surface correspond to $y = 0$ before deformation, and make the coordinate y indicate the height of departure from the neutral surface. For an arbitrary fibre element with length dx departing distance y from the neutral surface, its length becomes

dx' after deformation. The increment is $du_x = dx' - dx$, where u_x is the x-directional displacement.

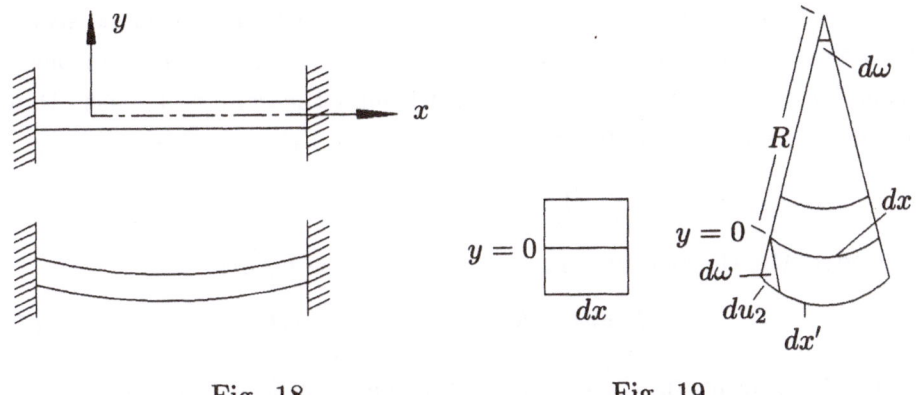

Fig. 18   Fig. 19

A proportional relationship can be obtained from Fig. 19 as

$$d\omega = -\frac{du_x}{y} = \frac{dx}{R},$$

hence

$$\varepsilon_{xx} = \frac{du_x}{dx} = \frac{-y}{R} = -K_z y. \tag{7.1}$$

Since $K_z > 0, \varepsilon_{xx} < 0$ means compression, with $y > 0$, while $\varepsilon_{xx} > 0$ means stretching, with $y < 0$. Hence the bending of beams is a kind of nonuniform pure stretching and compression. The corresponding normal stress based on Hooke's law is

$$\sigma_{xx} = E\varepsilon_{xx} = -EK_z y, \tag{7.2}$$

σ_{xx} stands for the force per unit area acting on the negative side from the positive side of the cross section S normal to the x-axis, and yields a moment of rotation about the z-axis, i.e., the z-directional bending moment

$$M_z = -\iint_S y\sigma_{xx}dS = EK_z \iint_S y^2 dS. \tag{7.3}$$

Since the beam is not subjected to longitudinal loads, the resultant force of the longitudinal stresses over every cross section vanishes, i.e.,

$$\iint_S \sigma_{xx}dS = -EK_z \iint_S ydS = 0.$$

Hence $\bar{y}$ is 0:

$$\bar{y} = \iint_S y\,dS \bigg/ \iint_S dS = 0,$$

where $\bar{y}$ is the y-coordinate of the center of mass of each cross section. It follows that the intersecting line of the cross section and the neutral surface $y = 0$, i.e., the z-axis, passes through the center of mass. Hence the inertial moment of the cross section rotated the z-axis is

$$I_z = \iint_S (y - \bar{y})^2\,dS = \iint_S y^2\,dS.$$

Consequently, the bending moment (7.3) can be expressed as

$$M_z = EI_z K_z = DK_z, \ D = EI_z. \tag{7.4}$$

In the bending problem for beams, one may take the point of view that the "strain" is described by the curvature K_z and the "stress" is described by the bending moment M_z. Hence the relation (7.4) indicates that the stress is proportional to the strain and can be called the Hooke's law for the beam bending problem, where the constant $D = EI_z$ is called the flexural rigidity of the cross section. The larger its magnitude, the more the bending can be resisted. Obviously, for a given kind of material and a given expenditure of material, i.e., a given cross sectional area, the farther the distribution of material in the cross section departs from the z-axis, the larger the magnitude of I_z, i.e., the larger the flexuous rigidity. Examples of this occur in daily life. For the beam of rectangular cross section shown in Fig. 20, the power of resistance to transverse bending is very different from the resistance to vertical bending.

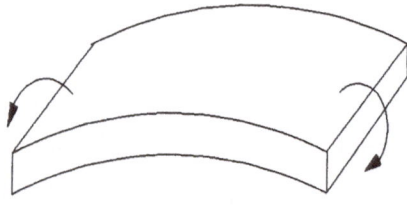 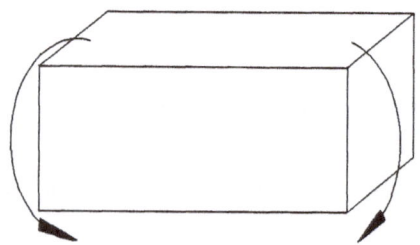

Fig. 20

Now let us try to express the curvature K_z in (7.4) in terms of the transverse (y-directional) displacement. Suppose every point on the neutral surface has a transverse displacement $u_y = u_y(x)$ which is generally called the deflection of the beam. Hence the angle of rotation $\omega_z = \omega_z(x)$ about the z-axis and the curvature $K_z = K_z(x)$ can be expressed as (Fig. 21)

$$\mathrm{tg}\,\omega_z = \frac{du_y}{dx},$$

$$K_z = \frac{d\omega_z}{dS}.$$

For this small deformation, the arc length $ds \approx dx, \mathrm{tg}\,\omega_z \approx \omega_z$, hence the curvature is

$$K_z = \frac{d\omega_z}{dx} = \frac{d^2 u_y}{dx^2}. \qquad (7.5)$$

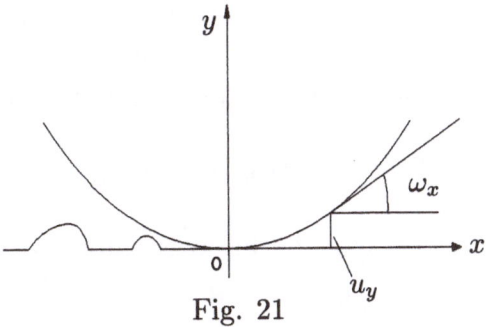

Fig. 21

The bending moment is

$$M_z = EI_z K_z = EI_z \frac{d^2 u_y}{dx^2}. \qquad (7.6)$$

7.2   Variational principles and equilibrium equations

For brevity, we will neglect the subscript which indicates the spatial direction, and denote the derivative $\dfrac{du}{dx}$ as u', i.e.,

$$u_y = u, \ \omega_z = \omega = u', \ K_z = K = u'',$$

$$M_z = M = EI_z K = EI_z u''.$$

Since the bending deformation is variety of nonuniform pure stretching and compression, the strain energy per unit volume of the beam can be obtained from (7.2) as

$$W = \frac{1}{2}\sigma_{xx}\varepsilon_{xx} = \frac{1}{2}E\varepsilon_{xx}^2 = \frac{1}{2}EK^2 y^2.$$

Hence the strain energy per unit length is

$$\overline{W} = \iint_S W\,dS = \frac{1}{2}EK^2 \iint_S y^2\,dS = \frac{1}{2}EI_z K^2$$

$$= \frac{1}{2}EI_z(u'')^2. \qquad (7.7)$$

Clearly,

$$M = EI_z K = \frac{\partial \overline{W}}{\partial K}, \ \overline{W} = \frac{1}{2}MK. \qquad (7.8)$$

Hence the total strain energy of the beam is

$$P(u) = \int_a^b \overline{W} dx = \frac{1}{2} \int_a^b M(u)K(u) dx$$

$$= \frac{1}{2} \int_a^b EI_z(u'')^2 dx = \frac{1}{2} D(u, u), \tag{7.9}$$

the corresponding functional of virtual work is

$$D(u, v) = \int_a^b M(u)K(v) dx = \int_a^b EI_z u'' v'' dx. \tag{7.10}$$

Suppose the beam is subjected to a transverse load. The force per unit length $f_y = f$ does work $\int_a^b f u dx$ in the direction of the transverse displacement u. Furthermore, suppose both the ends of the beam $x = a, b$ are not subjected to geometric constraints but subjected to transverse point forces g_a, g_b and z-directional point bending moments m_a, m_b, respectively. The work done by them in the directions of the transverse displacement u and in the z-directional angle of rotation $\omega = u'$ is

$$g_a u_a = g_b u_b + m_a u'_a + m_b u'_b.$$

Hence the total potential energy of external work of the beam is

$$-F(u) = -\left\{ \int_a^b f u dx + g_a u_a + g_b u_b + m_a u'_a + m_b u'_b \right\}, \tag{7.11}$$

consequently, the total potential energy is

$$J(u) = \frac{1}{2} D(u, u) - F(u)$$

$$= \frac{1}{2} \int_a^b EI_z(u'')^2 dx$$

$$- \int_a^b f u dx - g_a u_a - g_b u_b - m_a u'_a - m_b u'_b. \tag{7.12}$$

By the variational principle in §3, the displacement u of the equilibrium configuration satisfies

$$J(u) = \frac{1}{2} D(u, u) - F(u) = \text{Min} \tag{7.13}$$

which is also equivalent to

$$D(u, v) - F(v) = 0, \text{ for any } v. \tag{7.14}$$

The equilibrium equation and the equilibrium conditions of forces and moments at the two ends can still be derived from (7.14). To this end, we must perform integration by parts twice:

$$D(u,v) = \int_a^b EI_z u'' v'' dx$$

$$= -\int_a^b (EI_z u'')' v' dx + [EI_z u'' v']_a^b$$

$$= \int_a^b (EI_z u'')'' v dx - [(EI_z u'')' v]_a^b + [EI_z u'' v']_a^b.$$

Hence

$$D(u,v) - F(v) = \int_a^b [(EI_z u'')'' - f] v dx$$
$$+ [-(EI_z u'')'_b - g_b] v_b + [(EI_z u'')'_a - g_a] v_a$$
$$+ (EI_z u''_b - m_b) v'_b + (-EI_z u''_a - m_a) v'_a$$
$$= 0. \tag{7.15}$$

Because $v(x)$ is arbitrary, v_a, v_b, v'_a and v'_b, the above expression is equivalent to

$$\begin{cases}
(EI_z u'')'' - f = 0, & a < x < b, & (7.16) \\
(EI_z u'')'_a - g_a = 0, & x = a, & (7.17) \\
-(EI_z u'')_a - m_a = 0, & x = a, & (7.18) \\
-(EI_z u'')'_b - g_b = 0, & x = b, & (7.19) \\
(EI_z u'')_b - m_b = 0, & x = b. & (7.20)
\end{cases}$$

Now let us explain the mechanical meaning of (7.16)–(7.20). Take an arbitrary interval (x_1, x_2) on the beam. The transverse shearing force yielded by the shearing stresses σ_{yx} on each $x-$ cross section of coordinate x is

$$Q(x) = \iint_S \sigma_{yx} dS. \tag{7.21}$$

Moreover, the bending moment yielded by the normal stresses σ_{xx} on the cross section s is

$$M(x) = -\iint_S y\sigma_{xx} dS.$$

Considering the equilibrium of the y-directional forces on the beam element between (x_1, x_2) (See Fig. 22), we obtain

$$Q(x_2) - Q(x_1) + \int_{x_1}^{x_2} f dx = \int_{x_1}^{x_2} \left(\frac{dQ}{dx} + f\right) dx = 0.$$

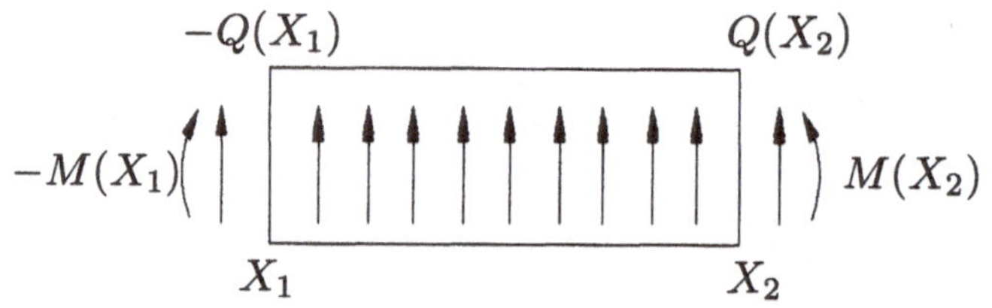

Fig. 22

As the above expression holds for arbitrary x_1, x_2, the equilibrium equation of shearing forces is

$$\frac{-dQ}{dx} = f. \tag{7.22}$$

Considering the equilibrium of the moment rotated the z-axis on the same interval, we obtain

$$(M(x_2) + x_2 Q(x_2)) - (M(x_1) + x_1 Q(x_1)) + \int_{x_1}^{x_2} xf/dx$$

$$= \int_{x_1}^{x_2} \left[\frac{dM}{dx} + \frac{d}{dx}(xQ) + xf\right] dx = 0.$$

The above expression holds for arbitrary x_1, x_2, hence

$$\frac{dM}{dx} + \frac{d}{dx}(xQ) + xf = 0.$$

Noting (7.22), the equilibrium equation of the moment is then obtained

$$\frac{dM}{dx} = -Q. \tag{7.23}$$

Substituting (7.23) into (7.22), the equilibrium equation of the beam is

$$\frac{d^2 M}{dx^2} = f.$$

This is exactly the equation (7.16) derived from the variational principle.

In addition, the shearing force at the left end $x = a$ is

$$Q_a = M'_a = (EI_z u'')'_a, \tag{7.24}$$

and the shearing force at the right end $x = b$ is

$$Q_b = -M_b' = -(EI_z u'')_b'. \tag{7.25}$$

Hence the boundary conditions of the differential equations (7.17) and (7.19) give the equilibrium of the shearing forces at the two ends. At the same time, the bending moment at the left end $x = a$ is

$$M_a = (EI_z u'')_a, \tag{7.26}$$

and the bending moment at the right end $x = b$ is

$$M_b = (EI_z u'')_b. \tag{7.27}$$

Hence the boundary conditions (7.18) and (7.20) denote the equilibrium of the moments at the two ends.

The equilibrium equation (7.16) is a fourth order differential equation of the displacement u. In order to determine unique solution, 4 boundary conditions are required. (7.17)–(7.20) give just the needed conditions. Since these 4 boundary conditions do not appear in equations (7.13) and (7.14) of the variational problem, they are the natural boundary conditions, giving the equilibrium of the forces and moments at the two ends and the mechanical boundary conditions as well. To sum up, in regard to the equilibrium of beams, we have

the angle of rotation: $\omega_z = \dfrac{du_y}{dx}$,

the curvature: $K_z = \dfrac{d\omega_z}{dx} = \dfrac{d^2 u_y}{dx^2}$,

the bending moment: $M_z = EI_z K_z = EI_z \dfrac{d^2 u_y}{dx^2}$,

the shearing force: $Q_y = -\dfrac{dM_z}{dx} = -\dfrac{d}{dx}\left(EI_z \dfrac{d^2 u_y}{dx^2}\right)$,

and the equilibrium equation: $-\dfrac{dQ_y}{dx} = f_y$, i.e.,

$$\frac{d^2 M_z}{dx^2} = \frac{d^2}{dx^2}\left(EI_z \frac{d^2 u_y}{dx^2}\right) = f_y.$$

7.3   Boundary conditions and interface conditions

Like rods in tension, there are three types of boundary conditions for the equilibrium problem of flexible beams.

1. Geometric constraints, dash the first kind of the boundary condition.

1.1   A fixed displacement

For example, the displacement is fixed at the left end point $x = a$

$$x = a : u_a = \bar{u}_a. \tag{7.28}$$

At that time, the displacement u of the equilibrium configuration corresponds to a minimum of the potential energy under the prescribed constraint, while the corresponding virtual displacement v satisfies the annihilating constraint

$$x = a : v_a = 0. \tag{7.29}$$

By the principle of virtual work (7.14), the displacement u of the equilibrium configuration satisfies the following equation,

$$D(u, v) - F(v) = 0, \text{ for any virtual displacement } v. \tag{7.30}$$

Using the boundary constraint (7.29), (7.15) becomes

$$D(u, v) - F(v) = \int_a^b [(EI_z u'')' - f] v dx + [-(EI_z u'')_b - g_b] v_b$$
$$+ [(EI_z u'')_b - m_b] v_b' + [-(EI_z u'')_a - m_a] v_a'$$
$$= 0.$$

Hence (7.30) is equivalent to the equilibrium equation (7.16) and the boundary conditions (7.18)–(7.20). Compared with the free boundary conditions at the two ends (7.17)–(7.20), it lacks the equilibrium condition of shearing forces (7.17) at the left end point $x = a$, which is replaced by the displacement constraint condition (7.28)

1.2   A fixed angle of rotation

If the angle of rotation is fixed at the left end point $x = a$

$$x = a : u_a' = \bar{\omega}_a, \tag{7.31}$$

then the corresponding virtual displacement v satisfies the annihilating constraint $v_a' = 0$. Thus, the equilibrium on bending moments at the end point a (7.18) is replaced by the constraint condition of the angle of rotation (7.31).

The case when the geometric constraint conditions (including the displacement and the angle of rotation) are applied at the right end is similar.

2. A prescribed load—dash the second kind of the boundary condition.

2.1   A prescribed point force (concentrated force) and free displacement.

2.2   Prescribed point moment (concentrated moment) and free angle of rotation.

These two cases have already been discussed above. For example, if a point force g_a is prescribed at the point $x = a$, then we have (7.17) as the equilibrium condition for shearing forces. Similarly, if a point moment m_a is prescribed at the point $x = a$, then we have (7.18) as the equilibrium condition for moments.

3. Elastic support—dash the third kind of the boundary condition.

3.1   The prescribed elastic reaction force and free displacement.

Suppose there is such an elastic support at the point a. Then we have an elastic reaction force proportional to the deviation of the displacements at that point

$$-c_a(u_a - \bar{u}_a) = -c_a u_a + g_a, \ g_a = c_a \bar{u}_a, \ c_a > 0.$$

In this case, a term $\frac{1}{2} c_a u_a^2$ must be added to the strain energy $\frac{1}{2} D(u, u)$, a term $-g_a u_a$ must be added to the potential energy of external work, and equation (7.17), giving the equilibrium condition for shearing forces, must be changed to

$$(EI_z u'')'_a = -Q_a = -c_a u_a + g_a. \tag{7.32}$$

3.2   Prescribed moment of elastic reaction and free angle of rotation.

Suppose that at point a there is an elastic support with a prescribed moment. Then we have an elastic reaction moment proportional to the deviation of the angles of rotation at that point

$$-c_a^*(u_a' - \bar{\omega}_a) = -c_a^* u_a' + m_a, \ m_a = c_a^* \bar{\omega}_a, \ c_a^* > 0.$$

In this case, a term $\frac{1}{2} c_a^* u_a'^2$ must be added to the strain energy $\frac{1}{2} D(u, u)$, and a term $-m_a u_a'$ must be added to the potential energy of the external work. Equation (7.18) giving the equilibrium condition for bending moments must be changed to

$$-(EI_z u'')_a = -M_a = -c_a^* u_a' + m_a. \tag{7.33}$$

At each end point, any two of the above six boundary conditions determine the solution of the equilibrium problem. However, the prescribed displacement and free displacement conditions cannot hold simultaneously; neither can the prescribed angle of rotation and free angle of rotation hold simultaneously. In the variational principle, only the first kind, i.e., the geometric boundary condition must be set out as a condition for determining the solution, which is the essential boundary condition. The second and the third kinds, i.e., the mechanical boundary conditions in the variational problem are the natural boundary conditions and need not be set out explicitly. This point is just the same as in the case of the variational principle for rods in tension.

In practice, the following three combinations of boundary condition, for flexible beams are encountered most often.

1° A fixed end: $u = 0$ (essential), $u' = 0$ (essential), at the end a shown in Fig. 23(a).

2° A pinned end: $u = 0$ (essential), $u'' = 0$ (natural), at the ends a and b shown in Fig. 23(b).

3° A free end: $(EI_z u'')' = 0$ (natural), $u'' = 0$ (natural), at the end b shown in Fig. 23(a).

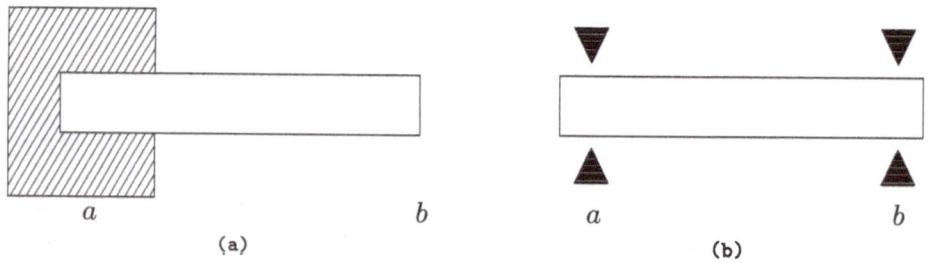

(a)   (b)

Fig. 23

A further case is that of a beam on an elastic foundation, i.e., a beam is elastically coupled with a foundation along its longitudinal direction. Suppose that within a certain interval $a' \leq x \leq b'(a \leq a' < b' \leq b)$ an elastic reaction force per unit length of $-c(u - \bar{u}) = -cu + \bar{f}, \bar{f} = c\bar{u}$, is applied, where c is the constant of elastic coupling. $\bar{f}$ may be absorbed into the load f and the elastic reaction force per unit length may be regarded as $-cu$. In the interval without elastic support, we may let $c = 0$. Then a term of $\dfrac{1}{2}\displaystyle\int_a^b cu^2 dx$ must be added to the total strain energy. The equilibrium equation (7.16) becomes

$$(EI_z u'')'' + cu = f. \tag{7.34}$$

Now we discuss interface conditions.

If the flexuous rigidity EI_z of the beam is discontinuous at $x = p$, e.g., because the cross section or the elastic modulus has a jump (Fig. 24) or there is a concentrated load r_p acting at $x = p$, then a term $-r_p u(p)$ must be added to the corresponding potential energy of the external work. In such a case, integration by parts must be carried out piecewise during the derivation of the equilibrium equation from the variational principle. The interface conditions at $x = p$ can be derived as follows:

$$-[Q]_{p-0}^{p+0} = [(EI_z u'')']_{p-0}^{p+0} = r_p, \tag{7.35}$$

$$[M]_{p-0}^{p+0} = [EI_z u'']_{p-0}^{p+0} = 0. \tag{7.36}$$

(7.36) indicates that the bending moment is always continuous whether there is a concentrated force or not. However, when there is a concentrated load at $x = p$, then (7.35) implies that the shearing force must have a jump at that point so as to produce an effective transverse point force to achieve equilibrium with the concentrated load. The interface conditions (7.35) and (7.36) are the natural boundary conditions, which can be derived automatically from the variational principle. In addition, it is still necessary to impose the continuity conditions of the displacement and of the angle of rotation,

$$[u]_{p-0}^{p+0} = 0, \qquad [u'] =_{p-0}^{p+0}= 0. \tag{7.37}$$

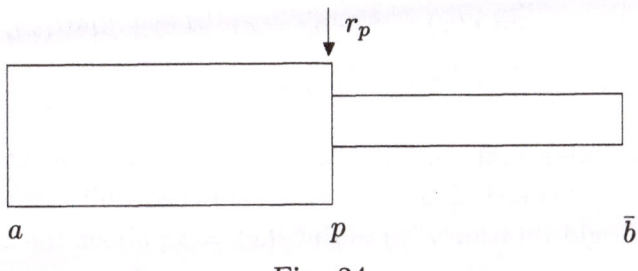

Fig. 24

It should be pointed out that, the energy functional contains only the first derivative of the displacement u' for the beam in tension, while the equilibrium equation containing u'' is of second order. The essential boundary condition containing u is of only one choice, while the natural boundary condition containing u' is also of only one choice. However, the energy functional contains the second derivative of the displacement u'' for the flexible beam, while the equilibrium equation containing u'''' is of fourth order. Thus, the essential boundary conditions fall into two classes, according to whether they contain u or u'. Likewise, the natural boundary conditions divide into those that contain u'' and those that condtain u'''. Furthermore, the boundary conditions may be combined in many ways. Hence, in comparison with the rod in tension, the situation here is quite complex. In the variational principle form, the complex natural boundary conditions (including the interface condition) can be omitted and the order of the derivative of u in the energy functional is lower two ranks than that in the equilibrium equation. Thus the superiority of the mathematical formulation in terms of the variational principle becomes much clearer of the beams bending than in the case of rods in tension during the actual solution process.

7.4   Strainless states

The strain energy of bending is obviously nonnegative, i.e., satisfies

$$D(v,v) = \frac{1}{2} \int_a^b EI_z(v'')^2 dx \geq 0, \text{ for any } v.$$

However it becomes degenerate when the boundary is not subject to any geometric constraints. This is because

$$D(v,v) = 0 \Longleftrightarrow v''(x) = 0$$
$$\Longleftrightarrow v(x) = \alpha + \beta x = av^{(1)}(x) + \beta v^{(2)}(x),$$
$$v^{(1)}(x) = 1, v^{(2)}(x) = x. \tag{7.38}$$

Hence the strainless state has has two degrees of freedom. One of them is a transverse rigid translation $v(x) = av^{(1)}(x) = \alpha$, while another one is an infinitesimal rigid rotation $v(x) = \beta v^{(2)}(x) = \beta x$ about the z-axis. It has already been proved in §3 that a necessary and sufficient condition for the existence of solutions f of the degenerative equilibrium problem (7.13) or

(7.14) is

$$F(v^{(i)}) = 0, \qquad i = 1, 2,$$

i.e.,

$$F(v^{(i)}) = \int_a^b f\,dx + g_a + g_b = 0, \tag{7.39}$$

$$F(v^{(2)}) = \int_a^b xf\,dx + a \cdot g_a + b \cdot g_b + m_a + m_b = 0. \tag{7.40}$$

These two expressions denote the equilibrium of the y-directional forces and the equilibrium of the moments of rotation about the z-axis of the external loads.

Under the prerequisites (7.39) and (7.40), there exists a solution of the equilibrium problem of rods in bending but the solution for the displacement is not unique and may differ by a rigid motion $\alpha + \beta x$. Since the rigid motion contributes nothing to the bending moment and to the shearing force, the solution for the stress is still unique.

We have already seen in §3 that introducing geometric constraints or elastic supports in a problem always reduces the number of degrees of freedom of the strainless state, i.e., to raise the positive definiteness of the strain energy for rods in tension. This is also true for bending beams. For example, if the angle of rotation u_0' is fixed at end point a, then the virtual displacement v satisfies the annihilating constraint condition $v'(a) = 0$. Hence, in the strainless virtual displacement $v(x) = \alpha + \beta x, \beta = v'(a) = 0$, and $v = \alpha$, which means that the strainless state must be a rigid translation. Thus, one degree of freedom is eliminated. Furthermore, if the displacement u_a at point a is fixed, then the virtual displacement v should also satisfy $v(a) = 0$, so that the strainless state is $v \equiv 0$. The strain energy $D(v, v)$ becomes positive definite and there exists a unique solution of the equilibrium problem.

The case of elastic supports is similar to the above. For example, when an elastic reaction force and a moment act simultaneously at the end point a, then the strain energy becomes

$$\frac{1}{2}D(v, v) = \frac{1}{2}\int_a^b EI_z(v'')^2\,dx + \frac{1}{2}c_a v_a^2 + \frac{1}{2}c_a^* v_a'^2,$$

$$c_a > 0, \quad c_a^* > 0.$$

Clearly,

$$D(v, v) = 0 \iff v'' = 0, v_a = 0, v'_a = 0 \iff v = 0.$$

Hence $D(v, v)$ is positive definite.

Furthermore, for the beam on elastic foundations, the strain energy is

$$\frac{1}{2}D(v, v) = \frac{1}{2}\int_a^b EI_z(v'')^2 dx + \frac{1}{2}\int_a^b cv^2 dx.$$

Hence the strainless virtual displacement must be $v \equiv 0$ provided that $c > 0$ over an arbitrary small interval.

Chapter 2

Static Elasticity

In Chapter 1, we learned the basic concepts of elastic displacements, strains, stresses, Hooke's law, strain energy, and variational principles of energy by discussing on some simple but typical elastic deformation modes. Now it is possible and also necessary to further systematize these fundamental concepts and rules and to organize a set of comparatively complete theories so as to solve the more complicated and more difficult problems of elastic structures, that we meet in engineering practice and in scientific experiments. We give a brief introduction to the fundamental theory of linear static elasticity will be given in this chapter.

§1   Displacements and Strains

1.1   Strains

Choose a set of rectangular coordinates in the three dimensional space. For brevity, name the coordinates $x = x_1$, $y = x_2$, and $z = x_3$. Suppose there is an elastic body Ω which becomes Ω' after a deformation. A mass point, whose position vector is $x = (x_1, x_2, x_3)^T$ in the internal body of Ω, moves to $x' = (x_1', x_2', x_3')^T$ after the deformation. Consequently, it undergoes a displacement $u = x' - x$, i.e.,

$$x' = x + u, \quad u = (u_1, u_2, u_3)^T, \quad u_i = u_1(x_1, x_2, x_3).$$

The essential point about elastic deformation is that the relative distances between various points in the body are changed, but any simple displacement of all points is not considered. When a body has made a rigid motion such as translation or rotation, its position has changed, but the relative distances between various points remain unchanged, i.e., there is no deformation. That is the strainless state mentioned several times before.

Take another point $x + dx$ in an infinitesimal neighbourhood of x, which changes to a point $x' + dx'$ in an infinitesimal neighbourhood of x' after deformation, $dx' = dx + du$. Because

$$du_j = \sum_{i=1}^{3} \frac{\partial u_j}{\partial x_i} dx_i, \quad i = 1, 2, 3,$$

the square of the distance between the two points x and $x + dx$

$$ds^2 = \sum_{j=1}^{3} dx_j^2,$$

$$ds'^2 = \sum_{j=1}^{3} dx_j'^2 = \sum_{j=1}^{3} \left(dx_j + \sum_{i=1}^{3} \frac{\partial u_j}{\partial x_i} dx_i \right)^2$$

$$= \sum_{j=1}^{3} dx_j^2 + \sum_{i,j=1}^{3} \left(\frac{\partial u_j}{\partial x_i} + \frac{\partial u_i}{\partial x_j} \right) dx_i dx_j$$

$$+ \sum_{i,j,k=1}^{3} \frac{\partial u_k}{\partial x_i} \frac{\partial u_k}{\partial x_j} dx_i dx_j$$

$$= ds^2 + 2 \sum_{i,j} \varepsilon_{ij} dx_i dx_j,$$

where

$$\varepsilon_{ij} = \frac{1}{2} \left(\frac{\partial u_j}{\partial x_i} + \frac{\partial u_i}{\partial x_j} + \sum_{k=1}^{3} \frac{\partial u_k}{\partial x_i} \frac{\partial u_k}{\partial x_j} \right), \quad i, j = 1, 2, 3.$$

In what follows, we always assume that the deformation is sufficiently small, hence the quadratic terms of the derivatives $\frac{\partial u_i}{\partial x_j}$ on the right hand of the above expressions are infinitesimals of higher order relative to the linear terms. This assumption is a fundamental prerequisite of the linear elasticity. Hence, neglecting the quadratic terms of the derivatives $\frac{\partial u_i}{\partial x_i}$, the strain tensor is defined as

$$\varepsilon_{ij} = \frac{1}{2} \left(\frac{\partial u_j}{\partial x_i} + \frac{\partial u_i}{\partial x_j} \right), \quad i, j = 1, 2, 3, \tag{1.1}$$

and we have

$$ds'^2 = ds^2 + 2 \sum_{i,j} \varepsilon_{ij} dx_i dx_j. \tag{1.2}$$

The strain tensor is obviously symmetric in i and j,

$$\varepsilon_{ij} = \varepsilon_{ji}, \quad i, j = 1, 2, 3. \tag{1.3}$$

Hence, among the nine components of ε_{ij}, there are only six essential components, which are enough to describe the change of relative distances in the elastic body and consequently can completely describe the elastic deformation. When $i = j$,

$$\varepsilon_{ij} = \frac{1}{2}\left(\frac{\partial u_i}{\partial x_i} + \frac{\partial u_i}{\partial x_i}\right) = \frac{\partial u_i}{\partial x_i}, \quad i = 1, 2, 3.$$

These are the normal strains which represent the rate of elongation of the displacements u_i in the x_i-directions. When $i \neq j$, $\varepsilon_{ij} = \varepsilon_{ji}$ are shear strains.

Suppose that the included angle of the local (x_i, x_j) axes, which is originally $\pi/2$, is decreased by an angle γ. The shear strains $\varepsilon_{ij} = \varepsilon_{ji}$ reprensente the half of the angle of shear $\gamma/2$ (see §5, Chapter 1).

1.2 Rotations

The rate of change inspace of a displacement field $u_i(x_1, x_2, x_3)$ is given by nine independent partial derivatives $\dfrac{\partial u_i}{\partial x_j}$, $i, j = 1, 2, 3$. The components ε_{ij} of the strain tensor field of u are symmetric combinations of the derivatives $\dfrac{\partial u_i}{\partial x_j}$. There are only six independent ε_{ij}'s. Hence, although the strain field is enough to describe the deformation, it is still not enough to completely describe the rate of change of the displacement. To describe completely the rate of change of the displacement, we must introduce the complementary antisymmetric combinations of $\dfrac{\partial u_i}{\partial x_j}$ as follows:

$$\omega_{ij} = \frac{1}{2}\left(\frac{\partial u_j}{\partial x_i} - \frac{\partial u_i}{\partial x_j}\right), \quad i, j = 1, 2, 3. \tag{1.4}$$

ω_{ij} is called the rotation tensor. Obviously it is antisymmetric:

$$\omega_{ij} = -\omega_{ji}, \omega_{ii} = 0, i, j = 1, 2, 3, \tag{1.5}$$

and we have

$$\frac{\partial u_j}{\partial x_i} = \varepsilon_{ij} + \omega_{ij}. \tag{1.6}$$

There are only three nonvanishing independent rotational components, e.g., ω_{12}, ω_{23}, and ω_{31}. Generally speaking, we assume

$$\omega_1 = \omega_{23}, \omega_2 = \omega_{31}, \omega_3 = \omega_{12}. \tag{1.7}$$

In the tensor analysis, for an arbitrary vector field $u = (u_1, u_2, u_3)^T$, we can define another vector field

$$(\text{rot } \boldsymbol{u})_1 = \frac{\partial u_3}{\partial x_2} - \frac{\partial u_2}{\partial x_3},$$

$$(\text{rot } \boldsymbol{u})_2 = \frac{\partial u_1}{\partial x_3} - \frac{\partial u_3}{\partial x_1},$$

$$(\text{rot } \boldsymbol{u})_3 = \frac{\partial u_2}{\partial x_1} - \frac{\partial u_1}{\partial x_2},$$

which is called the curl field rot $\boldsymbol{u}$.

Hence $\omega = (\omega_1, \omega_2, \omega_3)^T$, regarded as a vector field, is half the curl of the displacement field $u(x_1, x_2, x_3)$:

$$\omega = \frac{1}{2} \text{ rot } \boldsymbol{u}. \tag{1.8}$$

The rotation tensor ω_{ij} contributes nothing to the elastic deformation, i.e., makes no contribution to the change in relative distances. However, it has its own geometrical meaning. Let ox_1x_2 be the local (x_1, x_2) coordinate system which becomes $o'x_1'x_2'$ after deformation (projected again on the x_1x_2-plane). We may translate origins of coordinates o and o' to make them coincide. Suppose OM is the bisector before the deformation and OM' is the bisector after the deformation. The angle through which the bisector is rotated about the x_3-axis (Fig. 25) is

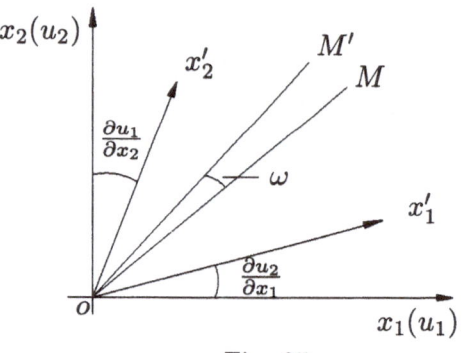

Fig. 25

$$\angle MOM' \approx \frac{1}{2}\left(\frac{\pi}{2} - \frac{\partial u_1}{\partial x_2} - \frac{\partial u_2}{\partial x_1}\right) + \frac{\partial u_2}{\partial x_1} - \frac{\pi}{4}$$

$$= \frac{1}{2}\left(\frac{\partial u_2}{\partial x_1} - \frac{\partial u_1}{\partial x_2}\right) = \omega_{12} = \omega_3.$$

Hence, $\omega_3 = \omega_{12}$ can be regarded as the angle of rotation of the bisector of the local (x_1, x_2) coordinate axes about the x_3-axis. $\omega_3 > 0$ denotes a positive rotation about the x_3-axis according to the right hand rule, while $\omega_3 < 0$ denotes a negative rotation. The meaning of ω_1 and ω_2 is similar to that of ω_3. Hence ω_{ij} is called the rotation tensor and ω_i are generally called the (infinitesimal) angles of rotation about the x_i-axes.

The strains ε_{ij} and the rotations ω_{ij} are independent of each other in themselves, but their derivatives are related. In fact,

$$\frac{\partial \omega_{ij}}{\partial x_k} = \frac{\partial}{\partial x_k}\frac{1}{2}\Big(\frac{\partial u_j}{\partial x_i} - \frac{\partial u_i}{\partial x_j}\Big) = \frac{1}{2}\Big(\frac{\partial^2 u_j}{\partial x_k \partial x_i} - \frac{\partial^2 u_i}{\partial x_k \partial x_j}\Big)$$

$$= \frac{1}{2}\Big(\frac{\partial^2 u_j}{\partial x_i \partial x_k} + \frac{\partial^2 u_k}{\partial x_i \partial x_j} - \frac{\partial^2 u_k}{\partial x_j \partial x_i} - \frac{\partial^2 u_i}{\partial x_j \partial x_k}\Big).$$

Hence

$$\frac{\partial \omega_{ij}}{\partial x_k} = \frac{\partial \varepsilon_{jk}}{\partial x_i} - \frac{\partial \varepsilon_{ik}}{\partial x_j}, \quad ij = 12, 23, 31, k = 1, 2, 3. \tag{1.9}$$

1.3   Strainless states and infinitesimal rigid displacements

We say that an elastic body is in a strainless state if the following equality holds everywhere.

$$\varepsilon_{ij} = \frac{1}{2}\Big(\frac{\partial u_j}{\partial x_i} + \frac{\partial u_i}{\partial x_j}\Big) = 0, \quad i, j = 1, 2, 3. \tag{1.10}$$

That means that under an infinitesmal deformation there is no elastic deformation, i.e., $ds'^2 = ds^2$. The consistency of this definition of the strainless state with the definition in terms of the vanishing of the strain energy given in Chapter 1, will be explained in §5 of this Chapter.

There are two kinds of infinitesimal rigid displacements, as they are called. One is the translation, i.e., $\boldsymbol{u} \equiv \boldsymbol{a} = (a_1, a_2, a_3)^T$ is a constant vector and another is the infinitesimal rotation, namely, $\boldsymbol{u} \equiv \boldsymbol{b} \wedge \boldsymbol{x}$, where $\boldsymbol{b} = (b_1, b_2, b_3)^T$ is a constant vector, where $\boldsymbol{b} \wedge \boldsymbol{x}$ stands for the cross product. Hence

$$u_l = b_2 x_3 - b_3 x_2, u_2 = b_3 x_1 - b_1 x_3, u_3 = b_1 x_2 - b_2 x_1.$$

The general form of infinitesimal rigid displacements is the superimposition of the above two forms: $\boldsymbol{u} = \boldsymbol{a} + \boldsymbol{b} \wedge \boldsymbol{x}$, i.e.,

$$\begin{cases} u_1 = a_1 + b_2 x_3 - b_3 x_2, \\ u_2 = a_2 + b_3 x_1 - b_1 x_3, \\ u_3 = a_3 + b_1 x_2 - b_2 x_1. \end{cases} \tag{1.11}$$

Since

$$\varepsilon_{11} = \frac{\partial u_1}{\partial x_1} = \frac{\partial}{\partial x_1}(a_1 + b_2 x_3 - b_3 x_2) \equiv 0,$$

$$\varepsilon_{12} = \frac{1}{2}\Big(\frac{\partial u_2}{\partial x_1} + \frac{\partial u_1}{\partial x_2}\Big) = \frac{1}{2}(b_3 - b_3) \equiv 0,$$

etc., $\varepsilon_{ij} \equiv 0$. It shows that any infinitesimal rigid displacement must be strainless.

Conversely, a strainless displacement field can certainly be expressed in the form of infinitesimal rigid displacements as (1.11). From (1.9), we can see that $\dfrac{\partial \omega_{ij}}{\partial x_k} = 0$, i.e., ω_{ij} are constants, if $\varepsilon_{ij} = 0$. Denote

$$b_1 = \omega_{23}, \quad b_2 = \omega_{31}, \quad b_3 = \omega_{12}.$$

On the other hand, since $\varepsilon_{ij} = 0$,

$$\varepsilon_{11} = \frac{\partial u_1}{\partial x_1} = 0, \quad \varepsilon_{22} = \frac{\partial u_2}{\partial x_2} = 0, \quad \varepsilon_{33} = \frac{\partial u_3}{\partial x_3} = 0$$

and

$$\frac{\partial u_i}{\partial x_j} = -\frac{\partial u_j}{\partial x_i}, \quad i \neq j,$$

it follows that

$$b_1 = \omega_{23} = \frac{1}{2}\left(\frac{\partial u_3}{\partial x_2} - \frac{\partial u_2}{\partial x_3}\right) = \frac{\partial u_3}{\partial x_2} = -\frac{\partial u_2}{\partial x_3},$$

$$b_2 = \omega_{31} = \frac{1}{2}\left(\frac{\partial u_1}{\partial x_3} - \frac{\partial u_3}{\partial x_1}\right) = \frac{\partial u_1}{\partial x_3} = -\frac{\partial u_3}{\partial x_1},$$

$$b_3 = \omega_{12} = \frac{1}{2}\left(\frac{\partial u_2}{\partial x_1} - \frac{\partial u_1}{\partial x_2}\right) = \frac{\partial u_2}{\partial x_1} = -\frac{\partial u_1}{\partial x_2}.$$

Thus, we obtain th following three systems of the first order differential equations related to u_1, u_2, and u_3.

$$\left\{\begin{array}{l} \dfrac{\partial u_1}{\partial x_1} = 0, \\[2mm] \dfrac{\partial u_1}{\partial x_2} = -b_3, \\[2mm] \dfrac{\partial u_1}{\partial x_3} = b_2, \end{array}\right. \quad \left\{\begin{array}{l} \dfrac{\partial u_2}{\partial x_1} = b_3, \\[2mm] \dfrac{\partial u_2}{\partial x_2} = 0, \\[2mm] \dfrac{\partial u_2}{\partial x_3} = -b_1, \end{array}\right. \quad \left\{\begin{array}{l} \dfrac{\partial u_3}{\partial x_1} = -b_2, \\[2mm] \dfrac{\partial u_3}{\partial x_2} = b_1, \\[2mm] \dfrac{\partial u_3}{\partial x_3} = 0. \end{array}\right.$$

From above three systems of the equations, we get

$$u_1 = a_1 - b_3 x_2 + b_2 x_3,$$

$$u_2 = a_2 + b_3 x_1 - b_1 x_3,$$

$$u_3 = a_3 - b_2 x_1 + b_1 x_2.$$

These are just the rigid displacements (1.11).

§2   Transformation of Principal Axes

and Principal Strains

2.1   Rotation of coordinate axes

For a special kind of deformation, i.e., the multi-directional stretching and compression discussed in §4 of Chapter 1, the strain has only normal components but no shear components and the strain tensor is expressed in diagonal form as

$$[\varepsilon_{ij}] = \begin{bmatrix} \varepsilon_1 & 0 & 0 \\ 0 & \varepsilon_2 & 0 \\ 0 & 0 & \varepsilon_3 \end{bmatrix}.$$

The elastic rule is quite simple at that case.

For the case of pure shear in §5 of Chapter 1, the strain tensor has nondiagonal form

$$[\varepsilon_{ij}] = \begin{bmatrix} 0 & \tau/2G & 0 \\ \tau/2G & 0 & 0 \\ 0 & 0 & 0 \end{bmatrix}.$$

However, rotating the coordinate axes x_1 and x_2 about the x_3-axis through 45°, the strain tensor becomes diagonal in the new system of coordinates x_1', x_2', and x_3':

$$[\varepsilon_{ij}'] = \begin{bmatrix} \tau/2G & 0 & 0 \\ 0 & -\tau/2G & 0 \\ 0 & 0 & 0 \end{bmatrix}.$$

The pureshear case is transformed into the case of multi-dimensional stretching and compression. Thus we derive the elastic rule for pure shear. This is a special example, but it has general meaning. All arbitrary strain tensors ε_{ij} can be transformed into diagonal form, at least locally, by a suitable rotation of coordinate axes. In order to illustrate this problem, let us examine first how the components of the displacement vector and of the strain tensor vary when the coordinate axes are rotated.

2.2   Strain tensors in the transformed and original systems of coordinates

Rotate the coordinate axes x_1, x_2, and x_3 to a new orthogonal system of coordinates x_1', x_2', and x_3'. The transformation relations between the two systems of coordinates are

$$\begin{cases} x_1' = x_1 \cos(x_1', x_1) + x_2 \cos(x_1', x_2) + x_3 \cos(x_1', x_3), \\ x_2' = x_1 \cos(x_2', x_1) + x_2 \cos(x_2', x_2) + x_3 \cos(x_2', x_3), \\ x_3' = x_1 \cos(x_3', x_1) + x_2 \cos(x_3', x_2) + x_3 \cos(x_3', x_3), \end{cases} \qquad (2.1)$$

where $\cos(x_i', x_1)$, $\cos(x_i', x_2)$, and $\cos(x_i', x_3)$ stand for the three direction cosines of the new x_i'-axes with respect to the original coordinate axes x_1, x_2, and x_3, i.e., the three components of the unit vectors e_i' on the x_i'-axes,

$$e_i' = (\cos(x_i', x_1), \cos(x_i', x_2), \cos(x_i', x_3))^T. \qquad (2.2)$$

The transformation of coordinates (2.1) can be written in matrix form:

$$x' = Ax, \quad A = [a_{ij}], \quad a_{ij} = \cos(x_i', x_j). \qquad (2.3)$$

Because the new coordinate axes x_1', x_2', and x_3' are orthogonal, the three unit vectors $e_i'(i = 1, 2, 3)$ are also orthogonal. Hence the coefficient matrix A of (2.3) is an orthogonal matrix, which satisfies $AA^T = I$. So we have $A^{-1} = A^T$. Then the following reciprocally inverse transformation relations hold between the new and the original coordinates

$$x_i' = \sum_{i=1}^{3} a_{ij} x_i, \quad x_i = \sum_{j=1}^{3} a_{ji} x_j'. \qquad (2.4)$$

As stated in §1, the displacement vector $u = x^* - x'$ is expressed as the difference of two position vectors before and after deformation of the elastic body. Hence, the relations between the new and the original coordinates u_i and u_i' of the displacement vector during the rotation of coordinate axes are the same as (2.4),

$$u_i' = \sum_{j=1}^{3} a_{ij} u_j, \quad u_i = \sum_{j=1}^{3} a_{ji} u_j'. \qquad (2.5)$$

The components of the strain tensors derived from the same displacement field in the new and in the original systems of coordinates are

$$\varepsilon_{ij} = \frac{1}{2} \left(\frac{\partial u_j}{\partial x_i} + \frac{\partial u_i}{\partial x_j} \right), \varepsilon_{ij}' = \frac{1}{2} \left(\frac{\partial u_j'}{\partial x_i'} + \frac{\partial u_i'}{\partial x_j'} \right),$$

respectively. Since

$$\frac{\partial u'_j}{\partial x'_i} = \sum_{k=1}^{3} \frac{\partial u'_j}{\partial x_k} \frac{\partial x_k}{\partial x'_i} = \sum_{k=1}^{3} a_{ik} \frac{\partial}{\partial x_k} \left(\sum_{l=1}^{3} a_{jl} u_l \right)$$

$$= \sum_{k,l=1}^{3} a_{ik} a_{jl} \frac{\partial u_l}{\partial x_k},$$

exchanging the subscripts i and j and then k and l in the above expression, we can obtain

$$\frac{\partial u'_i}{\partial x'_j} = \sum_{k,l=1}^{3} a_{jk} a_{il} \frac{\partial u_l}{\partial x_k} = \sum_{k,l=1}^{3} a_{jl} a_{ik} \frac{\partial u_k}{\partial x_l}.$$

Hence the relations between the components of the strain in the new and in the original systems of coordinates are

$$\varepsilon'_{ij} = \sum_{k,l=1}^{3} a_{ik} \varepsilon_{kl} a_{jl}. \tag{2.6}$$

This can be written in the matrix form. Let

$$\varepsilon = [\varepsilon_{ij}], \quad \varepsilon' = [\varepsilon'_{ij}],$$

Both are symmetric matrices. Then

$$\varepsilon' = A\varepsilon A^T. \tag{2.7}$$

Because A is an orthogonal matrix, the following inverse relation holds.

$$\varepsilon = A^{-1}\varepsilon' A = A^T \varepsilon' A, i.e., \varepsilon_{ij} = \sum_{k,l} a_{ki} \varepsilon'_{kl} a_{lj}. \tag{2.8}$$

2.3　Principal axes and principal strains

For symmetric matrices in linear algebra the following fundamental theorem holds: for an arbitrary real symmetric matrix ε, there always exists an appropriate real orthogonal matrix A such that

$$A\varepsilon A^T = \begin{bmatrix} \varepsilon_1 & 0 & 0 \\ 0 & \varepsilon_2 & 0 \\ 0 & 0 & \varepsilon_3 \end{bmatrix} \tag{2.9}$$

is diagonal. The above expression can be written as

$$\varepsilon A^T = A^T \begin{bmatrix} \varepsilon_1 & 0 & 0 \\ 0 & \varepsilon_2 & 0 \\ 0 & 0 & \varepsilon_3 \end{bmatrix},$$

i.e.,

$$
\begin{bmatrix} \varepsilon_{11} & \varepsilon_{12} & \varepsilon_{13} \\ \varepsilon_{21} & \varepsilon_{22} & \varepsilon_{23} \\ \varepsilon_{31} & \varepsilon_{32} & \varepsilon_{33} \end{bmatrix} \begin{bmatrix} a_{11} & a_{21} & a_{31} \\ a_{12} & a_{22} & a_{32} \\ a_{13} & a_{23} & a_{33} \end{bmatrix} = \begin{bmatrix} \varepsilon_1 a_{11} & \varepsilon_2 a_{21} & \varepsilon_3 a_{31} \\ \varepsilon_1 a_{12} & \varepsilon_2 a_{22} & \varepsilon_3 a_{32} \\ \varepsilon_1 a_{13} & \varepsilon_2 a_{23} & \varepsilon_3 a_{33} \end{bmatrix}.
$$

This means that ε_i are the eigenvalues of the matrix ε and the three column vectors of A^T are just the eigenvectors of the matrix ε.

Since the transformation relation between the strain tensors in the new and in the original systems of coordinates is

$$
\varepsilon' = A\varepsilon A^T,
$$

if the three eigenvectors of the strain tensor ε are taken as the new coordinate axes e_i', then according to the fundamental theorem, the strain tensor ε' is reduced to diagonal form in this new system of coordinates. Its diagonal entries ε_1, ε_2, and ε_3 are just the eigenvalues of the matrix ε, which are called the principal strains, while the eigenvectors e_1', e_2', and e_3' are called the principal axes, and the corresponding stresses are called the principal stresses. In other words, the strain tensor transformed to the principal axes is diagonal, i.e., it appears as multi-directional stretching and compression without shear component.

Suppose the characteristic polynomial of matrix ε is

$$
\varphi(\lambda) = \begin{vmatrix} \varepsilon_{11} - \lambda & \varepsilon_{12} & \varepsilon_{13} \\ \varepsilon_{21} & \varepsilon_{22} - \lambda & \varepsilon_{23} \\ \varepsilon_{31} & \varepsilon_{32} & \varepsilon_{33} - \lambda \end{vmatrix} = -\lambda^3 + I_1 \lambda^2 - I_2 \lambda + I_3. \quad (2.10)
$$

It is not difficult to show that

$$
I_1 = \varepsilon_{11} + \varepsilon_{22} + \varepsilon_{33},
$$

$$
I_2 = \begin{vmatrix} \varepsilon_{11} & \varepsilon_{12} \\ \varepsilon_{21} & \varepsilon_{22} \end{vmatrix} + \begin{vmatrix} \varepsilon_{22} & \varepsilon_{23} \\ \varepsilon_{32} & \varepsilon_{33} \end{vmatrix} + \begin{vmatrix} \varepsilon_{11} & \varepsilon_{13} \\ \varepsilon_{31} & \varepsilon_{33} \end{vmatrix},
$$

$$
I_3 = \begin{vmatrix} \varepsilon_{11} & \varepsilon_{12} & \varepsilon_{13} \\ \varepsilon_{21} & \varepsilon_{22} & \varepsilon_{23} \\ \varepsilon_{31} & \varepsilon_{32} & \varepsilon_{33} \end{vmatrix}.
$$

On the other hand, because ε_1, ε_2, and ε_3 are the three eigenvalues of the matrix ε,

$$
\varphi(\lambda) = (\varepsilon_1 - \lambda)(\varepsilon_2 - \lambda)(\varepsilon_3 - \lambda). \quad (2.11)
$$

Comparing (2.10) with (2.11), we obtain

$$I_1 = \varepsilon_1 + \varepsilon_2 + \varepsilon_3, \tag{2.12}$$

$$I_2 = \varepsilon_1\varepsilon_2 + \varepsilon_2\varepsilon_3 + \varepsilon_3\varepsilon_1, \tag{2.13}$$

$$I_3 = \varepsilon_1\varepsilon_2\varepsilon_3. \tag{2.14}$$

Although the components of the strain tensor are changed after an arbitrary rotation of coordinate axes (whether they are principal axes or not), we know from (2.12)–(2.14) that, the values of I_1, I_2 and I_3 are unchanged, which are the three invariants of the strain tensor. Among them, an invariant

$$I_1 = \varepsilon_{11} + \varepsilon_{22} + \varepsilon_{33} = \varepsilon_1 + \varepsilon_2 + \varepsilon_3 \tag{2.15}$$

is often used.

§3   Stresses

3.1   Components of stress

When an elastic body is deformed under the action of loads, the body is in a strained state. All parts of the body are affected by the stresses. The stress state at each point (x_1, x_2, x_3) in the body is given by the nine components of stresses σ_{ij}, $i, j = 1, 2, 3$.

Taking a cross section normal to the $+x_i$-axis at a point (x_1, x_2, x_3), the force per unit area exerted on the "negative side" from the positive side" is σ_j. The three components of the force are denoted by σ_{1j}, σ_{2j}, and σ_{3j}, respectively. σ_{jj} is the positive stress, i.e., the normal stress $\sigma_{jj} > 0$ indicates a tensile force, while $\sigma_{ij} < 0$ indicates a compressive force. When $i \neq j$, σ_{ij} are the shearing stresses tangential to the cross section. $\sigma = [\sigma_{ij}]$ is called the stress tensor, and the dimension of which is [force/area].

The importance of the components of stress σ_{ij} lies on the unit cross section, in which the normal direction at any point (x_1, x_2, x_3) is $\vec{n} = (n_1, n_2, n_3)^T$ and $n_1^2 + n_2^2 + n_3^2 = 1$, the three components σ_{1n}, σ_{2n} and σ_{3n} of the stress σ_n exerted on the "in side" from the "out side" can be expressed in terms of σ_{ij} and n_j, i.e.,

$$\sigma_{in} = \sum_{j=1}^{3} \sigma_{ij} n_j, \quad i = 1, 2, 3. \tag{3.1}$$

In fact, we need only take an infinitesimal tetrahedron V with height h (sufficiently small) and base area S (Fig. 26) at that point. The outward normal directions of the four surfaces S, S_1, S_2, and S_3 of V are $(n_1, n_2, n_3)^T$, $(-1, 0, 0)^T$, $(0, -1, 0)^T$, and $(0, 0, -1)^T$, respectively. Note that the volumne of the tetrahedron $V = \frac{1}{3}hS$ and $S_i = n_i S$. Suppose the density of the load acting on the interior of the elastic body is $f = (f_1, f_2, f_3)^T$. According to the equilibrium of body force and surface force in the three directions, we obtain

$$f_i V + \sigma_{in} S - \sigma_{i1} S_1 - \sigma_{i2} S_2 - \Sigma_{i3} S_3 = 0, \quad i = 1, 2, 3,$$

i.e.,

$$f_i \cdot \frac{1}{3}hS + \sigma_{in} S - \Big(\sum_{j=1}^{3} \sigma_{ij} n_j \Big) S = 0.$$

Because the outward normal direction n_j to the base of the tetrahedron S remains unchanged when the height h varys, (3.1) can be obtained immediately if we let $h \to 0$ in the above expression.

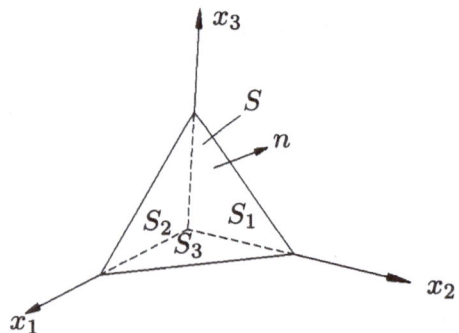

Fig. 26

It follows that, the projection σ_n of the stress on the cross section on any direction

$$m = (m_1, m_2, m_3)^T, \quad m_1^2 + m_2^2 + m_3^2 = 1$$

is

$$\sigma_{mn} = \sum_{i=1}^{3} \sigma_{in} m_i = \sum_{i,j=1}^{3} \sigma_{ij} m_i n_j, \tag{3.2}$$

where n is the normal direction of the cross section.

Especially, the normal component of the cross section stress σ_n is

$$\sigma_{mn} = \sum_{i,j=1}^{3} \sigma_{ij} n_i n_j. \tag{3.3}$$

3.2   Equilibrium equations

Suppose an elastic body Ω is subjected to a volume load with density $\boldsymbol{f} = (f_1, f_2, f_3)^T$. Consider an arbitrary subvolume $V \subset \Omega$, whose outward normal direction is $\boldsymbol{n} = (n_1, n_2, n_3)^T$. Through the boundary ∂V, an external force acts on the subvolume V. The x_j-directional component of the surface force exerted on the area element dS is

$$\sigma_{in} dS = \sum_{j=1}^{3} \sigma_{ij} n_j dS. \tag{3.4}$$

The x_i-directional component of the body force acting on a volume element dV within V is $f_i dV$. Assume that the x_i-directional components of force are in equilibrium. We obtain

$$\iint_{\partial V} \sum_{j=1}^{3} \sigma_{ij} n_j dS + \iiint_V f_i dV = 0, \quad i = 1, 2, 3, \tag{3.5}$$

$$\text{for any } V \subset \Omega.$$

This is the equilibrium equation in integral form.

According to Gauss's formula

$$\iint_{\partial V} \sum_{j=1}^{3} \sigma_{ij} n_i dS = \iiint_V \sum_{j=1}^{3} \frac{\partial \sigma_{ij}}{\partial x_j} dV,$$

we have

$$\iiint_V \left(\sum_{j=1}^{3} \frac{\partial \sigma_{ij}}{\partial x_j} + f_i \right) dV = 0, \quad i = 1, 2, 3,$$

$$\text{for any } V \subset \Omega.$$

Shrinking V to a point $(x_1, x_2, x_3) \in \Omega$, then we obtain an equation of elastic equilibrium in the differential form

$$\Omega : -\sum_{j=1}^{3} \frac{\partial \sigma_{ij}}{\partial x_j} = f_i, \quad i = 1, 2, 3. \tag{3.6}$$

Now let us consider the equilibrium of moments. Recall the definition of the moment: suppose a force $\boldsymbol{f} = (f_1, f_2, f_3)^T$ acts at the point $\boldsymbol{x} =$

$(x_1, x_2, x_3)^T$, the cross product $\boldsymbol{M} = \boldsymbol{x} \wedge \boldsymbol{f}$ is then the moment. The components of the three coordinate axes are, respectively,

$$M_1 = x_2 f_3 - x_2 f_2, \quad M_2 = x_3 f_1 - x_1 f_3, \quad M_3 = x_1 f_2 - x_2 f_1.$$

Then, the x_1-directional component of the stress moment of the area element $dS \in \partial V$ is

$$\left(x_2 \sum_j \sigma_{3j} n_j - x_3 \sum_j \sigma_{2j} n_j \right) \cdot dS,$$

while the x_I-directional component of the moment of the body forces on the volume element $dV \in V$ is $(x_2 f_3 - x_3 f_2) \cdot dV$. Hence we obtain the following equilibrium equation for moments in integral form:

x_1–axial :

$$\iint_{\partial V} \sum_{j=1}^{3} (x_2 \sigma_{3j} - x_3 \sigma_{2j}) n_j dS + \iiint_V (x_2 f_3 - x_3 f_2) dV = 0,$$

$$\text{for any } V \subset \Omega.$$

Similar equations in the x_2-and x_3-axial directions can be obtained by cyclic permutation of subscripts $1 \to 2 \to 3 \to 1$. Using Gauss's formula again and shrinking V to a point, we obtain the equilibrium equation of moments in differential form as

$x_1 -$ axial :

$$\sum_{j=1}^{3} \frac{\partial}{\partial x_j} (x_2 \sigma_{3j} - x_3 \sigma_{2j}) + x_2 f_3 - x_3 f_2$$

$$= x_2 \left(\sum_{j=1}^{3} \frac{\partial \sigma_{3j}}{\partial x_j} + f_3 \right) - x_3 \left(\sum_{j=1}^{3} \frac{\partial \sigma_{2j}}{\partial x_j} + f_2 \right) + (\sigma_{32} - \sigma_{23})$$

$$= 0.$$

Based on the equilibrium equation of forces (3.6), we have

$$\sigma_{32} - \sigma_{23} = 0, \quad i.e., \quad \sigma_{32} = \sigma_{23}.$$

Similarly,

$$\sigma_{13} = \sigma_{31}, \quad \sigma_{21} = \sigma_{12}.$$

Thus, the symmetry of the stress tensor can be derived from the equilibrium of forces and moments

$$\sigma_{ij} = \sigma_{ji}, \quad i, j = 1, 2, 3. \tag{3.7}$$

3.3 Principal stresses

We proved the symmetry of the stress tensor above. Let us now examine the relationship between the new and the original stress tensors under the rotation of coordinate axes.

Suppose the coordinate axes x_1, x_2 and x_3 are rotated to the new coordinate axes x_1', x_2', and x_3', and the unit vector on the x_i'-axis is

$$\boldsymbol{e}_j' = (a_{i1}, a_{i2}, a_{i3})^T$$

(see (2.2) of Chapter 2). In the new system of coordinates, the stress component σ_{ij}' is the x'-directional component of the stress on the cross section, the normal direction of which is $+x_i'$-axis. Then, in (3.2), let

$$\boldsymbol{n} = \boldsymbol{e}_j' = (a_{j1}, a_{j2}, a_{j3})^T, \quad \boldsymbol{m} = \boldsymbol{e}_i' = (a_{i1}, a_{i2}, a_{i3})^T,$$

we obtain

$$\sigma_{ij}' = \sum_{k,l=1}^{3} a_{ik}\sigma_{kl}a_{jl}. \tag{3.8}$$

Written inmatrix form

$$\sigma' = A\sigma A^T. \tag{3.9}$$

This equation gives the relation between stress tensors in the new and in the original systems of coordinates. The relation is the same as the transformation relation (2.7) between the new and the original strain tensors. Therefore, by a suitable rotation of coordinate axes, the stress tensor can be expressed in diagonal form, i.e., there are only direct stresses but no shearing stresses:

$$A\sigma A^T = \sigma' = \begin{bmatrix} \sigma_1 & 0 & 0 \\ 0 & \sigma_2 & 0 \\ 0 & 0 & \sigma \end{bmatrix}, A^T = A^{-1}, \tag{3.10}$$

where σ_1, σ_2, and σ_3 are called the principal stresses, which are also the eigenvalues of the symmetric matrix σ, while the vectors of principal axes $\boldsymbol{e}_1'$, $\boldsymbol{e}_2'$, and $\boldsymbol{e}_3'$ are then the eigenvectors of σ.

Note that the transformation matrix $A = [a_{ij}]$, which transforms the stress tensor into the principal axes, is dependent on $\sigma = [\sigma_{ij}]$. Generally, it is different from the transformation matrix which transforms the strain tensor into the principal axes.

§4   Hooke's Law and Strain Energy

4.1   Hooke's law

We saw in §2 that, through a suitable rotation of coordinate axes

$$x' = Ax, \tag{4.1}$$

the strain tensor $\varepsilon = (\varepsilon_{ij})$ can be transformed into diagonal form

$$A\varepsilon A^T = \varepsilon' = \begin{bmatrix} \varepsilon_1 & 0 & 0 \\ 0 & \varepsilon_2 & 0 \\ 0 & 0 & \varepsilon_3 \end{bmatrix}, \tag{4.2}$$

which becomes a pure three dimensional stretching and compression without shear strain.

Suppose that the medium is isotropic. We saw in §4 of Chapter 1 that, the corresponding stress tensor $\sigma' = [\sigma'_{ij}]$ also has no shear components, so σ' is diagonal, too:

$$\sigma' = \begin{bmatrix} \sigma_1 & 0 & 0 \\ 0 & \sigma_2 & 0 \\ 0 & 0 & \sigma_3 \end{bmatrix}.$$

Hooke's law is

$$\sigma_i = \frac{E}{1+\nu}\varepsilon_i + \frac{E\nu}{(1+\nu)(1-2\nu)}\sum_{k=1}^{3}\varepsilon_k, \quad i = 1, 2, 3. \tag{4.3}$$

Written in matrix form,

$$\sigma' = \frac{E}{1+\nu}\varepsilon' + \frac{E\nu}{(1+\nu)(1-2\nu)}\left(\sum_{k=1}^{3}\varepsilon_k\right)l. \tag{4.4}$$

Since the stress tensor becomes diagonal in the new system of coordinates, the new coordinate axes x'_i determined by (4.1) are the principal axes of both the strain tensor and the stress tensor, i.e.,

$$\sigma' = A\sigma A^T. \tag{4.5}$$

This shows that, for isotropic bodies, the stress tensor and the strain tensor can be simultaneously diagonalized, i.e., into the principal axes through a suitable rotation of coordinate axes.

Premultiplying both sides of (4.4) by A^T and postmultiplying them by A, since

$$\sigma = A^T\sigma'A, \quad \varepsilon = A^T\varepsilon'A, \quad A^TA = l,$$

we obtain

$$\sigma = \frac{E}{1+\nu}\varepsilon + \frac{E\nu}{(1+\nu)(1-2\nu)}\Big(\sum_{k=1}^{3}\varepsilon_k\Big)I.$$

Moreover, according to (2.15)

$$\sum_{k=1}^{3}\varepsilon_k = \sum_{k=1}^{3}\varepsilon_{kk},$$

so that

$$\sigma = \frac{E}{1+\nu}\varepsilon + \frac{E\nu}{(1+\nu)(1-2\nu)}\Big(\sum_{k=1}^{3}\varepsilon_{kk}\Big)I. \tag{4.6}$$

Written componentwise, this becomes

$$\sigma_{ij} = \frac{E}{1+\nu}\varepsilon_{ij} + \frac{E\nu}{(1+\nu)(1-2\nu)}\Big(\sum_{k=1}^{3}\varepsilon_{kk}\Big)\delta_{ij}, \tag{4.7}$$

$$i,j = 1,2,3,$$

where

$$\delta_{ij} = \begin{cases} 1, & i = j, \\ 0, & i \neq j. \end{cases}$$

(4.6) or (4.7) is just a general form of Hooke's law

Letting $i = j$ in (4.7) and superimposing the three expressions of $i = 1,2,3$, we obtain

$$\sum_{k=1}^{3}\sigma_{kk} = \frac{E}{1+\nu}\sum_{k=1}^{3}\varepsilon_{kk} + \frac{3E\nu}{(1+\nu)(1-2\nu)}\sum_{k=1}^{3}\varepsilon_{kk}$$

$$= \frac{E}{1-2\nu}\sum_{k=1}^{3}\varepsilon_{kk}.$$

Moreover, substituting the above expresssion into (4.6), we obtain the inverse relation

$$\varepsilon = \frac{1+\nu}{E}\sigma - \frac{\nu}{E}\Big(\sum_{k=1}^{3}\sigma_{kk}\Big)I, \tag{4.8}$$

or

$$\varepsilon_{ij} = \frac{1+\nu}{E}\sigma_{ij} - \frac{\nu}{E}\Big(\sum_{k=1}^{3}\sigma_{kk}\Big)\delta_{ij}, i,j = 1,2,3. \tag{4.9}$$

4.2   Strain energy

For an isotropic elastic body, the strain energy per unit volume with the principal axes is

$$W = \frac{1}{2} \sum_{i=1}^{3} \sigma_i \varepsilon_i. \tag{4.10}$$

Note that

$$\sigma'\varepsilon' = \begin{bmatrix} \sigma_1 & 0 & 0 \\ 0 & \sigma_2 & 0 \\ 0 & 0 & \sigma_3 \end{bmatrix} \begin{bmatrix} \varepsilon_1 & 0 & 0 \\ 0 & \varepsilon_2 & 0 \\ 0 & 0 & \varepsilon_3 \end{bmatrix} = \begin{bmatrix} \sigma_1\varepsilon_1 & 0 & 0 \\ 0 & \sigma_2\varepsilon_2 & 0 \\ 0 & 0 & \sigma_3\varepsilon_3 \end{bmatrix},$$

and

$$\sigma'\varepsilon' = A\sigma A^T A\varepsilon A^T = A\sigma\varepsilon A^T,$$

hence

$$\sum_{i=1}^{3} (\sigma'\varepsilon')_{ii} = \sum_{i=1}^{3} (\sigma\varepsilon)_{ii},$$

consequently,

$$W = \frac{1}{2} \sum_{i=1}^{3} \sigma_i \varepsilon_i = \frac{1}{2} \sum_{i=1}^{3} (\sigma'\varepsilon')_{ii} = \frac{1}{2} \sum_{i=1}^{3} (\sigma\varepsilon)_{ii} = \frac{1}{2} \sum_{i,j=1}^{3} \sigma_{ij}\varepsilon_{ij},$$

i.e.,

$$W = \frac{1}{2} \sum_{i,j=1}^{3} \sigma_{ij}\varepsilon_{ij}. \tag{4.11}$$

Substituting the linear stress-strain relation (4.7), obtained from Hooke's law, into the expression above, we obtain the volume density of strain energy

$$W = \frac{1}{2} \left\{ \frac{E}{1+\nu} \sum_{i,j=1}^{3} \varepsilon_{ij}^2 + \frac{E\nu}{(1+\nu)(1-2\nu)} \left(\sum_{k=1}^{3} \varepsilon_{kk} \right)^2 \right\}. \tag{4.12}$$

It is easy to verify that

$$\sigma_{ij} = \frac{\partial W}{\partial \varepsilon_{ij}}, i,j = 1,2,3,$$

so that

$$W = \frac{1}{2} \sum_{i,j=1}^{3} \frac{\partial W}{\partial \varepsilon_{ij}} \varepsilon_{ij}. \tag{4.13}$$

We see from (4.12) that W is quadratic and homogeneous in ε_{ij}. Therefore (4.13) is a special case of an n-th homogeneous expression as in Euler's theorem in algebra, with $n = 2$.

Since $E > 0$, $0 < \nu < \dfrac{1}{2}$,

$$\frac{E\nu}{(1+\nu)(1-2\nu)} > 0, \quad \frac{E}{(1+\nu)(1-2\nu)}\Big(\sum_{k=1}^{3}\varepsilon_{kk}\Big)^2 \geq 0,$$

hence

$$W = \frac{1}{2} = \Big\{ \frac{E}{1+\nu}\sum_{i,j=1}^{3}\varepsilon_{ij}^2 + \frac{E\nu}{(1+\nu)(1-2\nu)}\Big(\sum_{k=1}^{3}\varepsilon_{kk}\Big)^2 \Big\}$$

$$\geq \frac{1}{2}\frac{E}{1+\nu}\sum_{i,j=1}^{3}\varepsilon_{ij}^2.$$

That is to say, W is positive definite as a quadratic form in ε_{ij}, i.e., we always have $W \geq 0$ for any ε_{ij}, and

$$W = 0 \Longrightarrow \varepsilon_{ij} = 0, \quad i,j = 1,2,3.$$

This reflects the facts that the strain energy is positive definite in mechanics, and that solutions of the elastic equilibrium equation for stress exist and are unique.

In the literature of elasticity theory, two fundamental elastic moduli of isotropic bodies, called Lame's moduli,

$$\lambda = \frac{E\nu}{(1+\nu)(1-2\nu)}, \quad \mu = \frac{E}{2(1+\nu)} \tag{4.14}$$

are often used to replace the two conventional fundamental moduli E and ν in engineering. E and ν can be expressed in terms of λ and μ by

$$E = \frac{\mu(3\lambda + 2\mu)}{\lambda + \mu}, \quad \nu = \frac{\lambda}{2(\mu + \lambda)}. \tag{4.15}$$

Furthermore, the bulk modulus K and the modulus of the shear rigidity G can also be expressed in terms of λ and μ as

$$K = \frac{E}{3(1-2\nu)} = \lambda + \frac{2}{3}\mu, \tag{4.16}$$

$$G = \frac{E}{2(1+\nu)} = \mu. \tag{4.17}$$

Hooke's law and the strain energy for isotropic bodies can then be expressed in terms of λ and μ by

$$\sigma_{ij} = 2\mu\varepsilon_{ij} + \lambda\Big(\sum_{k=1}^{3}\varepsilon_{kk}\Big)\delta_{ij}, \quad i,j = 1,2,3, \tag{4.18}$$

$$W = \frac{1}{2}\Big\{2\mu\sum_{i,j=1}^{3}\varepsilon_{ij}^{2} + \lambda\Big(\sum_{k=1}^{3}\varepsilon_{kk}\Big)^{2}\Big\}. \tag{4.19}$$

§5   Variational Principles and Elastic Equilibrium

5.1   Variational principles

Suppose an elastic body Ω deformed elastically under the action of loads, including body and surface loads. That is to say, every point within the body Ω undergoes a displacement $u_i = u_i(x_1, x_2, x_3)$. This displacement induces a strain field

$$\varepsilon_{ij}(\boldsymbol{u}) = \frac{1}{2}\Big(\frac{\partial u_j}{\partial x_i} + \frac{\partial u_i}{\partial x_j}\Big)$$

and a stress field

$$\sigma_{ij}(\boldsymbol{u}) = \frac{E}{1+\nu}\varepsilon_{ij}(\boldsymbol{u}) + \frac{E\nu}{(1+\nu)(1-2\nu)}\Big(\sum_{k=1}^{3}\varepsilon_{kk}(\boldsymbol{u})\Big)\delta_{ij}.$$

A strain energy $p(\boldsymbol{u})$ is always stored when the elastic body undergoes a displacement u_i whether it is in equilibrium or not. According to (4.11) and (4.12),

$$p(\boldsymbol{u}) = \iiint_{\Omega} W\,dV = \frac{1}{2}\iiint_{\Omega}\sum_{i,j=1}^{3}\sigma_{ij}(\boldsymbol{u})\varepsilon_{ij}(\boldsymbol{u})\,dV$$

$$= \frac{1}{2}\iiint_{\Omega}\Big\{\frac{E}{1+\nu}\sum_{i,j=1}^{3}\varepsilon_{ij}^{2}(\boldsymbol{u}) + \frac{E\nu}{(1+\nu)(1-2\nu)}$$

$$\times\Big(\sum_{k=1}^{3}\varepsilon_{kk}(\boldsymbol{u})\Big)^{2}\Big\}dV = \frac{1}{2}D(\boldsymbol{u},\boldsymbol{u}). \tag{5.1}$$

From (5.1) we get

$$D(\boldsymbol{u},\boldsymbol{u}) = 0 \iff \varepsilon_{ij}(\boldsymbol{u}) \equiv 0, \quad i,j = 1,2,3.$$

This shows that the two definitions about the strainless states mentioned in Section 1.3 of the Chapter 2 are equivalent. Let

$$D(\boldsymbol{u}, \boldsymbol{v}) = \iiint_\Omega \sum_{i,j=1}^3 \sigma_{ij}(\boldsymbol{u})\varepsilon_{ij}(\boldsymbol{v})dV$$

$$= \iiint_\Omega \Big\{ \frac{E}{1+\nu} \sum_{i,j=1}^3 \varepsilon_{ij}(\boldsymbol{u})\varepsilon_{ij}(\boldsymbol{v})$$

$$+ \frac{E\nu}{(1+\nu)(1-2\nu)} \sum_{k=1}^3 \varepsilon_{kk}(\boldsymbol{u}) \cdot \sum_{k=1}^3 \varepsilon_{kk}(\boldsymbol{v}) \Big\}dV. \qquad (5.2)$$

Clearly, $D(\boldsymbol{u}, \boldsymbol{u})$ is a nonnegative quadratic functional of $\boldsymbol{u}$, while $D(\boldsymbol{u}, \boldsymbol{v})$ is a bilinear symmetric functional of $\boldsymbol{u}$ and $\boldsymbol{v}$, which is just the functional of virtual work.

Suppose the density of the body load acting on the interior of the body Ω is $\boldsymbol{f} = (f_1, f_2, f_3)^T$, and the density of the surface load acting on the boundary $\partial\Omega$ is $\boldsymbol{g} = (g_1, g_2, g_3)^T$. For the time being, we will not consider a displacement constraint or elastic support temporarily. Then the potential energy of external work produced by the body load f and the surface load g is

$$-F(\boldsymbol{u}) = -\Big\{ \iiint_\Omega \sum_{i=1}^3 f_i u_i dV + \iint_{\partial\Omega} \sum_{i=1}^3 g_i u_i dS \Big\} \qquad (5.3)$$

and the total potential energy of the system is

$$J(\boldsymbol{u}) = \frac{1}{2}D(\boldsymbol{u}, \boldsymbol{u}) - F(\boldsymbol{u}). \qquad (5.4)$$

By the variational principles described in Chapter 1, the following two problems are equivalent:

$$\text{problem 1,} \quad J(\boldsymbol{u}) = \frac{1}{2}D(\boldsymbol{u}, \boldsymbol{u}) - F(\boldsymbol{u}) = \text{Min}, \qquad (5.5)$$

$$\text{problem 2,} \quad D(\boldsymbol{u}, \boldsymbol{v}) - F(\boldsymbol{v}) = 0, \quad \text{for any } \boldsymbol{b}. \qquad (5.6)$$

For the strainless state v,

$$D(\boldsymbol{v}, \boldsymbol{v}) = 0 \longleftrightarrow \varepsilon_{ij}(\boldsymbol{v}) \equiv 0 \longleftrightarrow \boldsymbol{v} = \boldsymbol{a} + \boldsymbol{b} \wedge \boldsymbol{x} \qquad (5.7)$$

which is also an infinitesimal rigid displacement determined by (1.11) in §1 of Chapter 2. It has six degrees of freedom, and the corresponding six linearly independent rigid displacements are

$$
\boldsymbol{v}^{(1)} = \begin{bmatrix} 1 \\ 0 \\ 0 \end{bmatrix}, \quad
\boldsymbol{v}^{(2)} = \begin{bmatrix} 0 \\ 1 \\ 0 \end{bmatrix}, \quad
\boldsymbol{v}^{(3)} = \begin{bmatrix} 0 \\ 0 \\ 1 \end{bmatrix},
$$

$$
\boldsymbol{v}^{(4)} = \begin{bmatrix} 0 \\ -x_3 \\ x_2 \end{bmatrix}, \quad
\boldsymbol{v}^{(5)} = \begin{bmatrix} x_3 \\ 0 \\ -x_1 \end{bmatrix}, \quad
\boldsymbol{v}^{(6)} = \begin{bmatrix} -x_2 \\ x_1 \\ 0 \end{bmatrix}. \tag{5.8}
$$

Hence, the necessary and sufficient conditions for the existence of solution of the variational problems 1 or 2 are

$$
F(\boldsymbol{v}^{(i)}) = 0, \quad i = 1, 2, \cdots, 6.
$$

Substituting (5.8) into (5.3), we get

$$
\iiint_{\Omega} f_i dV + \iint_{\partial\Omega} g_i dS = 0, \quad i = 1, 2, 3, \tag{5.9}
$$

$$
\begin{cases}
\iiint_{\Omega} (x_2 f_3 - x_3 f_2) dV + \iint_{\partial\Omega} (x_2 g_3 - x_3 g_2) dS = 0, \\[2mm]
\iiint_{\Omega} (x_3 f_1 - x_1 f_3) dV + \iint_{\partial\Omega} (x_3 g_1 - x_1 g_3) dS = 0, \\[2mm]
\iiint_{\Omega} (x_1 f_2 - x_2 f_1) dV + \iint_{\partial\Omega} (x_1 g_2 - x_2 g_1) dS = 0.
\end{cases} \tag{5.10}
$$

These are equilibrium conditions of forces and moments of the external loads. Under the prerequisites of (5.9)–(5.10), there exists a solution of the variational problem. However the displacement solution is not unique and can differ by an infinitesimal rigid displacement, while the stress solution is unique. Because the rigid displacement has six degrees of freedom, six integral conditions such as

$$
\iiint_{\Omega} u_i dV = 0, \quad i = 1, 2, 3, \tag{5.11}
$$

$$
\iiint_{\Omega} (x_2 u_3 - x_3 u_2) dV = \iiint_{\Omega} (x_3 u_1 - x_1 u_3) dV
$$

$$
= \iiint_{\Omega} (x_1 u_2 - x_2 u_1) dV = 0 \tag{5.12}
$$

have to be added in order to eliminate these six degrees of freedom. Under these conditions, the displacement solution is unique.

5.2   Equilibrium equations

From the above variational problem 2, we can derive an equivalent problem:

Problem 3:

$$
\begin{cases}
\Omega : -\displaystyle\sum_{j=1}^{3} \frac{\partial \sigma_{ij}(\boldsymbol{u})}{\partial x_j} = f_i, \quad i = 1, 2, 3, & (5.13) \\[4mm]
\partial\Omega : \displaystyle\sum_{j=1}^{3} \sigma_{ij}(\boldsymbol{u})n_j = g_i, \quad i = 1, 2, 3. & (5.14)
\end{cases}
$$

The proof is as follows. By Gauss's integral formula and the symmetry of the stress tensor

$$
\begin{aligned}
D(\boldsymbol{u}, \boldsymbol{v}) &= \iiint_\Omega \sum_{i,j=1}^{3} \sigma_{ij}(\boldsymbol{u})\varepsilon_{ij}(\boldsymbol{v})dV \\
&= \iiint_\Omega \sum_{i,j=1}^{3} \sigma_{ij}(\boldsymbol{u}) \cdot \frac{1}{2}\left(\frac{\partial v_j}{\partial x_i} + \frac{\partial v_i}{\partial x_j}\right)dV \\
&= \iiint_\Omega \sum_{i,j=1}^{3} \sigma_{ij}(\boldsymbol{u})\frac{\partial v_i}{\partial x_j}dV \\
&= \iint_{\partial\Omega} \sum_{i,j=1}^{3} \sigma_{ij}(\boldsymbol{u})n_j v_i ds - \iiint_\Omega \sum_{i,j=1}^{3} \frac{\partial \sigma_{ij}(\boldsymbol{u})}{\partial x_j}v_i dV.
\end{aligned}
$$

Hence

$$
\begin{aligned}
D(\boldsymbol{u}, \boldsymbol{v}) - F(\boldsymbol{v}) &= -\iiint_\Omega \sum_{i=1}^{3}\left[\sum_{j=1}^{3} \frac{\partial \sigma_{ij}(\boldsymbol{u})}{\partial x_j} + f_i\right]v_i dV \\
&= \iint_{\partial\Omega} \sum_{i=1}^{3}\left[\sum_{j=1}^{3} \sigma_{ij}(\boldsymbol{u})n_j - g_i\right]v_i dS = 0. \quad (5.15)
\end{aligned}
$$

Since v_i is arbitrary, the terms included in the brackets of the above two integrals must vanish:

$$
\begin{cases}
\displaystyle\sum_{j=1}^{3} \frac{\partial \sigma_{ij}(\boldsymbol{u})}{\partial x_j} + f_i = 0, \quad i = 1, 2, 3, \ \text{in } \Omega, \\[4mm]
\displaystyle\sum_{j=1}^{3} \sigma_{ij}(\boldsymbol{u})n_j - g_i = 0, \quad i = 1, 2, 3, \ \text{on } \partial\Omega.
\end{cases}
$$

These are the two systems of equations of Problem 3. We know from Section 3.2 of Chapter 2 that these two systems of equations are the equilibrium equations inside the elastic body Ω and on its boundary $\partial\Omega$. Thus, we have derived the equilibrium equations directly from the variational problem.

Conversely, if we want to derive Problem 2 from Problem 3, we need only note that, provided $\boldsymbol{u}$ is a solution of equations (5.13) and (5.14), (5.15) still holds by the Gauss integral formula. Hence for arbitrary $\boldsymbol{v}$, we always have

$$D(\boldsymbol{u}, \boldsymbol{v}) - F(\boldsymbol{v}) = 0.$$

Thus, we have proved the equivalence of the three mathematical formulations of the elastic equilibrium problem. Since

$$
\begin{aligned}
\sigma_{ij}(\boldsymbol{u}) &= \frac{E}{1+\nu}\varepsilon_{ij}(\boldsymbol{u}) + \frac{E\nu}{(1+\nu)(1-2\nu)}\Big(\sum_{k=1}^{3}\varepsilon_{kk}(\boldsymbol{u})\Big)\delta_{ij} \\
&= \frac{E}{2(1+\nu)}\Big(\frac{\partial u_j}{\partial x_i} + \frac{\partial u_i}{\partial x_j}\Big) \\
&\quad + \frac{E\nu}{(1+\nu)(1-2\nu)}\Big(\sum_{k=1}^{3}\frac{\partial u_k}{\partial x_k}\Big)\delta_{ij},
\end{aligned}
\tag{5.16}
$$

the equilibrium equations (5.13) are three second order elliptic partial differential equations for the three unknown functions u_1, u_2, and u_3. In order to determine a unique solution, we must prescribe three boundary conditions on the boundary $\partial\Omega$ in addition to the boundary conditions are given by (5.14).

5.3   Boundary conditions and interface conditions

In §5.2, the equilibrium equations inside the elastic body and the boundary conditions, which are derived from the variational principle, are obtained based on the essumption that the whole boundary $\partial\Omega$ is subjected to surface loads. Now we will discuss the general case, i.e., according to the different boundary mechanisms, the whole boundary $\partial\Omega$ is divided into several parts,

$$\partial\Omega = \Gamma_1 + \Gamma_2 + \Gamma_3,\tag{5.17}$$

with a different boundary condition on each part.

1. The first kind of the boundary condition: fixed supports. Geometric constraints conditions are prescribed on Γ_1, e.g., the fixed displacements are known,

$$\Gamma_1 : u_i = \bar{u}_i, \quad i = 1, 2, 3. \tag{5.18}$$

2.   The second kind of the boundary condition: loading supports. Surface loads with density $g = (g_1, g_2, g_3)^T$ are exerted on Γ_2.

3.   The third kind of the boundary conditon: elastic supports. The boundary is coupled with an external elastic body on Γ_3. A given unit area is subjected to an elastic reaction force, "which is proportional" to the local displacement $-\sum_{j=1}^{3} c_{ij} u_j + g_i$, $i = 1, 2, 3$. Here, $C = (c_{ij})$ is a symmetric positive definite matrix indicating the elastic coefficients of the support. This is a direct generalization of the ease of a one dimensional elastic support stated in Chapter 1, where c is a positive coefficient, while in the present case C is a positive definite coefficient matrix. Obviously, the boundary condition on Γ_2 can be considered as a special case of that on Γ_3, corresponding to $c_{ij} = 0$.

In addition to these three kinds of boundaries, we further assume that there is an interface Γ' in the interior of Ω, that the elastic media on its two sides consist of different materials, and that the elastic modulus has the discontinuity or jump on the interface Γ'. Define a positive normal $\nu = (\nu_1, \nu_2, \nu_3)^T$. We call the side in the positive direction Ω^+, and the side in the negative direction $\Omega^- : \Omega = \Omega^+ + \Omega^-$, and

$$E^+ \neq E^-, \quad \nu^+ \neq \nu^-. \tag{5.19}$$

Note that the displacement u on the interface Γ' is still continuous, i.e.,

$$u^+ = u^-, \quad \text{on } \Gamma'. \tag{5.20}$$

Let us now write down the strain energy and the potential energy of external work of this system. The strain energy is

$$\frac{1}{2} D(u, u) = \frac{1}{2} \iiint_{\Omega^+ + \Omega^-} \sum_{i,j=1}^{3} \sigma_{ij}(u) \varepsilon_{ij}(u) dV$$

$$+ \frac{1}{2} \iint_{\Gamma_3} \sum_{i,j=1}^{3} c_{ij} u_i u_j dS. \tag{5.21}$$

The potential energy of external work is

$$-F(u) = -\left\{ \iiint_{\Omega^+ + \Omega^-} \sum_{i=1}^{3} f_i u_i dV + \iint_{\Gamma_2 + \Gamma_3} \sum_{i=1}^{3} g_i u_i dS \right\}. \tag{5.22}$$

The functional of virtual work is

$$D(\boldsymbol{u}, \boldsymbol{v}) = \iiint_{\Omega^+ + \omega^-} \sum_{i,j=1}^{3} \sigma_{ij}(\boldsymbol{u})\varepsilon_{ij}(\boldsymbol{v})dV$$

$$+ \iint_{\Gamma_3} \sum_{i,j=1}^{3} c_{ij}u_i v_j dS. \tag{5.23}$$

Since the displacement $\boldsymbol{u}$ on Γ_1 has geometric constraints (5.18), the virtual displacement $\boldsymbol{v}$ in the variational problem must satisfy the corresponding annihilating constraints

$$\Gamma_1 : v_i = 0, \quad i = 1, 2, 3. \tag{5.24}$$

Moreover, since the displacement $\boldsymbol{u}$ is continuous on the interface Γ', the virtual displacement $\boldsymbol{v}$ must also be continuous on Γ', i.e.,

$$\Gamma' : v_i^+ = v_i^-, \quad i = 1, 2, 3. \tag{5.25}$$

Hence the complete statement of variational problem 2 is

$$\begin{cases} D(\boldsymbol{u}, \boldsymbol{v}) - F(\boldsymbol{v}) = 0 \text{ for any virtual displacement } \boldsymbol{v} \\[2mm] \text{satisfying constraints (5.24)–(5.25)}, \\[2mm] \Gamma_1 : u_i = \bar{u}_i, \\[2mm] \Gamma' : u_i^+ = u_i^-, \quad i = 1, 2, 3. \end{cases} \tag{5.26}$$

Due to the discontinuity of the medium, we use Gauss's integral formula piecewise on Ω^+ and Ω^-. Noting the continuity of v_i on the interface Γ' and the symmetry of σ_{ij} and c_{ij}, we obtain

$$D(\boldsymbol{u}, \boldsymbol{v}) = -\iiint_{\Omega^+ + \Omega^-} \sum_{i,j=1}^{3} \frac{\partial \sigma_{ij}(\boldsymbol{u})}{\partial x_j} v_i dV + \iint_{\Gamma_3} \sum_{i,j=1}^{3} c_{ij}u_j v_i dS$$

$$+ \iint_{\partial\Omega} \sum_{i,j=1}^{3} \sigma_{ij}(\boldsymbol{u})n_j v_i dS + \int_{\Gamma'} \sum_{i=1}^{3} \Big[\sum_{j=1}^{3} \sigma_{ij}^-(\boldsymbol{u})v_j - \sum_{j=1}^{3} \sigma_{ij}^+(\boldsymbol{u})v_j\Big]v_i dS,$$

where σ_{ij}^+ and σ_{ij}^- denote the stress components in Ω^+ and Ω^-, respectirely. The material constants in Ω^+ are E^+, v^+ and the material constants in Ω^- are E^-, v^-. Further, using the constraint conditions on v_i (5.24), we have

$$D(\boldsymbol{u}, \boldsymbol{v}) - F(\boldsymbol{v}) = - \iiint_{\Omega^+ + \Omega^-} \sum_{i=1}^{3} \left[\sum_{j=1}^{3} \frac{\partial \sigma_{ij}(\boldsymbol{u})}{\partial x_j} + f_i \right] v_i dV$$

$$+ \iint_{\Gamma_3} \sum_{i=1}^{3} \left[\sum_{j=1}^{3} \sigma_{ij}(\boldsymbol{u}) n_j - g_i \right] v_i dS$$

$$+ \iint_{\Gamma_3} \sum_{i=1}^{3} \left[\sum_{j=1}^{3} \sigma_{ij}(\boldsymbol{u}) n_j + \sum_{j=1}^{3} C_{ij} u_j - g_i \right] v_i dS$$

$$+ \iint_{\Gamma'} \sum_{i=1}^{3} \left[\sum_{j=1}^{3} \sigma_{ij}^-(\boldsymbol{u}) u v j - \sum_{j=1}^{3} \sigma_{ij}^+(\boldsymbol{u}) v_j \right] v_i dS = 0.$$

Because the v_i are arbitrary, the terms included in the brackets under the above integral signs must vanish. Hence

$$\begin{cases} -\sum_{j=1}^{3} \dfrac{\partial \sigma_{ij}(\boldsymbol{u})}{\partial x_j} = f_i, \quad i = 1,2,3, \quad \Omega^+ + \Omega^-; & (5.27) \\[2mm] u_i = \bar{u}_i, \quad i = 1,2,3, \quad \Gamma_1; & (5.28) \\[2mm] \sum_{j=1}^{3} \sigma_{ij}(\boldsymbol{u}) n_j = g_i, \quad i = 1,2,3, \quad \Gamma_2; & (5.29) \\[2mm] \sum_{j=1}^{3} \sigma_{ij}(\boldsymbol{u}) n_j + \sum_{j=1}^{3} c_{ij} u_j = g_i, \quad i = 1,2,3, \quad \Gamma_3; & (5.30) \\[2mm] u_i^+ = u_i^-, \quad i = 1,2,3, \quad \Gamma'; & (5.31) \\[2mm] \sum_{j=1}^{3} \sigma_{ij}^-(\boldsymbol{u}) \nu_i - \sum_{j=1}^{3} \sigma_{ij}^+(\boldsymbol{u}) \nu_j = 0, \quad i = 1,2,3. & (5.32) \end{cases}$$

Substituting the formula for the stress $\sigma_{ij}(\boldsymbol{u})$ (5.16), in which $\sigma_{ij}(\boldsymbol{u})$ is represented by the first partial derivatives of the displacements $\boldsymbol{u}_i$ into the above expressions, we obtain the following results:

Corresponding to (5.27), within $\Omega^+ + \Omega^-$, we get a system of second order partial differential equation satisfied by the displacement u_i of the equilibrium configuration;

Corresponding to (5.28), on Γ_1 we get geometric boundary conditions involving only the displacement;

Corresponding to (5.29) and (5.30), on Γ_2 and Γ_3, we get mechanical boundary conditions, which involve the first order of partial derivatives of the displacements;

There are also two interface conditions on the interface Γ' within the medium Ω: One is the displacement continuity condition corresponding to (5.31). The other is the stress equilibrium condition involving the first order partial derivatives, which corresponds to (5.32). This stress equilibrium condition is also a kind of complete "geometrical" mathematical formulation of the elastic equilibrium problem, expressed as a boundary value problem of the system of second order partial differential equations in three unknown functions u_1, u_2, and u_3.

5.4   Strainless states

We mentioned in Chapter 1 that adding geometric constraints or elastic supports always reduces the number of degrees of freedom of the strainless state and raises the positive definiteness of the strain energy. This is just as it was for the equilibrium problem of the three-dimensional elastic body.

As stated in 5.3, if the whole boundary $\partial\Omega$ has the second kind of the boundary condition loading support, then the strain energy is degenerat, and the strainless state is an infinitesimal rigid displacement, which is six degrees of freedom. If there is a geometric constraint on a subset Γ_1 of $\partial\Omega$, then the original stress equilibrium condition is replaced by the geometric constraint condition on Γ_1, and the strainless virtual displacement v must satisfy the annihilating constraint condition on Γ_1 in addition to being an infinitesimal rigid displacement. Hence, we must have $v \equiv 0$ provided there are three points in Γ_1 not on the same straight line. Thus, under such a constraint, the strain energy becomes positive definite, and the equilibrium problem has a unique displacement solution. If Γ_1 is a straight line, or only one or two components of the displacement are prescribed. Accordingly, the strain energy may still not be positive definite, but the degrees of freedom of the strainless state are decreased.

On the other hand, if there has an elastic support on another part Γ_3 of the boundary $\partial\Omega$, then the strainless state v satisfies $\sum_{i,j=1}^{3} c_{ij}v_i v_j = 0$ on Γ_3 as well, where c_{ij} is a symmetric positive definite matrix. So we must have $v_i = 0$ on Γ_3. As in the case of a geometric constraint, if three points in Γ_3 do not lie on the same straight line, then we must have $v \equiv 0$. Consequently, the strain energy becomes positive definite as well.

Furthermore, we have seen that, in the equilibrium problem of a three-dimensional elastic body, the mechanical boundary conditions (5.29) and (5.30) and the stress equilibrium conditions on the interface (5.32) are all natural boundary conditions. Only the geometrical boundary conditions (5.28) and the displacement continuity conditions (5.31) on the interface are essential boundary conditions.

5.5   On variational principles and finite element methods

We have shown above in detail that, an elastic equilibrium problem may have mathematical formulations which are different in form but equivalent in essence. One mathematical formulation is the variational problem which is the problem of minimizing the energy functional as required by the variational principle. Another is to state the problem as a boundary value problem of the system of second order elliptic equations given by the equilibrium equation.

The different mathematical formulations have inspired different ways of solving the problem in practice. These different ways have different effects. If we want to solve the problem analytically, i.e., to find a solution in closed form, the way via equilibrium equations is comparatively straightforward. However, it is well known that, only a few model problems, for which both geometrical and physical conditions of the problem are extremely regular and simple, can be solved in closed form. In practice both geometrical and physical conditions are complicated. Generally the difficulty of the analytic method makes it necessary to find solutions numerically. This makes the method based on the variational principle becomes more advantageous. The reason is that the variational principle states the problem in a comparatively simple and compact form: the energy expression only involves lower derivatives, and only the comparatively simple essential boundary condition are retained as constraint conditions while the comparatively complicated natural boundary conditions (including the interface condition of the medium) are omitted. This character of the variational principle becomes more prominent when the geometrical and the behaviour of the material of the elastic body and the load conditions become complicated. To simplify calculations and its statement, we have expressed the variational principle in a comparatively abstract form. A more important reason for doing so is to express all static elasticity problems encountered henceforth in such a unified form that only the energy expression and the constraint condition differ in concrete problems.

Such unity in mathematical form reflects the unity of the fundamental laws of mechanics, and it can greatly reduce the tediousness inherent in elasticity problems.

Recently, on the basis of productive practice both in China and in other countries, a complete and systematic set of numerical methods, knows as the difference-variational method or the finite element method, has been developed for elliptical equation problems, elasticity problems including the problem in elasticity. It is just based on the variational principle with the cooperation of the approximation of geometric subdivision that the potential superiority of the variational principle can be brought into full play as a result, especially it suitable to both geometrically and physically complicated problems, This numerical method possesses high currency and flexibility, which is suitable to computer-calculation, and has the advantages of strong intuitive properties on both geometry and physics. The method itself is also easy to be grasped. This set of methods has stood extensive tests in practice, met with great success, occupied a leading position in the field of numerical methods of the problem in both elasticity and structural mechanics, and promoted the widespread application of computers to engineering technical problems. A brief introduction of the finite element method will be given in Chapter 5 of the present book using a number of typical elastic equilibrium problems discussed in Chapters 1 to 4.

§6   Geometrical Compatibility

6.1   Integrability conditions of vector fields and topological properties of domain

As is well known in calculus that, a differentiable function in a single variable $f = f(x)$ has its derivative $\dfrac{df}{dx}$. Conversely, an integrable function $g(x)$ must have a primitive function $f(x)$ such that $\dfrac{df}{dx} = g$. The general form of the primitive function is

$$f(x) = f(x_0) + \int_{x_0}^{x} g(x)dx,$$

where $f(x_0)$ is an arbitrary constant of integration.

A differentiable function $f = f(x_1, x_2)$ with two variables has two first partial derivatives $\frac{\partial f}{\partial x_1}$ and $\frac{\partial f}{\partial x_2}$. Conversely, for two arbitrary double-variate functions of g_1 and $g_2{}^{1)}$, the problem whether there exists a primitive function $f = f(x_1, x_2)$ such that

$$\frac{\partial f}{\partial x_1} = g_1, \quad \frac{\partial f}{\partial x_2} = g_2$$

is called the problem of integrability of g_1 and g_2. Obviously it is different from the one dimensional case and a primitive function need not exist. In fact, if they are integrable, then

$$\frac{\partial^2 f}{\partial x_1 \partial x_2} = \frac{\partial^2 f}{\partial x_2 \partial x_1}.$$

Hence, g_1 and g_2 must satisfy

$$\frac{\partial g_2}{\partial x_1} - \frac{\partial g_1}{\partial x_2} = 0, \tag{6.1}$$

this is a necessary condition for integrability.

There is a similar integrability problem as well for the three dimensional case, of three variables. Assume g_1, g_2, and g_3 are three functions with three variables. The problem of integrability of g_1, g_2 and g_3 is whether there exists a primitive function $f = f(x_1, x_2, x_3)$ such that

$$\frac{\partial f}{\partial x_i} = g_i, \quad i = 1, 2, 3.$$

Speaking in the language of vector analysis, for an arbitrary vector field

$$g = (g_1, g_2, g_3)^T,$$

the problem of integrability is whether g can be expressed as a gradient field of one scalar quantity f such that

$$\text{grad } f = \left(\frac{\partial f}{\partial x_1}, \frac{\partial f}{\partial x_2}, \frac{\partial f}{\partial x_3} \right)^T = g.$$

It is obvious that a primitive need not exist, since

$$\frac{\partial^2 f}{\partial x_i \partial x_j} = \frac{\partial^2 f}{\partial x_j \partial x_i}.$$

Necessary conditions for integrability are

$$\frac{\partial g_3}{\partial x_2} - \frac{\partial g_2}{\partial x_3} = 0, \quad \frac{\partial g_1}{\partial x_3} - \frac{\partial g_3}{\partial x_1} = 0, \quad \frac{\partial g_2}{\partial x_1} - \frac{\partial g_1}{\partial x_2} = 0, \tag{6.2}$$

---

[1] Here we assume that f has continuous second partial derivatives and both g_1 and g_1 have continuous first partial derivatives. We do not want to go into these continuity conditions deeply but only want to investigate the integrability condition.

or

$$\text{Rot } \boldsymbol{g} = 0.$$

Thus $\boldsymbol{g}$ must be an irrotational field.

The conditions (6.1) or (6.2) are necessary conditions of integrability. Are these also sufficient conditions? If not, what additional conditions are needed? The answer to these questions depends on the topological properties of the domain. In what follows, we shall mainly discuss the three dimensional problem.

Lemma. Suppose a domain Ω is connected, i.e., any two points in Ω can be connected by an arc in Ω. The g_1, g_2, and g_3 are integrable over Ω if there a exists a single-valued function $f = f(x_1, x_2, x_3)$ such that

$$\frac{\partial f}{\partial x_k} = g_k, \quad k = 1, 2, 3, \ \Omega. \tag{6.3}$$

A necessary and sufficient condition for integrability is:

$$\oint_L \sum_{k=1}^{3} g_k dx_k = 0, \quad \text{for any loop } l \subset \Omega. \tag{6.4}$$

When g_1, g_2 and g_3 are integrable, any primitive function f satisfies

$$f(P) = f(P_0) + \int_{P_0}^{P} \sum_{k=1}^{3} g_k dx_k, \tag{6.5}$$

where $P_0 = (x_1^{(0)}, x_2^{(0)}, x_3^{(0)})$ is an arbitrary initial point in Ω, $P = (x_1, x_2, x_3)$ is an arbitrary point in Ω, and the path of integration $P_0 P$ can be chosen arbitrarily in Ω.

Proof. Necessity. If there exists a singlevalued function f such that (6.3) holds, then

$$\oint_L \sum_{k=1}^{3} g_k dx_k = \oint_L \sum_{k=1}^{3} \frac{\partial f}{\partial x_k} dx_k = \oint_L df = 0, \quad \text{for any loop } L \subset \Omega,$$

i.e., (6.4) holds. But since the domain is connected, any two points P_0 and P can be connected by an arc in Ω. Consequently,

$$f(P) = f(P_0) + \int_{P_0}^{P} \sum_{k=1}^{3} \frac{\partial f}{\partial x_k} dx_k = f(P_0) + \int_{P_0}^{P} \sum_{k=1}^{3} g_k dx_k,$$

hence (6.5) also holds.

Sufficiency. If (6.4) holds, then the integral at the right side of the expression above is independent of the selection of the path of P_0P. Hence a single-valued function $f(x_1, x_2, x_3) = f(P)$, where $f(x_1^{(0)}, x_2^{(0)}, x_3^{(0)}) = f(P_0)$ is a constant of integration chosen arbitrarily, can be defined in Ω. Obviously, such a function f satisfies (6.3). Q.E.D.

Stokes integral formula is well-known in calculus

$$\oint_{\partial S} \sum_{k=1}^{3} g_k dx_k = \iint_S \left(\frac{\partial g_3}{\partial x_2} - \frac{\partial g_2}{\partial x_3} \right) dx_2 dx_3 + \iint \left(\frac{\partial g_1}{\partial x_3} - \frac{\partial g_3}{\partial x_1} \right) dx_3 dx_1$$

$$+ \iint \left(\frac{\partial g_2}{\partial x_1} - \frac{\partial g_1}{\partial x_2} \right) dx_1 dx_2. \tag{6.6}$$

From now on, the curved surface S, the curves ∂S and L, etc. will all be considered as oriented. The orientation of a curved surface S is given by a direction n normal to it. If a direction t tangential to the curved surface S is defined on ∂s so as to form a right-handed system $\{v, t, n\}$ where v is the outward normal direction to the contour ∂S (Fig. 27), then the direction of ∂S is defined to be the tangent t.

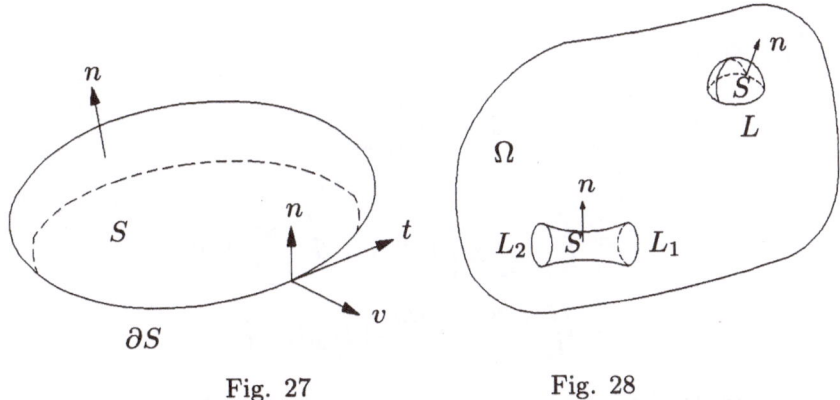

Fig. 27   Fig. 28

By (6.2), irrotationality is the necessary condition of integrability. By the lemma and Stokes formula, irrotationality is also a sufficient condition for integrability if an arbitrary loop L in Ω can be expressed as the contour $\partial S(= L)$ of some curved surface S in Ω. A domain Ω with this property is said to be simply connected. To investigate simple connectivity, we introduce the concept of homology of loops.

Definition.

1° For an oriented loop $L \subset \Omega$, if there exists an oriented curved surface $S \subset \Omega$ such that $\partial S = L$, then we say that L is homologous to 0 in Ω. We write this as $L \sim 0$.

2° For two oriented loops L_1 and L_2 in Ω, if $L_1 - L_2 \sim 0$, then L_1 and L_2 are called homologous in Ω (Fig. 28).

Roughly speaking, a loop homologous to 0 in Ω is just a loop which can be shrunk to a point via a continuous deformation in Ω. Mutually homologous loops are loops which can be transformed from one loop to the other loop via a continuous deformation in Ω. It is convenient to introduce the foregoing terminology and notation because we shall mainly discuss the loop integral, called the circulation, of irrotational fields. Irrotationality and Stokes formula imply that if $L \sim 0$ then

$$
\oint_L \sum_{k=1}^{3} g_k dx_k = \oint_{\partial S} \sum_{k=1}^{3} g_k dx_k
$$
$$
= \iint_S \left(\frac{\partial g_3}{\partial x_2} - \frac{\partial g_2}{\partial x_3} \right) dx_2 dx_3 + \left(\frac{\partial g_1}{\partial x_3} - \frac{\partial g_3}{\partial x_1} \right) dx_3 dx_1
$$
$$
+ \left(\frac{\partial g_2}{\partial x_1} - \frac{\partial g_1}{\partial x_2} \right) dx_1 dx_2
$$
$$
= 0. \tag{6.7}
$$

If $L_1 \sim L_2$, then

$$
\oint_{L_1} \sum_{k=1}^{3} g_k dx_k - \oint_{L_2} \sum_{k=1}^{3} g_k dx_k
$$
$$
= \oint_{L_1 - L_2} \sum_{k=1}^{3} g_k dx_k = \oint_{\partial S} \sum_{k=1}^{3} g_k dx_k
$$
$$
= \iint_S \left(\frac{\partial g_3}{\partial x_2} - \frac{\partial g_2}{\partial x_3} \right) dx_2 dx_3 + \cdots = 0. \tag{6.8}
$$

That is to say, for irrotational fields the circulation of a loop homologous to 0 equals 0, and the circulations of homologous loops are equal.

Let us now illustrate the concepts of simply connectivity and multiply connectivity. If every loop in a connected domain Ω is homologous to 0, then Ω is said to be simply connected, otherwise it is said to be multiply connected.

In the two dimensional case, any domains without a hole is simply connected, while any domain with a hole is multiply-connected because a loop around a hole is not homologous to 0.

In three dimensional space, any domain without a hole is simply connected, while a domain with holes can be either simply connected or multiply connected. For example, the domain between two concentric spheres is still simply connected because any loop L in it can be shrunk to a point via continuous deformation keeping away from the center hole. Consequently, a cap-like curved surface S (Fig. 29) lies in the domain with L as its contour. Typical multiply connected domain is the torus a sphere with a hole through it. The center axis of a torus cannot be expressed in the form ∂S for S a surface contained in the torus. There are domains with higher degrees of multiple connectivity, such as the double torus, multiple torus, etc. (Fig. 30).

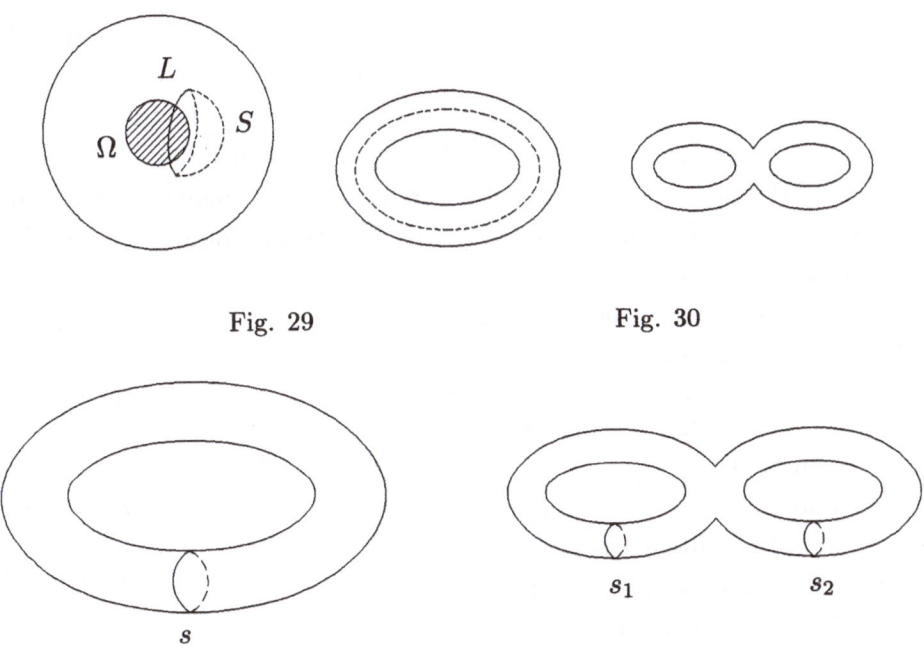

Fig. 29   Fig. 30

Fig. 31

Let us now proceed to investigate multiply connected domains. By cutting an arbitrary multiply connected domain Ω by one or more suitably chosen crosscuts, we can always make the separated domain Ω' simply connected. Here, the "'crosscuts" $S_1, S_2, \cdots, S_p$ are cross sections in Ω, of which all the boundaries $\partial S_1 \partial S_2, \cdots, \partial S_p$ lie on the boundary surface $\partial\Omega$ of Ω. Although the selection of cross sections which make Ω' simply connected is quite arbitrary, the number p of the cross sections is always the same. This number ρ is a topological invariant called the connectivity number of Ω. When Ω is simply connected, $p = 0$. The connectivity number p of the torus is one, and the crosscut S_1 can be any meridian cross section perpendicular to the axis of the torus. For the double torus, $p = 2$, and the crosscuts S_1 and S_2 can be taken as that shown in Fig. 31.

The situation is much simpler for domains in the plane. The crosscut is a cross line in Ω and the connectivity number p just equals the number of holes in the domain.

Now let Ω be a multiply connected domain, whose connectivity number is p. Ω becomes a simply connected domain Ω' after being cut by cross sections $S_1, \cdots, S_p$.

Each cross section S_i divides Ω locally into two parts, one positive and the other negative which side is positive and which negative may be difined. Hence, the cross section S_i may viewed as having a positive side S_i^+ and a negative side S_i^-, which are two parts of the boundary of the separated simply connected domain Ω' but have a coincident geometrical location. Arbitrarily choose a point Ω_i on each S_i. We may also to assume that Ω_i is divided into one positive point Ω_i^+ and one negative point Ω_i^-, which belong separately to the two sides S_i^+ and S_i^- but are coincident.

Draw an arbitrary oriented curve L_i from Ω_i^- to Ω_i^+ in Ω'. In Ω', L_i is not a loop (a closed curve), but it seems in Ω, that L_i is an oriented loop which crosses S_i (from the positive to the negative side) at point Ω_i. A set of such loops $L_1, \cdots, L_p$, can be selected in Ω (Fig. 32).

Although choice of the set of loops $\{L_1, \cdots, L_p\}$ is quite arbitrary, it always has the following fundamental properties[1] independent of the selection method:

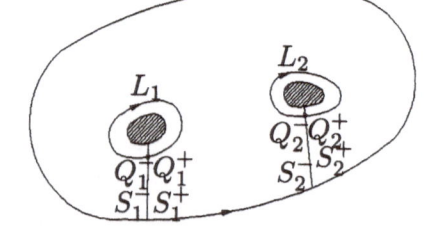

Fig. 32

For each oriented loop L in Ω, there exists a uniquely defined set of integers (positive, negative or 0, depending on L), $n_1, \cdots, n_p$, such that

$$L \sim \sum_{i=1}^{p} n_i L_i. \tag{6.9}$$

This shows that up to homology equivalence, the set of loops $\{L_1, \cdots, L_p\}$

---

[1] We do not want to prove this conclusion.

generates all loops in Ω. The uniqueness of the coefficients $n_1, \cdots, n_p$ indicates that the loops are independent of one another. Hence, $\{L_1, \cdots L_p\}$ called a fundamental set of loops, which constitute a basis set for the multiply connected domain Ω.

For plane domains, any set containing an arbitrary simple closed curve around each hole constitutes a fundamental set of loops.

On the basis of the above discussion on simply and multiply connected domains, we can answer the question about the sufficient condition for integrability.

Proposition 2.1   For a simply connected domain Ω, a necessary and sufficient condition for integrability of a vector field $g = (g_1, g_2, g_3)^T$ is

$$\text{rot } g = 0.$$

The primitive f satisfies the equation

$$f(P) = f(P_0) + \int_{P_0}^{P} \sum_{k=1}^{3} g_k dx_k,$$

where the path of integration P_0P can be chosen arbitrarily.

Proof. Sufficiency. Since any loop $L \sim 0$ in a simply connected domain, from (6.7), the circulation of an irrotational field g satisfies

$$\oint_{L} \sum_{k=1}^{3} g_k dx_k = 0.$$

By the lemma, g is integrable and the above equation for f holds.

Necessity. The necessary condition for integrability rot $g = 0$ was proved in (6.2) and it is omitted here.

Proposition 2.2   On a multiply connected domain Ω, necessary and sufficient conditions for the integrability of a vector field $g = (g_1, g_2, g_3)^T$ are

1. rot $g = 0$,

2. $\oint_{L_i} \sum_{k=1}^{3} g_k dx_k = 0, \quad i = 1, \cdots, p,$

where $\{L_1, \cdots, L_p\}$ constitute a fundamental set of loops in Ω, and the primitive f satisfies the equation

$$f(P) = f(P_0) + \int_{P_0}^{P} \sum_{k=1}^{3} g_k dx_k,$$

where the path of integration P_0P can be chosen arbitrarily.

Proof. Sufficiency. Because any loop $L \subset \Omega$ in a multiply connected domain satisfies

$$L \sim \sum_{i=1}^{p} n_i L_i,$$

by (6.8), the circulation of g is

$$\oint_L \sum_{k=1}^{3} g_k dx_k = \sum_{i=1}^{p} n_i \oint_{L_i} \sum_{k=1}^{3} g_k dx_k.$$

The right hand side of the above expression vanishes by condition 2, hence the left hand side vanishes,

$$\oint_L \sum_{k=1}^{3} g_k dx_k = 0, \quad \text{for any loop } L \subset \Omega.$$

By the lemma. The vector field g is integrable.

Necessity.

1. We proved earlier. that rot $g = 0$.

2. Because g is integrable, there exists a single valued function f such that $\dfrac{\partial f}{\partial x_k} = g_k$, $k = 1, 2, 3$. Therefore, for any loop $L \subset \Omega$, we always have

$$\int_L df = 0.$$

Particularly, taking $L = L_i$ (a fundamental loop), we have

$$\int_{L_i} df = \int_{L_i} \sum_{k=1}^{3} \frac{\partial f}{\partial x_k} dx_k = \int_{L_i} \sum_{k=1}^{3} g_k dx_k = 0, \quad i = 1, \cdots, p,$$

That is the condition 2. Q.E.D.

6.2   Equations of geometric compatibility and conditions for integrability

Suppose a displacement field

$$u_i = u_i(x_1, x_2, x_3), \quad i = 1, 2, 3$$

is defined in a space domain Ω and has first partial derivatives $\dfrac{\partial u_j}{\partial x_i}$, $i, j = 1, 2, 3$. Making a symmetric combination of these partial derivatives, we obtain the strain field

$$\varepsilon_{ij} = \frac{1}{2}\left(\frac{\partial u_j}{\partial x_i} + \frac{\partial u_i}{\partial x_j}\right), \quad \varepsilon_{ij} = \varepsilon_{ji}. \tag{6.10}$$

Making an antisymmetric combination, we obtain the rotation field

$$\omega_{ij} = \frac{1}{2}\left(\frac{\partial u_j}{\partial x_j} - \frac{\partial u_i}{\partial x_j}\right), \quad \omega_{ij} = -\omega_{ji}.$$

Thus, we have

$$\frac{\partial u_j}{\partial x_i} = \varepsilon_{ij} + \omega_{ij}, \tag{6.11}$$

$$\frac{\partial \omega_{ij}}{\partial x_k} = \frac{\partial \varepsilon_{jk}}{\partial x_i} - \frac{\partial \varepsilon_{ik}}{\partial x_j}, \quad k = 1, 2, 3. \tag{6.12}$$

We know that the elastic deformation depends only on the strain field; the following two questions concern the inverse dependence:

(1) Is an arbitrary symmetric tensor field ε_{ij} always the strain field of some displacement field, i.e., do there always exist u_i, $i = 1, 2, 3$ on Ω that make (6.10) true? If not, then under what condition does the displacement field exist?

(2) How is the displacement field u_i expressed in terms of the corresponding strain tensor ε_{ij}?

These two problems are similar to the problem of integrability, i.e., the problem of relationship between the primitive function and its partial derivatives discussed above. However, the problems here are more complicated than that of integrability. The reason is that here we require three primitive functions u_j, each required to have three of the nine $\frac{\partial u_j}{\partial x_i} = \varepsilon_{ij} + \omega_{ij}$, expressed in terms of the six symmetric combinations of ε_{ij}.

The answer to question (1) can be expressed in terms of the following proposition.

Proposition 2.3   If the domain Ω is simply connected, a necessary and sufficient condition for a symmetric tensor field ε_{ij} to be a strain field that it satisfy Saint Venant's six equations of geometric compatibility:

$$
\begin{cases}
\dfrac{\partial^2 \varepsilon_{22}}{\partial x_l^2} + \dfrac{\partial^2 \varepsilon_{11}}{\partial x_2^2} - 2\dfrac{\partial^2 \varepsilon_{12}}{\partial x_1 \partial x_2} = 0, \quad 1 \to 2 \to 3 \to 1, & (6.13) \\[3mm]
-\dfrac{\partial^2 \varepsilon_{23}}{\partial x_l \partial x_2} + \dfrac{\partial^2 \varepsilon_{12}}{\partial x_2 \partial x_3} - \dfrac{\partial^2 \varepsilon_{13}}{\partial x_2^2} - \dfrac{\partial^2 \varepsilon_{22}}{\partial x_1 \partial x_3} = 0, & (6.14) \\[3mm]
\qquad\qquad 1 \to 2 \to 3 \to 1.
\end{cases}
$$

If Ω is multiply connected, a necessary and sufficient condition is given by adding to Saint Venant's equations the $6p$ Volterra integral conditions:

$$
\begin{cases}
b_{ij}(L_m) = \oint_{L_m} \sum_{k=1}^{3} \left(\dfrac{\partial \varepsilon_{jk}}{\partial x_i} - \dfrac{\partial \varepsilon_{ik}}{\partial x_j} \right) dx_k = 0, \\[2mm]
\qquad ij = 12, 23, 31; \, m = 1, \cdots, p, \\[4mm]
a_j(L_m) = \oint_{L_m} \sum_{k=1}^{3} \left[\varepsilon_{kj} - \sum_{i=1}^{3} \left(\dfrac{\partial \varepsilon_{jk}}{\partial x_i} - \dfrac{\partial \varepsilon_{ik}}{\partial x_j} \right) x_i \right] dx_k = 0, \\[2mm]
\qquad j = 1, 2, 3; \; m = 1, \cdots, p,
\end{cases}
$$

$$(6.15)$$

$$(6.16)$$

where L_m, $m = 1, \cdots, p$ constitute a fundamental set of loops of the multiply connected domain Ω, and the connectivity number of Ω is p.

Proof. The proof divides into three steps. 1. Find first a necessary and sufficient condition for the existence of ω_{ij} satisfying (6.12), given ε_{ij}. 2. Find then a necessary and sufficient conditon for the existence of u_j satisfying (6.11) given ε_{ij} and w_{ij}. 3. Prove that u_j is a displacement field iducing the symmetric tensor field ε_{ij} as its strain field. We carry out these steps as follows.

1. By propositions 2.1 and 2.2 of Section 6.1, necessary and sufficient conditions for the existence of single-valued functions ω_{ij}, $ij = 12, 23, 31$ satisfying

$$
\frac{\partial \omega_{ij}}{\partial x_k} = \frac{\partial \varepsilon_{jk}}{\partial x_i} - \frac{\partial \varepsilon_{ik}}{\partial x_j}, \quad k = 1, 2, 3 \tag{6.12}
$$

are that

$$
\frac{\partial}{\partial x_l} \left(\frac{\partial \varepsilon_{jk}}{\partial x_i} - \frac{\partial \varepsilon_{ik}}{\partial x_j} \right) = \frac{\partial}{\partial x_k} \left(\frac{\partial \varepsilon_{jl}}{\partial x_i} - \frac{\partial \varepsilon_{jl}}{\partial x_j} \right),
$$

$$
ij = 12, 23, 31; \quad k, l = 1, 2, 3. \tag{6.17}
$$

When the domain Ω is multiply connected, we require $3p$ further integral conditions

$$
\oint_{L_m} \sum_{k=1}^{3} \left(\frac{\partial \varepsilon_{jk}}{\partial x_i} - \frac{\partial \varepsilon_{ik}}{\partial x_j} \right) dx_k = 0, ij = 12, 23, 31; \quad m = 1, \cdots, p. \tag{6.18}
$$

Let us investigate the conditions in (6.17). Obviously, only the equations with subscripts $ij, kl = 12, 23, 31$ need be discussed here. There are nine equations in all, but it follows easily from the symmetry of ε_{ij} that, if we interchange the subscripts i and k or j and l, then the equations remain unaltered. Hence, there are really only six distinct equations among these nine. Letting $ij = kl = 12, 23, 31$, we can obtain three different equations.

(6.13) in Proposition 2.3 is the case of $ij = kl = 12$, and the other two equations can be obtained through the cyclic permutation of subscripts $1 \to 2 \to 3 \to 1$. Furthermore, letting

$$ij = 12, kl = 23; ij = 23, kl = 31; ij = 31, kl = 12;$$

we obtain three more distinct equations. (6.14) in Proposition 2.3 is the case of $ij = 12$, $kl = 23$, and the other two equations follow similarly by the cyclic permutation of subscripts, too.

Thus, (6.17), the first part of the necessary and sufficient conditions in 1, is just Saint Venant's equations of geometric compatibility, and obviously, (6.18), the second part of the necessary and sufficient conditions is just Volterra's integral condition (6.15).

2. From the single-valued functions ω_{ij}, $ij = 12, 23, 31$ determined in 1, define an antisymmetric tensor

$$\omega_{ji} = -\omega_{ij}, \quad ij = 12, 23, 31, \quad \omega_{ii} = 0, \quad i = 1, 2, 3.$$

By Propositions 2.1 and 2.2, given for the symmetric tensor field ε_{ij} and the antisymmetric tensor field ω_{ij} in Ω, necessary and sufficient conditions for the existence of single-valued functions u_j, $j = 1, 2, 3$ satisfying

$$\frac{\partial u_j}{\partial x_i} = \varepsilon_{ij} + \omega_{ij}, \quad i, j = 1, 2, 3 \tag{6.19}$$

are

$$\frac{\partial}{\partial x_l}(\varepsilon_{kj} + \omega_{kj}) = \frac{\partial}{\partial x_k}(\varepsilon_{lj} + \omega_{lj}), \quad k, l = 1, 2, 3. \tag{6.20}$$

When the domain Ω is multiply connected, we need to add $3p$ more integral conditions

$$\oint_{L_m} \sum_{k=1}^{3} (\varepsilon_{kj} + \omega_{kj}) dx_k = 0, \quad j = 1, 2, 3, \quad m = 1, \cdots, p. \tag{6.21}$$

By the definition ω_{ij} in 1 and the antisymmetry stipulated here in 2, (6.12) holds. Hence

$$\frac{\partial \omega_{kj}}{\partial x_l} = \frac{\partial \varepsilon_{jl}}{\partial x_k} - \frac{\partial \varepsilon_{kl}}{\partial x_j}, \quad \frac{\partial \omega_{lj}}{\partial x_k} = \frac{\partial \varepsilon_{jk}}{\partial x_l} - \frac{\partial \varepsilon_{lk}}{\partial x_j}.$$

By substituting the two expressions above into (6.20), the two sides are identical. Hence the necessary and sufficient conditions are in fact automatically satisfied.

Let us now investigate the second part of the necessary and sufficient conditions as shown in (6.21). These conditions are equivalent to

$$\oint_L \sum_{k=1}^3 (\varepsilon_{kj} + \omega_{kj}) dx_k = 0, \quad \text{for any loop } L \subset \Omega.$$

Choose an arbitrary point $P_0 = (x_1(P_0), x_2(P_0), x_3(P_0))$ as the initial point of integration, then for an arbitrary point P in Ω, the integral

$$\int_{P_0}^P \sum_{k=1}^3 (\varepsilon_{kj} + \omega_{ki}) dx_k$$

is independent of the path of integration. Integrating by parts, we get

$$\int_{P_0}^P \sum_{k=1}^3 (\varepsilon_{kj} + \omega_{kj}) dx_k = \int_{P_0}^P \sum_{k=1}^3 \varepsilon_{kj} dx_k + \int_{P_0}^P \sum_{k=1}^3 \omega_{kj} dx_k$$

$$= \int_{P_0}^P \sum_{k=1}^3 \varepsilon_{kj} dx_k + \sum_{k=1}^3 \omega_{kj}(P) x_k(P) - \sum_{k=1}^3 \omega_{kj}(P_0) x_k(P_0)$$

$$- \int_{P_0}^P \sum_{k=1}^3 \left(\sum_{i=1}^3 \frac{\partial \omega_{kj}}{\partial x_i} dx_i \right) x_k$$

$$= \sum_{k=1}^3 \omega_{kj}(P) x_k(P) - \sum_{k=1}^3 \omega_{kj}(P_0) x_k(P_0)$$

$$+ \int_{P_0}^P \sum_{k=1}^3 \left[\varepsilon_{kj} - \sum_{i=1}^3 \frac{\partial \omega_{ij}}{\partial x_k} x_i \right] dx_k$$

$$= \sum_{k=1}^3 \omega_{kj}(P) x_k(P) - \sum_{k=1}^3 \omega_{kj}(P_0) x_k(P_0)$$

$$+ \int_{P_0}^P \sum_{k=1}^3 \left[\varepsilon_{kj} - \sum_{i=1}^3 \left(\frac{\partial \varepsilon_{jk}}{\partial x_i} - \frac{\partial \varepsilon_{ik}}{\partial x_j} \right) x_i \right] dx_k. \qquad (6.22)$$

Due to the monodromy of ω_{ij}, for any closed curve L in Ω, we have

$$\oint_L \sum_{k=1}^3 (\varepsilon_{kj} + \omega_{kj}) dx_k = \oint_L \sum_{k=1}^3 \left[\varepsilon_{kj} - \sum_{i=1}^3 \left(\frac{\partial \varepsilon_{jk}}{\partial x_i} - \frac{\partial \varepsilon_{ik}}{\partial x_j} \right) x_i \right] dx_k = 0,$$

$$j = 1, 2, 3,$$

which is equivalent to

$$\oint_{L_m} \sum_{k=1}^3 \left[\varepsilon_{kj} - \sum_{i=1}^3 \left(\frac{\partial \varepsilon_{jk}}{\partial x_i} - \frac{\partial \varepsilon_{ik}}{\partial x_j} \right) x_i \right] dx_k = 0.$$

$$j = 1, 2, 3; \quad m = 1, \cdots, p.$$

These are the Volterra integral conditions (6.16) in Proposition 2.3.

3. By (6.19), we have

$$\frac{\partial u_j}{\partial x_i} = \varepsilon_{ij} + \omega_{ij}, \quad \frac{\partial u_i}{\partial x_j} = \varepsilon_{ji} + \omega_{ji}.$$

Adding and subtracting these two equations and using the symmetry of ε_{ij} and the antisymmetry of ω_{ij}, we obtain

$$\varepsilon_{ij} = \frac{1}{2}\left(\frac{\partial u_j}{\partial x_i} + \frac{\partial u_i}{\partial x_j}\right)$$

and

$$\omega_{ij} = \frac{1}{2}\left(\frac{\partial u_j}{\partial x_i} - \frac{\partial u_j}{\partial x_j}\right).$$

This shows that u_j actually is a displacement field whose strain field is the given symmetric field ε_{ij} and whose rotational field is ω_{ij}.

To sum up 1 and 2, the necessary and sufficient conditions to determine u_j from ε_{ij} are the six Saint Venant's equations (6.13) and (6.14), plus the $6p$ Volterra's integral conditions (6.15) and (6.16) when Ω is multiply connected. Q.E.D.

On the basis of Proposition 2.3, it is very easy to solve the second problem, i.e., how to express the displacement fields in terms of the strain fields. Because

$$\frac{\partial u_j}{\partial x_i} = \varepsilon_{ij} + \omega_{ij},$$

it follows that

$$u_j(P) = u_j(P_0) + \int_{P_0}^{P} \sum_{k=1}^{3} \frac{\partial u_j}{\partial x_k} dx_k = u_j(P_0) + \int_{P_0}^{P} \sum_{k=1}^{3} (\varepsilon_{kj} + \omega_{kj}) dx_k.$$

By (6.22), this expression can be written

$$u_j(P) = u_j(P_0) + \sum_{i=1}^{3} \omega_{ij}(P) x_i(P) - \sum_{i=1}^{3} \omega_{ij}(P_0) x_i(P_0)$$

$$+ \int_{P_0}^{P} \sum_{k=1}^{3} \left[\varepsilon_{kj} - \sum_{i=1}^{3}\left(\frac{\partial \varepsilon_{jk}}{\partial x_i} - \frac{\partial \varepsilon_{jk}}{\partial x_j}\right) x_i\right] dx_k. \tag{6.23}$$

The right hand side still contains a rotational component $\omega_{ij}(P)$ independent of the strain component ε_{ij} in the expression. But integrating (6.12), we get

$$\omega_{ij}(P) = \omega_{ij}(P_0) + \int_{P_0}^{P} \sum_{k=1}^{3} \left(\frac{\partial \varepsilon_{ik}}{\partial x_i} - \frac{\partial \varepsilon_{jk}}{\partial x_j}\right) dx_k.$$

Substituting this into (6.23), we obtain a final form

$$u_j(P) = u_j(P_0) + \sum_{i=1}^{3} \omega_{ij}(P_0)(x_i(P) - x_i(P_0))$$

$$+ \int_{P_0}^{P} \sum_{k=l}^{3} \left[\varepsilon_{kj} + \sum_{i=1}^{3} \left(\frac{\partial \varepsilon_{jk}}{\partial x_i} - \frac{\partial \varepsilon_{ik}}{\partial x_j} \right) \times (x_i(P) - x_i) \right] dx_k.$$

This expression contains a total of six constants of integration $u_j(P_0)$ and $\omega_{ij}(P_0)$, while the integral on the right side is independent of the path of integration.

§7   Thermal Effects

7.1   Hooke's law and strain energy

Suppose the unloaded state of an elastic body at uniform temperature T_0 is a strainless state. Now this body is heated to a temperature $T = T(x_1, x_2, x_3)$ and is free to expand. Suppose the body is isotropic and every element in the body will expand at equal amount in each direction without shear. Then no stress is produced in the body and the thermal strains ε_{ij}^* are given by

$$\varepsilon_{ji}^* = \alpha(T - T_0), \quad i = 1, 2, 3, \tag{7.1}$$

$$\varepsilon_{ij}^* = 0, \quad i \neq j. \tag{7.2}$$

(7.1) is the primary law of thermal expansion indicating that the relative elongation is proportional to the temperature rise, where α is the coefficient of linear thermal expansion of the material. For the sake of convenience, we denote the temperature rise by τ, i.e.,

$$T - T_0 = \tau = \tau(x_1, x_2, x_3). \tag{7.3}$$

Then (7.1) and (7.2) can be combined and written

$$\varepsilon_{ij}^* = \alpha \tau \delta_{ij}. \tag{7.4}$$

If the surface of the body is restrained so as not to be able to expand freely, or if the heating is nonuniform or there are external loads, then stresses σ_{ij} and strains ε_{ij} will be produced in the interior of the body. In that case, the strains will be the superposition of the free thermal strains ε_{ij}^* mentioned above and the strains ε_{ij}^{**} produced by the stresse

σ_{ij} without increase in temperature. That is,

$$\varepsilon_{ij} = \varepsilon_{ij}^* + \varepsilon_{ij}^{**}. \tag{7.5}$$

According to the conventional Hooke's law

$$\varepsilon_{ij}^{**} = \frac{1+\nu}{E}\sigma_{ij} - \frac{\nu}{E}\left(\sum_{k=1}^{3}\sigma_{kk}\right)\delta_{ij},$$

hence

$$\varepsilon_{ij} = \frac{1+\nu}{E}\sigma_{ij} - \frac{\nu}{E}\left(\sum_{k=1}^{3}\sigma_{kk}\right)\delta_{ij} + \alpha\tau\delta_{ij}. \tag{7.6}$$

Letting $i = j = 1, 2, 3$ in the above expression and then adding, we obtain

$$\sum_{k=1}^{3}\varepsilon_{kk} = \frac{1-2\nu}{E}\sum_{k=1}^{3}\sigma_{kk} + 3\alpha\tau,$$

so

$$\sum_{k=1}^{3}\sigma_{kk} = \frac{E}{1-2\nu}\sum_{k=1}^{3}\varepsilon_{kk} - \frac{3E\alpha}{1-2\nu}\tau.$$

Hence we obtain an inverse relation

$$\sigma_{ij} = \frac{E}{1+\nu}\varepsilon_{ij} + \frac{E\nu}{(1+\nu)(1-2\nu)}\left(\sum_{k=1}^{3}\varepsilon_{kk}\right)\delta_{ij} - \frac{E\alpha}{1-2\nu}\tau\delta_{ij}. \tag{7.7}$$

This is the Hooke's law under thermal effects. The difference from the conventional Hooke's law without thermal effects is that the relation between the stress and the strain is no longer a linear homogeneous expression but a linear nonhomogeneous expression including a term for zero order induced by thermal effects. Owing to the appearance of the term for thermal effects, there may exist stresses when the strain is absent, and may exist strains when the stress is absent as well.

Hooke's law, (7.6) or (7.7), can be written in matrix form as

$$\varepsilon = \frac{1+\nu}{E}\sigma - \frac{\nu}{E}\left(\sum_{k=1}^{3}\sigma_{kk}\right)I + \alpha\tau I, \tag{7.8}$$

$$\sigma = \frac{E}{1+\nu}\varepsilon + \frac{E\nu}{(1+\nu)(1-2\nu)}\left(\sum_{k=1}^{3}\varepsilon_{kk}\right)I - \frac{E\alpha}{1-2\nu}\tau I. \tag{7.9}$$

In order to determine the volume density of the strain energy under thermal effects, we must perform the principal transformation of coordinates following the method of §2–4. It is clear, as in §4, that the strains and the stresses can be transformed onto the principal axes simultaneously, i.e.,

$$\varepsilon' = A\varepsilon A^T = \begin{pmatrix} \varepsilon_1 & 0 & 0 \\ 0 & \varepsilon_2 & 0 \\ 0 & 0 & \varepsilon_3 \end{pmatrix}, \quad \sigma' = A\sigma A^T = \begin{pmatrix} \sigma_1 & 0 & 0 \\ 0 & \sigma_2 & 0 \\ 0 & 0 & \sigma_3 \end{pmatrix}.$$

Premultiplying by A and postmultiplying by A^T on both the sides of (7.9) and writing

$$\sum_{k=1}^{3} \varepsilon_{kk} = \sum_{k=1}^{3} \varepsilon_k,$$

we obtain

$$\sigma' = \frac{E}{1+\nu}\varepsilon' + \frac{E\nu}{(1+\nu)(1-2\nu)}\Big(\sum_{k=1}^{3}\varepsilon_k\Big)I - \frac{E\alpha}{1-2\nu}\tau I,$$

i.e.,

$$\sigma_i = \frac{E}{1+\nu}\varepsilon_i + \frac{E\nu}{(1+\nu)(1-2\nu)}\sum_{k=1}^{3}\varepsilon_k - \frac{E\alpha}{1-2\nu}\tau, i = 1,2,3.$$

Letting $\varepsilon_i' = t\varepsilon_i$, we get

$$\sigma_i' = t\Big(\frac{E}{1+\nu}\varepsilon_i + \frac{E\nu}{(1+\nu)(1-2\nu)}\sum_{k=1}^{3}\varepsilon_k\Big) - \frac{E\alpha}{1-2\nu}\tau.$$

ε_i' varies from 0 to ε_i and σ_i' varies from $-\dfrac{E\alpha}{1-2\nu}\tau$ to σ_i as t varies from 0 to 1. Hence the strain energy per unit volume is

$$W = \sum_{i=1}^{3}\int_0^{\varepsilon_j} \sigma_i' d\varepsilon_i'$$

$$= \sum_{i=1}^{3}\int_0^{1}\Big\{t\Big(\frac{E}{1+\nu}\varepsilon_i + \frac{E\nu}{(1+\nu)(1-2\nu)}\sum_{k=1}^{3}\varepsilon_k\Big) - \frac{E\alpha}{1-2\nu}\tau\Big\}\varepsilon_i dt$$

$$= \frac{1}{2}\Big\{\frac{E}{1+\nu}\sum_{k=1}^{3}\varepsilon_k^2 + \frac{E\nu}{(1+\nu)(1-2\nu)}\Big(\sum_{k=1}^{3}\varepsilon_k\Big)^2\Big\} - \frac{E\alpha}{1-2\nu}\tau\sum_{k=1}^{3}\varepsilon_k.$$

Note that the factor $\dfrac{1}{2}$ appears only in the first term but not in the second term!

Returning to the original system of coordinates, since

$$\sum_{k=1}^{3}\varepsilon_k = \sum_{k=1}^{3}\varepsilon_{kk}, \quad \sum_{k=1}^{3}\varepsilon_k^2 = \sum_{i,j=1}^{3}\varepsilon_{ij}^2,$$

we obtain the volume density of the strain energy under thermal effects

$$W = \frac{1}{2}\left\{\frac{E}{1+\nu}\sum_{i,j=1}^{3}\varepsilon_{ij}^{2} + \frac{E\nu}{(1+\nu)(1-2\nu)}\Big(\sum_{k=1}^{3}\varepsilon_{kk}\Big)^{2}\right\} - \frac{E\alpha}{1-2\nu}\tau\sum_{k=1}^{3}\varepsilon_{kk}.$$
$$(7.10)$$

Comparing with (4.12), the equation for strain energy without thermal effects, we see that here W is no longer a quadratic homogeneous expression of ε_{ij} but a nonhomogeneous expression which includes a first order thermal effect term.

It is easy to verify that the relation

$$\sigma_{ij} = \frac{\partial W}{\partial \varepsilon_{ij}}$$

still holds, while the relation $W = \frac{1}{2}\sum_{i,j=1}^{3}\sigma_{ij}\varepsilon_{ij}$ no longer holds, because here W is not a quadratic homogeneous expression of ε_{ij} and Euler's theorem is not applicable.

7.2   Variational principles and equilibrium equations

Whether or not we consider thermal effects, the relation between strains and displacements is always

$$\varepsilon_{ij} = \frac{1}{2}\Big(\frac{\partial u_j}{\partial x_i} + \frac{\partial u_i}{\partial x_j}\Big).$$

In addition, the relations of the geometric compatibility which must be satisfied by the strain components, i.e., Saint Venant's equations and Volterra's integral conditions, still hold because these relations are purely geometric properties and independent of thermal effects.

From (7.10), the strain energy of the elastic body is

$$\iiint_{\Omega}W\,dV = \frac{1}{2}\iiint_{\Omega}\Big\{\frac{E}{1+\nu}\sum_{i,j=1}^{3}\varepsilon_{ij}^{2} + \frac{E\nu}{(1+\nu)(1-2\nu)}$$

$$\times\Big(\sum_{k=1}^{3}\varepsilon_{kk}\Big)^{2}\Big\}dV - \iiint_{\Omega}\frac{E\alpha}{1-2\nu}\tau\Big(\sum_{k=1}^{3}\varepsilon_{kk}\Big)dV. \qquad (7.11)$$

Suppose the elastic body is subjected to body loads inside Ω, with densities f_i, $i = 1,2,3$, and surface loads, on the boundary $\partial\Omega$, with densities g_i, $i = 1,2,3$. Then the potential energy of external work is

$$-\iiint_{\Omega}\sum_{i=1}^{3}f_i u_i\,dV - \iint_{\partial\Omega}\sum_{i=1}^{3}g_i u_i\,dS.$$

The total potential energy of the system is

$$J(\boldsymbol{u}) = \iiint_\Omega W(\boldsymbol{u})dV - \iiint_\Omega \sum_{i=1}^3 f_i u_i dV - \iint_{\partial\Omega} \sum_{i=1}^3 g_i u_i dS. \qquad (7.12)$$

This is different from before, because there is a thermal effects term. The strain energy (7.11) is no longer a quadratic functional of the displacement $\boldsymbol{u}$ and cannot be considered as $\dfrac{1}{2}D(\boldsymbol{u}, \boldsymbol{u})$. For this reason, we introduce the following notation

$$\sigma'_{ij} = \frac{E}{1+\nu}\varepsilon_{ij} + \frac{E\nu}{(1+\nu)(1-2\nu)}\Big(\sum_{k=1}^3 \varepsilon_{kk}\Big)\delta_{ij}$$

$$= \frac{E}{2(1+\nu)}\Big(\frac{\partial u_j}{\partial x_i} + \frac{\partial u_i}{\partial x_j}\Big) + \frac{E\nu}{(1+\nu)(1-2\nu)}\Big(\sum_{k=1}^3 \frac{\partial u_k}{\partial x_k}\Big)\delta_{ij}. \qquad (7.13)$$

Note that σ'_{ij} is not the stress but only a constituent part of the stress. From (7.7), the stress

$$\sigma_{ij} = \sigma'_{ij} - \frac{E\alpha}{1-2\nu}\tau\delta_{ij}. \qquad (7.14)$$

When there is no thermal effect, i.e., $\tau = 0$, we have $\sigma'_{ij} = \sigma_{ij}$.

Let

$$D(\boldsymbol{u}, \boldsymbol{u}) = \iiint_\Omega \sum_{i,j=1}^3 \sigma'_{ij}(\boldsymbol{u})\varepsilon_{ij}(\boldsymbol{u})dV,$$

$$D(\boldsymbol{u}, \boldsymbol{v}) = \iiint_\Omega \sum_{i,j=1}^3 \sigma'_{ij}(\boldsymbol{u})\varepsilon_{ij}(\boldsymbol{v})dV,$$

$$F(\boldsymbol{u}) = \iiint_\Omega \frac{E\alpha}{1-2\nu}\tau \sum_{k=1}^3 \varepsilon_{kk}(\boldsymbol{u})dV + \iiint_\Omega \sum_{i=1}^3 f_i u_i dV$$

$$+ \iint_{\partial\Omega} \sum_{i=1}^3 g_i u_i dS.$$

Clearly, $D(\boldsymbol{u}, \boldsymbol{u})$ is a nonnegative quadratic functional of $\boldsymbol{u}$, $D(\boldsymbol{u}, \boldsymbol{v})$ is a bilinear symmetric functional of $\boldsymbol{u}$ and $\boldsymbol{v}$, and $F(\boldsymbol{u})$ is a linear functional of $\boldsymbol{u}$. From (7.12), the total potential energy is

$$J(\boldsymbol{u}) = \frac{1}{2}D(\boldsymbol{u}, \boldsymbol{u}) - F(\boldsymbol{u}).$$

Thus according to the variational principle,

$$J(\boldsymbol{u}) = \text{Min}$$

is equivalent to
$$D(\boldsymbol{u}, \boldsymbol{v}) - F(\boldsymbol{v}) = 0, \quad \text{for all } \boldsymbol{v}.$$

The equivalent equilibrium equations can be derived from above as follows:

$$D(\boldsymbol{u}, \boldsymbol{v}) - F(\boldsymbol{v})$$

$$= \iiint_\Omega \sum_{i,j=1}^3 \sigma'_{ij}(\boldsymbol{u}) \varepsilon_{ij}(\boldsymbol{v}) dV - \iiint_\Omega \frac{E\alpha}{1-2\nu} \tau \sum_{k=1}^3 \varepsilon_{kk}(\boldsymbol{v}) dV$$

$$- \iiint_\Omega \sum_{i=1}^3 f_i v_i dV - \iint_{\partial\Omega} \sum_{i=1}^3 g_i v_i dS$$

$$= - \iiint_\Omega \left\{ \sum_{j=1}^3 \frac{\partial \sigma'_{ij}(\boldsymbol{u})}{\partial x_j} - \frac{\partial}{\partial x_i} \times \left(\frac{E\alpha}{1-2\nu}\tau \right) + f_i \right\} v_i dV$$

$$+ \iint_{\partial\Omega} \left\{ \sum_j \sigma'_{ij}(\boldsymbol{u}) n_j - \frac{E\alpha}{1-2\nu}\tau n_i - g_i \right\} v_i ds = 0,$$

hence

$$\begin{cases} \Omega: \quad -\sum_{j=1}^3 \frac{\partial \sigma'_{ij}}{\partial x_j} = f_i - \frac{\partial}{\partial x_i}\left(\frac{E\alpha}{1-2\nu}\tau \right), \\[4mm] \partial\Omega: \quad \sum_{i=1}^3 \sigma_{ij} n_j = g_i + \frac{E\alpha}{1-2\nu}\tau n_i, \quad i = 1, 2, 3. \end{cases} \quad (7.15)$$

If the σ'_{ij} in the above expression are replaced by the stresses σ_{ij}, then the equilibrium equations of the stresses in the body and on the boundary are respectively,

$$\begin{cases} -\sum_{j=1}^3 \frac{\partial \sigma_{ij}}{\partial x_j} = f_i, \quad i = 1, 2, 3, \quad \Omega, \\[4mm] \sum_{j=1}^3 \sigma_{ij} n_j = g_i, \quad i = 1, 2, 3, \quad \partial\Omega. \end{cases} \quad (7.16)$$

Formally (7.16) is identical to (5.13) and (5.14). Substituting (7.13) and (7.14), which are stresses represented by the partial derivatives of displacements, into (7.16), we obtain

$$\Omega: -\sum_{j=1}^3 \frac{\partial}{\partial x_j}\left\{ \frac{E}{2(1+\nu)}\left(\frac{\partial u_j}{\partial x_j} + \frac{\partial u_i}{\partial x_j} \right) \right\} + \frac{\partial}{\partial x_i} \frac{E\nu}{(1+\nu)(1-2\nu)} \sum_{k=1}^3 \frac{\partial u_k}{\partial x_k}$$

$$= f_i - \frac{\partial}{\partial x_i}\left(\frac{E\alpha}{1-2\nu}\tau \right), \quad i = 1, 2, 3,$$

$$(7.17)$$

$$\partial\Omega : \sum_{j=1}^{3} \left\{ \frac{E}{2(1+\nu)} \left(\frac{\partial u_j}{\partial x_i} + \frac{\partial u_i}{\partial x_j} \right) + \frac{E\nu}{(1+\nu)(1-2\nu)} \times \left(\sum_{k=1}^{3} \frac{\partial u_k}{\partial x_k} \right) \delta_{ij} \right\} n_j$$

$$= g_i + \frac{E\alpha}{1-2\nu} \tau n_i.$$

$$(7.18)$$

Compared to the case without thermal effects, a temperature gradient term $-\dfrac{\partial}{\partial x_i} \left(\dfrac{E\alpha}{1-2\gamma} \tau \right)$ has been added to the right hand side of (7.17). This term plays the part of a load, and may as well called the thermal body load. Similarly, a thermal surface load term $\dfrac{E\alpha}{1-2\gamma} \tau n_i$ has been added to the right hand side of the boundary condition (7.18).

Using the same method as in this section, we can discuss other kinds of boundary conditions and interface conditions of the medium.

Chapter 3

Typical Problems of Elastic Equilibrium

The classical linearized theory of the threedimensional elastic body has been introduced in Chapter 2, where we emphasized the view-point of energy and variational principles. Since all elastic bodies are three-dimensional, they can in principle be solved by directly using that general three-dimensional theory. In practice, this way of solving problems is used more and more, especially when finding numerical solutions based on the finite element method. However, there are many typical elastic structures in engineering which have their own special geometric and mechanical features, such as thin plates and shells, slender beams and rods, and bodies with various symmetries.

The task of this chapter is apply the fundamental theory in Chapter 2 to the concrete case from the general view-point. Some typical two-dimensional and one-dimensional modes are examined and the corresponding variational principles are obtained based on the geometrical and mechanical characteristics of the modes. Thus, not only are simplified methods of solution found, but the theory is also enriched and developed. We shall return again to the problems introduced in Chapter 1, and discuss them more completely.

§1   Plane Elastic Problems

The plane elastic problem is a general term for the plane strain problem and plane stress problem. Their common characteristic is that, the displacement u depends only on two coordinates x_1 and x_2, but is independent of the third coordinate of x_3:

$$u_i = u_i(x_1, x_2), \quad i = 1, 2, 3. \tag{1.1}$$

The so-called plane strain problems should further satisfy

$$\varepsilon_{13} = \varepsilon_{23} = \varepsilon_{33} = 0, \tag{1.2}$$

while the plane stress problem should satisfy

$$\sigma_{13} = \sigma_{23} = \sigma_{33} = 0. \tag{1.3}$$

For the plane strain problem, from (1.1) and (1.2), we have

$$\varepsilon_{13} = \frac{1}{2}\left(\frac{\partial u_3}{\partial x_i} + \frac{\partial u_i}{\partial x_3}\right) = \frac{1}{2}\frac{\partial u_3}{\partial x_i} = 0, \quad i = 1, 2, 3,$$

so that u_3 is constant. We may let $u_3 = 0$. This corresponds to, the displacement being constrained in one of the directions (the x_3−direction). Conversely, for the plane stress problem, (1.3) implies that there is no load in the x_3−direction, while the displacement is absolutely free.

For a uniform cylinder with infinite length, whose the axial direction is x_3, its elastic deformation belongs to the plane strain problem when the load distribution is independent of x_3 and the component of the load in the x_3−direction vanishes. If the cylinder with infinite length is transformed into that with finite length, e.g., if a segment included between two parallel rigid cross sections is to be treated, whose displacement in the x_3−direction is restrained, then this is also a plane strain problem. In general, a straight and long dam or tunnel, a long roller in roller bearings, a long tube under internal and external pressures, etc., can all be considered as cases of the plane strain problem.

The plane stress problem is mainly intended to describe in-plane deformation of thin plates, i.e., the so-called thin plate in stretching. In that case, the loads are parallel to the plane of the plate and distributed uniformly over the thickness of the plate. The two surfaces of the plate are completely free. Let the plane be normal to the x_3 axis. On the plane of the plate there is the relation $\sigma_{i3} = 0$ holds, and it also holds in the plate because the plate is very thin. The plane stress problem also applies to the bending of beams. If the width of the beam is comparable to the length of the beam in the plane of bending, the so-called deep beam or retaining wall, the beam bending problem can also be considered to be a plane stress problem.

Although the mechanical background of the plane strain and of the plane stress problems are entirely different, we will see that they have a unified mathematical form. We shall discuss the general case containing thermal effects. The conclusion in the absence of thermal efffects can be obtained immediately by letting the temperature rise $\tau = 0$.

1.1   Plane strain problems

Substituting the conditions

$$\varepsilon_{13} = \varepsilon_{23} = \varepsilon_{33} = 0$$

into Hooke's law (7.7) and formula (7.10) for the volume density of strain energy of a three-dimensional elastic body with thermal effects in §7 of Chapter 2, we get

$$\sigma_{ii} = \frac{E}{1+\nu}\varepsilon_{ii} + \frac{E\nu}{(1+\nu)(1-2\nu)}\sum_{k=1}^{2}\varepsilon_{kk} - \frac{E\alpha}{1-2\nu}\tau$$

$$= \frac{E}{1+\nu}\frac{\partial u_i}{\partial x_i} + \frac{E\nu}{(1+\nu)(1-2\nu)}\left(\frac{\partial u_1}{\partial x_1} + \frac{\partial u_2}{\partial x_2}\right)$$

$$- \frac{E\alpha}{1-2\nu}\tau, \quad i = 1,2, \tag{1.4}$$

$$\sigma_{12} = \sigma_{21} = \frac{E}{1+\nu}\varepsilon_{12} == \frac{E}{2(1+\nu)}\left(\frac{\partial u_2}{\partial x_1} + \frac{\partial u_1}{\partial x_2}\right), \tag{1.5}$$

$$\sigma_{13} = \sigma_{23} + 0,$$

$$\sigma_{33} = \frac{E\nu}{(1+\nu)(1-2\nu)}(\varepsilon_{11} + \varepsilon_{22}) - \frac{E\alpha}{1-2\nu}\tau.$$

Note that σ_{33} does not vanish in general. All the stress components $\sigma_{ij}(i,j = 1,2,3)$ can be expressed in terms of the three strain components $\varepsilon_{ij}(i,j = 1,2)$.

The volume density of the strain energy is

$$W = W' - \frac{E\alpha}{1-2\nu}\tau\sum_{k=1}^{2}\varepsilon_{kk}, \tag{1.6}$$

where

$$W' = \frac{1}{2}\left\{\frac{E}{1+\nu}\sum_{i,j=1}^{2}\varepsilon_{ij}^2 + \frac{E\nu}{(1+\nu)(1-2\nu)} \times \left(\sum_{k=1}^{2}\varepsilon_{kk}\right)^2\right\}. \tag{1.7}$$

W is a nonhomogeneous quadratic form of $\varepsilon_{ij}(i,j = 1,2)$, whose homogeneous quadratic part W' is positive definite. Clearly , we have

$$\sigma_{ij} = \frac{\partial W}{\partial \varepsilon_{ij}}, \quad i,j = 1,2.$$

but

$$W \neq \frac{1}{2} \sum_{i,j=1}^{2} \sigma_{ij}\varepsilon_{ij}.$$

Let Ω denote a standard cross section of the original elastic body and $\partial\Omega$ denote the intersecting line of the original boundary and this cross section. Suppose that $\partial\Omega = \Gamma_1 + \Gamma_3$; that there are fixed displacements $u_i = \bar{u}_i$, $i = 1,2$ on Γ_1; and that there are elastic supportings on Γ_3, whose elastic reaction forces are $-\sum_{j=1}^{2} c_{ij}u_j + g_i$, $i = 1,2$, where $[c_{ij}]$ is a symmetric positive definite matrix. Furthermore, suppose that the densities of external loads on Ω are f_i, $i = 1,2$, and that there exists an interface line Γ' of the medium in the interior of Ω. Hence, the strain energy is

$$\iint_{\Omega} W(\boldsymbol{u})dS = \iint_{\Omega} W'(\boldsymbol{u})dS - \iint_{\Omega} \frac{E\alpha}{1-2\nu}\tau \sum_{k=1}^{2} \varepsilon_{kk}(\boldsymbol{u})dS$$

$$+\frac{1}{2}\int_{\Gamma_3} \sum_{i,j=1}^{2} c_{ij}u_iu_jdl, \tag{1.8}$$

where

$$\iint_{\Omega} W'(\boldsymbol{u})ds = \frac{1}{2}\iint_{\Omega} \left\{ \frac{E}{1+\nu} \sum_{i,j=1}^{2} \varepsilon_{ij}^2(\boldsymbol{u}) \right.$$

$$\left. +\frac{E\nu}{(1+\nu)(1-2\nu)}\left(\sum_{k=1}^{2} \varepsilon_{kk}(\boldsymbol{u})\right)^2 \right\}dS.$$

The potential energy of the external work is

$$-\iint_{\Omega} \sum_{i=1}^{2} f_iu_idS - \int_{\Gamma_3} \sum_{i=1}^{2} g_iu_idl. \tag{1.9}$$

Let

$$D(\boldsymbol{u},\boldsymbol{v}) = \iint_{\Omega} \left\{ \frac{E}{1+\nu} \sum_{i,j=1}^{2} \varepsilon_{ij}(\boldsymbol{u})\varepsilon_{ij}(\boldsymbol{v}) \right.$$

$$\left. +\frac{E\nu}{(1+\nu)(1-2\nu)} \sum_{k=1}^{2} \varepsilon_{kk}(\boldsymbol{u}) \sum_{k=1}^{2} \varepsilon_{kk}(\boldsymbol{v}) \right\}dS$$

$$+\int_{\Gamma_3} \sum_{i,j=1}^{2} c_{ij}u_iv_jdl, \tag{1.10}$$

$$F(\boldsymbol{u}) = \iint_\Omega \frac{E\alpha}{1-2\nu}\tau \sum_{k=1}^{2}\varepsilon_{kk}(\boldsymbol{u})dS + \iint_\Omega \sum_{i=1}^{2}f_i u_i dS$$

$$+ \int_{\Gamma_3} \sum_{i=1}^{2} g_i u_i dl. \tag{1.11}$$

Then the principle of minimum potential energy is

$$\begin{cases} \dfrac{1}{2}D(\boldsymbol{u},\boldsymbol{u}) - F(\boldsymbol{u}) = \text{Min}, \\[2mm] \Gamma_1 : u_i = \bar{u}_i, \quad i = 1,2, \end{cases} \tag{1.12}$$

which is equivalent to the principle of virtual work

$$\left.\begin{cases} D(\boldsymbol{u},\boldsymbol{v}) - F(\boldsymbol{v}) = 0, \quad \text{for any virtual displacement } \boldsymbol{v} \\[2mm] \Gamma_1 : u_i = \bar{u}_i, \quad i = 1,2. \end{cases}\right\}. \tag{1.13}$$

The equilibrium equations are

$$\begin{cases} \Omega - \Gamma' : -\displaystyle\sum_{j=1}^{2}\frac{\partial\sigma_{ij}}{\partial x_j} = f_i, \quad i = 1,2, \\[4mm] \Gamma_1 : u_i = \bar{u}_i, \quad i = 1,2, \\[4mm] \Gamma_3 : \displaystyle\sum_{j=1}^{2}\sigma_{ij}n_j + \sum_{j=1}^{2}c_{ij}u_j = g_i, \quad i = 1,2, \\[4mm] \Gamma' : \displaystyle\sum_{j=1}^{2}\sigma_{ij}'\nu_j - \sum_{j=1}^{2}\sigma_{ij}^+\nu_j = 0, \quad i = 1,2. \end{cases} \tag{1.14}$$

Here, (ν_1,ν_2) denotes a normal direction prescribed arbitrarily on the interface line Γ', σ_{ij}^- and σ_{ij}^+ denote the stress components evaluated according to the different material constants on the two sides of Γ', respectively. Substituting (1.4) and (1.5) into (1.14), we get a boundary value problem consisting of two second order partial differential equations relating the two unknown functions u_1 and u_2 on the plane domain Ω.

1.2   Plane stress problems

Substituting the conditions

$$\sigma_{13} = \sigma_{23} = \sigma_{33} = 0$$

into (7.7) of Chapter 2 $(j = 3)$, we obtain

$$\varepsilon_{13} = \varepsilon_{23} = 0, \quad \varepsilon_{33} = -\frac{\nu}{1-\nu}\sum_{k=1}^{2}\varepsilon_{kk} + \frac{1+\nu}{1-\nu}\alpha\tau,$$

so that

$$\sum_{k=1}^{3} \varepsilon_{kk} = \frac{1-2\nu}{1-\nu} \sum_{k=1}^{2} \varepsilon_{kk} + \frac{1+\nu}{1-\nu}\alpha\tau.$$

Substituting the foregoing expression into (7.7) of Chapter 2 ($j = 1, 2$) again, we obtain

$$\sigma_{ii} = \frac{E}{1+\nu}\varepsilon_{ii} + \frac{E\nu}{1-\nu^2} \sum_{k=1}^{2} \varepsilon_{kk} - \frac{E\alpha}{1-\nu}\tau$$

$$= \frac{E}{1+\nu}\frac{\partial u_i}{\partial x_i} + \frac{E\nu}{1-\nu^2}\left(\frac{\partial u_1}{\partial x_1} + \frac{\partial u_2}{\partial x_2}\right) - \frac{E\alpha}{1-\nu}\tau, \qquad (1.15)$$

$$\sigma_{12} = \sigma_{21} = \frac{E}{1+\nu}\varepsilon_{12} = \frac{E}{2(1+\nu)}\left(\frac{\partial u_2}{\partial x_1} + \frac{\partial u_1}{\partial x_2}\right). \qquad (1.16)$$

The stress components $\sigma_{ij}(i, j = 1, 2, 3)$ can all be expressed in terms of the three strain components $\varepsilon_{ij}(i, j = 1, 2)$. Note that ε_{33} does not vanish in general.

The volume density of the strain energy is

$$W = W' - \frac{E\alpha}{1-\nu}\tau \sum_{k=1}^{2} \varepsilon_{kk}, \qquad (1.17)$$

where

$$W' = \frac{1}{2}\left\{ \frac{E}{1+\nu} \sum_{i,j=1}^{2} \varepsilon_{ij}^2 + \frac{E\nu}{1-\nu^2}\left(\sum_{k=1}^{2} \varepsilon_{kk}\right)^2 \right\}. \qquad (1.18)$$

As in plane strain problem, W is a nonhomogeneous quadratic form of $\varepsilon_{ij}(i, j = 1, 2)$, whose homogeneous quadratic part W' is positive definite. We have

$$\sigma_{ij} = \frac{\partial W}{\partial \varepsilon_{ij}}, \quad i, j = 1, 2.$$

As in the plane strain problem, we can now write down the variational principle and the equivalent equilibrium equations for the plane stress problem.

1.3   Comparisons

As with the two dimensional stress and strain tensors σ_{ij}, $\varepsilon_{ij}(i, j = 1, 2)$, the Hooke's law and the strain energy expressions of the plane strain problem and of the plane stress problem are identical in form, differing only in their coefficients. From (1.4), the three coeffficients of the Hooke's law in the plane strain problem are

$$\frac{E}{1+\nu}, \quad \frac{E\nu}{(1+\nu)(1-2\nu)}, \quad \frac{E\alpha}{1-2\nu}.$$

From (1.15), the three corresponding coefficients of the plane stress problem are

$$\frac{E}{1+\nu}, \quad \frac{E\nu}{1-\nu^2}, \quad \frac{E\alpha}{1-\nu}.$$

If we let

$$E' = \frac{E(1+2\nu)}{(1+\nu)^2}, \quad \nu' = \frac{\nu}{1+\nu}, \quad \alpha' = \frac{\alpha(1+\nu)}{1+2\nu},$$

then we have

$$\frac{E'}{1+\nu'} = \frac{E}{1+\nu}, \quad \frac{E'\nu'}{(1+\nu')(1-2\nu')} = \frac{E\nu}{1-\nu^2}, \quad \frac{E'\alpha'}{1-2\nu'} = \frac{E\alpha}{1-\nu}.$$

Hence the plane stress problem, whose material constants are E, ν and α, is identical to the plane strain problem, whose material constants are E', ν', and α'.

Conversely, if let

$$E'' = \frac{E}{1-\nu^2}, \quad \nu'' = \frac{\nu}{1-\nu}, \quad \alpha'' = (1+\nu)\alpha,$$

then we have

$$\frac{E''}{1+\nu''} = \frac{E}{1+\mu},$$

$$\frac{E''\nu''}{(1+\nu'')(1-\nu'')} = \frac{E\nu}{(1+\nu)(1-2\nu)},$$

$$\frac{E''\alpha''}{1-\nu''} = \frac{E\alpha}{1-2\nu}.$$

Hence the plane strain problem, whose material constants are E, ν, and α, is identical to the plane stress problem, whose material constants are E'', ν'', and α''.

Comparing again the strain energy density of the plane strain problem with that of the plane stress problem without thermal effects, i.e., $\tau = 0$, we have $W = W'$. Since

$$\frac{E\nu}{(1+\nu)(1-2\nu)} > \frac{E\nu}{1-\nu^2},$$

from (1.7) and (1.8), we obtain

$$
W'_{\text{plane strain}} = \frac{1}{2}\left\{ \frac{E}{1+\nu}\sum_{i,j=1}^{2}\varepsilon_{ij}^2 + \frac{E\nu}{(1+\nu)(1-2\nu)}\left(\sum_{k=1}^{2}\varepsilon_{kk}\right)^2 \right\}
$$

$$
> \frac{1}{2}\left\{ \frac{E}{1+\nu}\sum_{i,j=1}^{2}\varepsilon_{ij}^2 + \frac{E\nu}{1-\nu^2}\left(\sum_{k=1}^{2}\varepsilon_{kk}\right)^2 \right\} = W'_{\text{plane stress}}
$$

Therefore, for the same material under the same strain, the potential energy stored in the plane strain state is larger than that in the plane stress state. This means that, for the same material, the power of resistance in the plane strain problem, in which displacement in the direction normal to the plane is restrained, is larger than that in the plane stress problem, in which the displacement in the normal direction is free. We obtained the same conclusion for one dimensional rods in §4 of Chapter 1.

Without thermal effects, the strainless states of plane problems are just infinitesimal rigid displacements in the plane. Because $\varepsilon_{ij} \equiv 0$, $i, j = 1, 2$, we immediately get

$$
v_1 = a_1 - b_3 x_2, \quad v_2 = a_2 + b_3 x_1,
$$

or

$$
\begin{bmatrix} v_1 \\ v_2 \end{bmatrix} = \begin{bmatrix} a_1 \\ a_2 \end{bmatrix} + \begin{bmatrix} 0 & -b_3 \\ b_3 & 0 \end{bmatrix} \begin{bmatrix} x_1 \\ x_2 \end{bmatrix}
$$

where b_3 is an infinitesimal rotation angle of $(x_1 - x_2)$ plane about the x_3-axis.

1.4   One dimensional problems

We can reduce the plane elastic problem still further to an one-dimensional elastic problem, in which the displacement vector depends only on one of the coordinates x_1 and is independent of x_2 and x_3: $u_i = u_i(x)$. Consequently, the problem of plane strain or of plane stress is reduced to a problem of one dimensional strains or one dimensional stresses, respectively.

1. The one dimensional strain problem

$$
\varepsilon_{13} = \varepsilon_{23} = \varepsilon_{33} = 0, \quad \text{which corresponds to } u_3 = 0,
$$

$$
\varepsilon_{12} = \varepsilon_{22} = \varepsilon_{32} = 0, \quad \text{which corresponds to } u_2 = 0.
$$

Hence

$$\sigma_{11} = \frac{E(1-\nu)}{(1+\nu)(1-2\nu)}\varepsilon_{11} - \frac{E\alpha}{1-2\nu}\tau,$$

$$\sigma_{22} = \sigma_{33} = \frac{E\nu}{(1+\nu)(1-2\nu)}\varepsilon_{11} - \frac{E\alpha}{1-2\nu}\tau,$$

$$\sigma_{12} = \sigma_{13} = \sigma_{23} = 0.$$

Note that, there are no displacements ($u_2 = u_3 = 0$) but there are still direct stresses σ_{22} and σ_{33} in the x_2- and x_3-directions.

The strain energy density is

$$W = W' - \frac{E\alpha}{1-2\nu}\tau\varepsilon_{11},$$

$$W' = \frac{1}{2}\frac{E(1-\nu)}{(1+\nu)(1-2\nu)}\varepsilon_{11}^2,$$

$$\sigma_{11} = \frac{\partial W}{\partial \varepsilon_{11}}.$$

2. The one dimensional stress problem

$$\sigma_{13} = \sigma_{23} = \sigma_{33} = 0, \quad \text{which corresponds to } u_3 \text{ being free,}$$

$$\sigma_{12} = \sigma_{22} = \sigma_{32} = 0, \quad \text{which corresponds to } u_2 \text{ being free.}$$

Hence

$$\sigma_{11} = E\varepsilon_{11} - E\alpha\tau, \quad \varepsilon_{12} = \varepsilon_{13} = \varepsilon_{23} = 0,$$

$$\varepsilon_{22} = \varepsilon_{33} = -\nu\varepsilon_{11} + (1+\nu)\alpha\tau.$$

In this case there are deformations in the x_2-and x_3-directions.

The strain energy density is

$$W = W' - E\alpha\tau\varepsilon_{11}, \quad W' = \frac{1}{2}E\varepsilon_{11}^2, \quad \sigma_{11} = \frac{\partial W}{\partial \varepsilon_{11}}.$$

Let

$$E' = \frac{E(1-\nu)}{(1+\nu)(1-2\nu)}, \quad \nu' = \nu, \quad \alpha' = \frac{1+\nu}{1-\nu}\alpha.$$

The problem of one dimensional strains, whose material constants are E, ν, and α, is then identical to the problem of one dimensional stresses, whose material constants are E', ν', and α'.

The transversely free or fixed rod in tension discussed in §3 and §4 of Chapter 1 is just the mode of one dimensional stresses or one dimensional strains, respectively.

§2   Plane Geometric Compatibility and Stress Function

2.1   Plane geometric compatibility

In the plane elastic problem, whether it is a plane strain or plane stress mode, the elastic deformation can be fully determined by a two dimensional displacement vector field

$$u_i = u_i(x_1, x_2), \quad i, j = 1, 2$$

and has the corresponding two dimensional strain tensor

$$\varepsilon_{ij} = \varepsilon_{ji} = \frac{1}{2}\left(\frac{\partial u_j}{\partial x_i} + \frac{\partial u_i}{\partial x_j}\right), i, j = 1, 2. \tag{2.1}$$

ε_{ij} have only three essential components

$$\varepsilon_{11} = \frac{\partial u_1}{\partial x_1}, \quad \varepsilon_{22} = \frac{\partial u_2}{\partial x_2}, \quad \varepsilon_{12} = \varepsilon_{21} = \frac{1}{2}\left(\frac{\partial u_2}{\partial x_1} + \frac{\partial u_1}{\partial x_2}\right).$$

Correspondingly, the two dimensional rotation tensor

$$\omega_{ij} = -\omega_{ji} = \frac{1}{2}\left(\frac{\partial u_j}{\partial x_i} - \frac{\partial u_i}{\partial x_j}\right), \quad i, j = 1, 2 \tag{2.2}$$

has only one essential component

$$\omega_{11} = \omega_{22} = 0,$$

$$\omega_{12} = -\omega_{21} = \omega_3 = \frac{1}{2}\left(\frac{\partial u_2}{\partial x_1} - \frac{\partial u_1}{\partial x_2}\right),$$

and

$$\frac{\partial u_j}{\partial x_i} = \varepsilon_{ij} + \omega_{ij}, \quad i, j = 1, 2. \tag{2.3}$$

As in §1 of Chapter 2, the derivatives of ε_{ij} and ω_{ij}, are in the relations

$$\frac{\partial \omega_{ij}}{\partial x_k} = \frac{\partial \varepsilon_{jk}}{\partial x_i} - \frac{\partial \varepsilon_{ik}}{\partial x_j}, \quad ij = 12, k = 1, 2, \tag{2.4}$$

i.e.,

$$\frac{\partial \omega_{12}}{\partial x_1} = \frac{\partial \varepsilon_{21}}{\partial x_1} - \frac{\partial \varepsilon_{11}}{\partial x_2}, \quad \frac{\partial \omega_{12}}{\partial x_2} = \frac{\partial \varepsilon_{22}}{\partial x_1} - \frac{\partial \varepsilon_{12}}{\partial x_2}.$$

Since

$$\frac{\partial}{\partial x_1}\frac{\partial \omega_{12}}{\partial x_2} = \frac{\partial}{\partial x_2}\frac{\partial \omega_{12}}{\partial x_1},$$

the two dimensional Saint-Venant geometric compatibility equation can be obtained from (2.4).

$$\frac{\partial^2 \varepsilon_{11}}{\partial x_2^2} + \frac{\partial^2 \varepsilon_{22}}{\partial x_1^2} = 2 \frac{\partial^2 \varepsilon_{12}}{\partial x_1 \partial x_2}. \tag{2.5}$$

Note that, in the case of three dimensions, there are six geometric compatibility equations in all (see (6.13)–(6.14) in §6 of Chapter 2), which are reduced to only one equation in the case of two dimensions.

For a simply-connected domain Ω, the geometric compatibility equation (2.5) is a necessary and sufficient condition of making ε_{ij} as a strain tensor (i.e., there exist such u_1 and u_2 as to make (2.1) valid).

Now consider a multiply-connected domain Ω, whose connectivity number, which is just the number of holes in Ω, is p. Taking a simple closed loop around each hole, we obtain $L_1, \cdots, L_p$, which constitute fundamental a set of loops of Ω. According to Proposition 2.3 in §6 of Chapter 2, we must add to the necessary and sufficient condition for making ε_{ij} a strain tensor $3p$ more Volterra integral conditions

$$\begin{cases} \oint_{L_m} \sum_{k=1}^{2} \left(\frac{\partial \varepsilon_{2k}}{\partial x_1} - \frac{\partial \varepsilon_{1k}}{\partial x_2} \right) dx_k = 0, \quad m = 1, 2, \cdots, p, \tag{2.6} \\ \oint_{L_m} \sum_{k=1}^{2} \left[\varepsilon_{kj} - \sum_{i=1}^{2} \left(\frac{\partial \varepsilon_{jk}}{\partial x_i} - \frac{\partial \varepsilon_{ik}}{\partial x_j} \right) x_i \right] dx_k = 0, \tag{2.7} \\ \qquad j =, 2; \quad m = 1, \cdots, p \end{cases}$$

in addition to the Saint-Venant geometric compatibility equation (2.5). These conditions imply the monodromy of the curl ω_{12} and of the displacements u_1 and u_2, i.e., for any loop L in Ω,

$$\oint_L d\omega_{12} = 0, \quad \oint_L du_1 = 0, \quad \oint_L du_2 = 0, \tag{2.8}$$

respectively.

2.2 Stress function

In §1 of the present chapter, we introduced a mathematical formulation and solution method for plane elasticity problems that uses u_1 and u_2 as fundamental variables. This is called the displacement method. It is also possible to take the stresses σ_{ij} as fundamental variables. That is called the force method. In the plane situation, the force method can be simplified in some measure by introducing the stress function.

How to analyze an elastic problem by the force method, depends as a rule, on the topological properties of the domain of the elastic body. Let us first consider the case of simply-connected domains.

Suppose we have a uniform, elastic plane body Ω, whose material moduli E, ν, α are all constants, and that there are no body loads: $f_1 = f_2 \equiv 0$. In the equilibrium state, the stress field σ_{ij} satisfies the homogeneous equilibrium equations

$$\Omega : \begin{cases} \dfrac{\partial \sigma_{11}}{\partial x_1} + \dfrac{\partial \sigma_{21}}{\partial x_2} = 0, \\[3mm] \dfrac{\partial \sigma_{21}}{\partial x_1} + \dfrac{\partial \sigma_{22}}{\partial x_2} = 0. \end{cases} \tag{2.9}$$

Since Ω is simply-connected, these two conditions respectively guarantee (Proposition 2.1 in §6 of Chapter 2) the existence of singlevalued functions ϕ_1 and ϕ_2 on Ω such that

$$\frac{\partial \phi_1}{\partial x_1} = -\sigma_{12}, \qquad \frac{\partial \phi_1}{\partial x_2} = \sigma_{11}, \tag{2.10}$$

$$\frac{\partial \phi_2}{\partial x_1} = -\sigma_{22}, \qquad \frac{\partial \phi_2}{\partial x_2} = \sigma_{21}. \tag{2.11}$$

Moreover, since $\sigma_{12} = \sigma_{21}$,

$$\frac{\partial \phi_1}{\partial x_1} + \frac{\partial \phi_2}{\partial x_2} = 0. \tag{2.12}$$

This guarantees the existence of a single-valued function φ on Ω such that

$$\frac{\partial \varphi}{\partial x_1} = -\varphi_2, \qquad \frac{\partial \varphi}{\partial x_2} = \phi_1. \tag{2.13}$$

Then

$$\sigma_{11} = \frac{\partial^2 \varphi}{\partial x_2^2}, \qquad \sigma_{22} = \frac{\partial^2 \varphi}{\partial x_1^2}, \qquad \sigma_{12} = \sigma_{21} = -\frac{\partial^2 \varphi}{\partial x_1 \partial x_2}. \tag{2.14}$$

This function φ is called the Airy stress function.

Suppose there are two stress functions φ and $\bar{\varphi}$ corresponding to the same stress field, then the difference $\varphi - \bar{\varphi}$ satisfies

$$\frac{\partial^2 (\varphi - \bar{\varphi})}{\partial x_1^2} = \frac{\partial^2 (\varphi - \bar{\varphi})}{\partial x_2^2} = \frac{\partial^2 (\varphi - \bar{\varphi})}{\partial x_1 \partial x_2} \equiv 0.$$

So that

$$\varphi - \bar{\varphi} = a_1 x_1 + a_2 x_2 + b. \tag{2.15}$$

Thus, the stress function corresponding to a given stress field depends on three constants of integration. These three constants of integration, can be taken as the values of $\dfrac{\partial \varphi}{\partial x_1}$, $\dfrac{\partial \varphi}{\partial x_2}$, and φ at a specified point P_0.

Up to a polynomial of degree 1, the stress function is determined uniquely by the stress field. Thus, the stress tensor field σ_{ij} can be replaced by a scalar field φ satisfying (2.14). Since

$$\frac{\partial \sigma_{11}}{\partial x_1} + \frac{\partial \sigma_{12}}{\partial x_2} = \frac{\partial^3 \varphi}{\partial x_1 \partial x_2^2} - \frac{\partial^3 \varphi}{\partial x_2 \partial x_1 \partial x_2} = 0,$$

$$\frac{\partial \sigma_{21}}{\partial x_1} + \frac{\partial \sigma_{22}}{\partial x_2} = \frac{-\partial^3 \varphi}{\partial x_1 \partial x_1 \partial x_2} + \frac{\partial^3 \varphi}{\partial x_2 \partial x_1^2} = 0,$$

the equilibrium equations (2.9) are satisfied automatically.

According to Hooke's law, the strain field ε_{ij} can also be expressed in terms of the stress function φ. For the sake of clarity, considering a plane stress mode. We have

$$\begin{cases} \varepsilon_{11} = \frac{1}{E}\sigma_{11} - \frac{\nu}{E}\sigma_{22} + \alpha\tau = \frac{1}{E}\frac{\partial^2 \varphi}{\partial x_2^2} - \frac{\nu}{E}\frac{\partial^2 \varphi}{\partial x_1^2} + \alpha\tau, \\[2mm] \varepsilon_{22} = \frac{1}{E}\sigma_{22} - \frac{\nu}{E}\sigma_{11} + \alpha\tau = \frac{1}{E}\frac{\partial^2 \varphi}{\partial x_1^2} - \frac{\nu}{E}\frac{\partial^2 \varphi}{\partial x_2^2} + \alpha\tau, \\[2mm] \varepsilon_{12} = \varepsilon_{21} = \frac{1+\nu}{E}\sigma_{12} = -\frac{1+\nu}{E}\frac{\partial^2 \varphi}{\partial x_1 \partial x_2}. \end{cases} \qquad (2.16)$$

Now, the geometric compatibility equation (2.5) satisfied by ε_{ij} can be expressed as

$$\Omega : \frac{1}{E}\Delta^2 \varphi = \alpha\Delta\tau, \qquad (2.17)$$

where

$$\Delta = \frac{\partial^2}{\partial x_1^2} + \frac{\partial^2}{\partial x_2^2}, \quad \Delta^2 = \frac{\partial^4}{\partial x_1^4} + 2\frac{\partial^4}{\partial x_1^2 \partial x_2^2} + \frac{\partial^4}{\partial x_2^4}.$$

Hence the stress function φ satisfies a nonhomogeneous biharmonic equation with a term $\alpha\Delta\tau$ on its right hand side, which is a fourth order elliptic equation. When no thermal effect is taken into account, i.e., $\tau \equiv 0$, or when there is a steady sourceless temperature distribution, i.e., $\Delta\tau \equiv 0$, the right hand of (2.17) vanishes. Then φ is a biharmonic function:

$$\Omega : \Delta^2 \varphi = 0. \qquad (2.18)$$

To sum up, we can see that, in the force method based on the stress function φ, the stresses and the strains are expressed in terms of second order derivatives of φ as in equations (2.14) and (2.16), respectively, while the geometric compatibility equation, which is expressed as a fourth order elliptic equation (2.17) in φ, plays the role of the fundamental equation. For simply-connected domains, such an equation guarantees that, the ε_{ij} determined from (2.16) are exactly the strains induced by a certain displacement field, and the σ_{ij} determined from (2.14) are exactly the stresses. If necessary, it is also possible to obtain the required displacements by integrating the strains ε_{ij}. The equilibrium equations (2.9) can be neglected because they are satisfied automatically.

Conversely, when a problem is solved by the displacement method, taking the displacements u_1 and u_2 as the unknown functions, the stresses and the strains are expressed in terms of the first derivatives of the displacements. The equilibrium conditions, which are expressed as two second order elliptic equations in u_1 and u_2, are the fundamental equations, and the displacements u_1 and u_2 can be found by solving it. And consequently, the strains and the stresses can be obtained without having to consider the geometric compatibility equation, because it is satisfied automatically once the displacements have been obtained.

2.3   Boundary conditions

Since the geometric compatibility equation (2.17) satisfied by the stress function φ is a fourth order elliptic equation, in order to determine a solution on Ω, it is necessary to prescribe two boundary conditions. Suppose the whole boundary is subject to given surface loads

$$\partial\Omega : \sigma_{11}n_1 + \sigma_{12}n_2 = g_1, \quad \sigma_{21}n_1 + \sigma_{22}n_2 = g_2, \tag{2.19}$$

where $\boldsymbol{n} = (n_1, n_2)^T$ is the outward normal direction of $\partial\Omega$.

From §5 of Chapter 2, we see that in order to guarantee that there exist a unique equilibrating stress field in Ω, the resultant force and the resultant moment of the prescribed external loads must vanish, i.e.,

$$\oint_{\partial\Omega} g_1 dl = 0, \quad \oint_{\partial\Omega} g_2 dl = 0,$$

$$\oint_{\partial\Omega} (x_1 g_2 - x_2 g_1) dl = 0. \tag{2.20}$$

When the stresses σ_{ij} are expressed in terms of the stress function φ, (2.19) becomes two boundary conditions involving the second derivatives of φ

$$\partial\Omega : \begin{cases} n_1 \dfrac{\partial^2 \varphi}{\partial x_2^2} - n_2 \dfrac{\partial^2 \varphi}{\partial x_1 \partial x_2} = g_1, \\[4mm] -n_1 \dfrac{\partial^2 \varphi}{\partial x_1 \partial x_2} + n_2 \dfrac{\partial^2 \varphi}{\partial x_1^2} = g_2. \end{cases} \tag{2.21}$$

These boundary conditions determine a unique solution of (2.17). However, the form of (2.21) is still not the standard form boundary conditions for biharmonic equations. It is disirable to transform them into the form only involving φ and $\dfrac{\partial \varphi}{\partial n}$.

Since Ω is assumed to be simply-connected, $\partial \Omega$ is a loop. Choose an arbitrary reference point $P_0 = (x_1^{(0)}, x_2^{(0)})$ on $\partial \Omega$. Define a tangential direction $\boldsymbol{t} = (t_1, t_2)^T$ along the loop that forms a right hand system $(\boldsymbol{n}, \boldsymbol{t})$ (Fig. 33). That is, the tangent points in the anticlockwise direction. Then we have

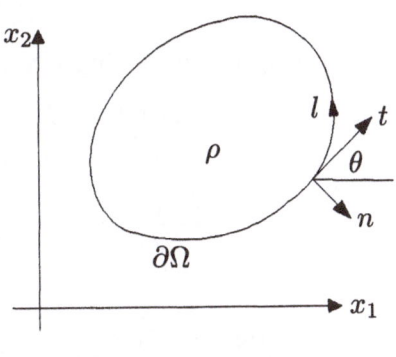

Fig. 33

$$t_1 = -n_2 = \frac{dx_2}{dl} = \cos\theta,$$

$$t_2 = n_1 = \frac{dx_1}{dl} = \sin\theta. \tag{2.22}$$

From the loads $g_1(l)$ and $g_2(l)$ on $\partial \Omega$, we can define three other functions:

$$G_1(P) = \int_{P_0}^{P} g_1 dl, \quad G_2(P) = \int_{P_0}^{P} g_2 dl,$$

$$M(P) = \int_{P_0}^{P} (x_1 g_2 - x_2 g_1) dl. \tag{2.23}$$

Here and below, integration always proceeds from the reference point P_0 to P in the positive direction on $\partial \Omega$. From (2.20) we see that, $G_1(P)$, $G_2(P)$, and $M(P)$ are all single-valued functions on $\partial \Omega$. Clearly,

$$G_1(P_0) = G_2(P_0) = M(P_0) = 0, \tag{2.24}$$

$$\frac{dG_1}{dl} = g_1, \quad \frac{dG_2}{dl} = g_2, \quad \frac{dM}{dl} = x_1 g_2 - x_2 g_1. \tag{2.25}$$

Furthermore, the stress function φ depends on three constants of integration which can be chosen arbitrarily and may be taken as

$$\left.\frac{\partial \varphi}{\partial x_1}\right|_{P_0} = A_1, \quad \left.\frac{\partial \varphi}{\partial x_2}\right|_{P_0} = A_2, \quad \varphi(P_0) = A_3. \tag{2.26}$$

Using to (2.22), the boundary conditions (2.21) can be rewritten as

$$n_1 \frac{\partial^2 \varphi}{\partial x_2^2} - n_2 \frac{\partial^2 \varphi}{\partial x_1 \partial x_2} = \frac{\partial}{\partial x_2}\left(\frac{\partial \varphi}{\partial x_2}\right)t_2 + \frac{\partial}{\partial x_1}\left(\frac{\partial \varphi}{\partial x_2}\right)t_1 = \frac{d}{dl}\left(\frac{\partial \varphi}{\partial x_2}\right) = g_1,$$

$$-n_1 \frac{\partial^2 \varphi}{\partial x_1 \partial x_2} + n_2 \frac{\partial^2 \varphi}{\partial x_1^2} = -\left[\frac{\partial}{\partial x_2}\left(\frac{\partial \varphi}{\partial x_1}\right)t_2 + \frac{\partial}{\partial x_1}\left(\frac{\partial \varphi}{\partial x_1}\right)t_2\right]$$

$$= -\frac{d}{dl}\left(\frac{\partial \varphi}{\partial x_1}\right) = g_2.$$

Integrating the foregoing two expression again, we obtain

$$\frac{\partial \varphi(P)}{\partial x_2} = \frac{\partial \varphi}{\partial x_2}\Big|_{P_0} + \int_{P_0}^{P} g_1 dl = A_2 + G_1(P),$$

$$\frac{\partial \varphi(P)}{\partial x_1} = \frac{\partial \varphi}{\partial x_1}\Big|_{P_0} - \int_{P_0}^{P} g_2 dl = A_1 - G_2(P).$$

Hence

$$\frac{\partial \varphi}{\partial n} = \frac{\partial \varphi}{\partial x_1}n_1 + \frac{\partial \varphi}{\partial x_2}n_2 = (A_1 - G_2(P))n_1 + (A_2 + G_1(P))n_2. \quad (2.27)$$

On the other hand, since

$$\frac{d\varphi}{dl} = \frac{\partial \varphi}{\partial x_1}t_1 + \frac{\partial \varphi}{\partial x_2}t_2 = (A_1 - G_2(P))t_1 + (A_2 + G_1(P))t_2,$$

integrating both sides and noting (2.24) and (2.25), we obtain

$$\varphi(P) = \varphi(P_0) + \int_{P_0}^{P} [(A_1 - G_2)t_1 + (A_2 + G_1)t_2]dl$$

$$= A_3 + \int_{P_0}^{P} \left[(A_1 - G_2)\frac{dx_1}{dl} + (A_2 + G_1)\frac{dx_2}{dl}\right]dl$$

$$= A_3 + \left[x_1(A_1 - G_2) + x_2(A_2 + G_1)\right]_{P_0}^{P}$$

$$\quad - \int_{P_0}^{P} \left[x_1\frac{d}{dl}(A_1 - G_2) + x_2(A_2 + G_1)\right]dl$$

$$= A_3 + x_1(A_1 - G_2(P)) + x_2(A_2 + G_1(P))$$

$$\quad -[X_1^{(0)}(A_1 - G_2(P_0)) + x_2^{(0)}(A_2 + G_1(P_0))]$$

$$\quad + \int_{P_0}^{P} (x_1 g_2 - x_2 g_1)dl$$

$$= A_3 + x_1(A_1 - G_2(P)) + X_2(A_2 + G_1(P))$$

$$\quad -(x_1^{(0)}A_1 + x_2^{(0)}A_2) + M(P). \quad (2.28)$$

Letting the constants of integration be $A_1 = A_2 = A_3 = 0$, we obtain two boundary conditions

$$\partial\Omega : \begin{cases} \varphi = -x_1 G_2 + x_2 G_1 + M, \\ \dfrac{\partial\varphi}{\partial n} = -n_1 G_2 + n_2 G_1 \end{cases} \tag{2.29}$$

which are equivalent to (2.21). The functions on the right hand of (2.29) are all given on $\partial\Omega$. These are the boundary conditions of the standard form of biharmonic equations.

2.4   Multiply-connected domains

The stress function and its boundary conditions discussed above all apply to simply-connected domains. Now we shall discuss the case of multiply-connected domains. Suppose Ω contains p holes. Then the boundary is decomposed into $p + 1$ loops

$$\partial\Omega = L_0 + L_1 + \cdots + L_p,$$

where, L_0 is the external boundary, and $L_1, \cdots, L_p$ are internal boundaries, i.e., the boundaries of the holes. Owing to the multiply-connectivity, the problem becomes complicated in two respects:

(1) In order to guarantee that ε_{ij} are the geometric compatibility condition of the strains, it is still necessary to add $3p$ Volterra integral conditions (2.6) and (2.7) in addition to the Saint-Venant equations (2.5). Therefore, the stress function φ must satisfy $3p$ integral conditions, besides satisfying the biharmonic equation (2.17) or (2.18). When E, ν, and α are constants, substituting (2.16) into (2.6) and (2.7), we obtain

$$\begin{cases} \displaystyle\oint_{L_m} -\frac{\partial F(\varphi)}{\partial x_2} dx_1 + \frac{\partial F(\varphi)}{\partial x_1} dx_2 = 0, \\[2mm] \displaystyle\oint_{L_m} \left(F(\varphi) - \frac{1+\nu}{E}\frac{\partial^2\varphi}{\partial x_1^2} - x_2\frac{\partial F(\varphi)}{\partial x_2} \right) dx_1 \\[2mm] \qquad + \left(-\frac{1+\nu}{E}\frac{\partial^2\varphi}{\partial x_1 \partial x_2} + x_2\frac{\partial F(\varphi)}{\partial x_1} \right) dx_2 = 0, \\[2mm] \displaystyle\oint_{L_m} \left(-\frac{1+\nu}{E}\frac{\partial^2\varphi}{\partial x_1 \partial x_2} + x_1\frac{\partial F(\varphi)}{\partial x_2} \right) dx_1 \\[2mm] \qquad + \left(F(\varphi) - \frac{1+\nu}{E}\frac{\partial^2\varphi}{\partial x_2^2} - x_1\frac{\partial F(\varphi)}{\partial x_1} \right) dx_2 = 0, \\[2mm] \qquad\qquad\qquad m = 1, \cdots, p, \end{cases} \tag{2.30}$$

where

$$F(\varphi) = \frac{1}{E}\Delta\varphi + \alpha\tau.$$

(2) In order to guarantee the existence of single-valued ψ_1 and ψ_2 satisfying (2.10) and (2.11) valid σ_{ij} must satisfy (2.9) and

$$\begin{cases} \oint_{L_m} -\sigma_{12}dx_1 + \sigma_{11}dx_2 = 0, \\ \\ \oint_{L_m} -\sigma_{22}dx_2 + \sigma_{21}dx_2 = 0, \end{cases} \qquad m = 1, 2, \cdots, p. \qquad (2.31)$$

Since

$$dx_1 = t_1 dl = -n_2 dl, \quad dx_2 = t_2 dl = n_1 dl,$$

(2.31) can be rewritten as

$$\begin{cases} \oint_{L_m} (\sigma_{11}n_1 + \sigma_{12}n_2)dl = 0, \\ \\ \oint_{L_m} (\sigma_{21}n_1 + \sigma_{22}n_2)dl = 0, \end{cases} \qquad m = 1, 2, \cdots, p. \qquad (2.32)$$

Thus, in order to guarantee the further existence of a single-valued φ satisfying (2.13), ψ_i must satisfy

$$\oint_{L_m} -\psi_2 dx_1 + \psi_1 dx_2 = \oint_{L_m} \left(\psi_1 \frac{dx_2}{dl} - \psi_2 \frac{dx_1}{dl}\right)dl$$

$$= \oint_{L_m} \left(x_1 \frac{d\psi_2}{dl} - x_2 \frac{d\psi_1}{dl}\right)dl$$

$$= \oint_{L_m} \left[x_1\left(\frac{\partial\psi_2}{\partial x_1}t_1 + \frac{\partial\psi_2}{\partial x_2}t_2\right)\right.$$

$$\left. -x_2\left(\frac{\partial\psi_1}{\partial x_1}t_1 + \frac{\partial\psi_1}{\partial x_1}t_2\right)\right]dl$$

$$= 0, \qquad m = 1, 2, \cdots, p$$

in addition to (2.12). The second of the equalities above uses the monodromy of ψ_1 and ψ_2. From (2.10), (2.11) and (2.22), the foregoing expressions become

$$\oint_{L_m} [x_1(\sigma_{21}n_1 + \sigma_{22}n_2) - x_2(\sigma_{11}n_1 + \sigma_{12}n_2)]dl = 0,$$

$$m = 1, 2, \cdots, p. \qquad (2.33)$$

Now suppose that surface loads on the whole boundary $\partial\Omega = L_0 + L_1 + \cdots + L_p$ are prescribed, i.e., the boundary conditions are of the form of (2.19), where L_0 stands for the external boundary and $L_1, \cdots, L_p$

stand for the internal boundaries. We can see by suubstituting them into (2.32) and (2.33) that, in order to guarantee the existence of a single-valued stress function φ in the multiply-connected domain Ω satisfying the relations (2.14), for the stress boundary condition (2.19), in addition to (2.20), these are the boundary conditions on the internal boundaries

$$\oint_{L_m} g_i dl = 0, \quad \oint_{L_m} g_2 dl = 0, \quad \oint_{L_m} (x_2 g_1 - x_1 g_2) dl = 0, \qquad (2.34)$$

where $m = 1, 2, \cdots, p$. Note that, (2.20) and (2.34) certainly imply

$$\oint_{L_0} g_1 dl = 0, \quad \oint_{L_0} g_2 dl = 0, \quad \oint_{L_0} (x_2 g_1 - x_1 g_2) dl = 0. \qquad (2.35)$$

This is to say, for a multiply-connected domain, the resultant force and the resutant moment on each loop of the internal and external boundaries must vanish.

Now let us consider how to transform the boundary conditions (2.21) into the standard form under the conditions (2.34) and (2.35). Following the discussions about the simply-connected domain in Section 2.3, choose a reference point $P_m = (x_1^{(m)}, x_2^{(m)})$ on each boundary $L_m (m = 0, 1, \cdots, p)$ and define three single-valued functions

$$G_1^{(m)}(P) = \int_{P_m}^{P} g_1 dl, \quad G_2^{(m)}(P) = \int_{P_m}^{P} g_2 dl,$$

$$M^{(m)}(P) = \int_{P_m}^{P} (x_1 g_2 - x_2 g_1) dl.$$

Note that the tangential direction t on the external boundary L_0 is anti-clockwise while t on the boundaries of holes $L_1, \cdots, L_p$ are clockwise, so that, on each boundary it is always the case that

$$t_1 = -n_2, \quad t_2 = n_1.$$

Moreover, define the constants $A_1^{(m)}, A_2^{(m)}, A_3^{(m)}, m = 1, \cdots, p$, by

$$\frac{\partial \varphi}{\partial x_1}\Big|_{P_m} = A_1^{(m)}, \quad \frac{\partial \varphi}{\partial x_2}\Big|_{P_m} = A_2^{(m)}, \quad \varphi(P_m) = A_3^{(m)}.$$

Then, as in (2.27) and (2.28), the boundary conditions for the stress fuunction φ can be obtained as follows:

$$\begin{cases} \dfrac{\partial \varphi}{\partial n} = n_1(A_1^{(m)} - G_2^{(m)}(P)) + n_2(A_2^{(m)} + G_1^{(m)}(P)), \\[2mm] \varphi = A_3^{(m)} - (x_1^{(m)} A_1^{(m)} + x_2^{(m)} A_2^{(m)}) + x_1(A_1^{(m)} \\[2mm] \qquad - G_2^{(m)}(P)) + x_2(A_2^{(m)} + G_1^{(m)}(P)) \\[2mm] \qquad + M^{(m)}(P), \quad m = 0, 1, \cdots, p. \end{cases} \qquad (2.36)$$

The three constants of integration on the external boundary may taken to be $A_1^{(0)} = A_2^{(0)} = A_3^{(0)} = 0$. There are $3p$ coefficients $A_1^{(m)}$, $A_2^{(m)}$, $A_3^{(m)}(m = 1, 2, \cdots, p)$ to be determined. The stress function φ satisfies the biharmonic equation (2.17) in the domain Ω and $3p$ integral conditions (2.30) on the loops.

§3   Torsion of Cylinders

3.1   Deformation modes

The preliminary investigation of the torsion of circular rods has been performed in §5 of Chapter 1. In the present section, we will further discuss the torsion of cylinders on the basis of the theory of three dimensional elasticity.

Consider the torsion of a cylinder with arbitrary cross sections, uniform material, and torque loads applied at the two ends, but without body and surface loads. We adopt the convention that, x_3 denotes the axial direction of the cylinder, the origin lies in the cross section at one end, and there is no thermal effect. Then, as for the circular rod, each cross section Ω of the cylinder has an infinitesimal rigid rotation with respect to the adjacent cross sections. Suppose the rotation angle is $\omega_3(x_3)$. Then the two displacement components of the cross section Ω are

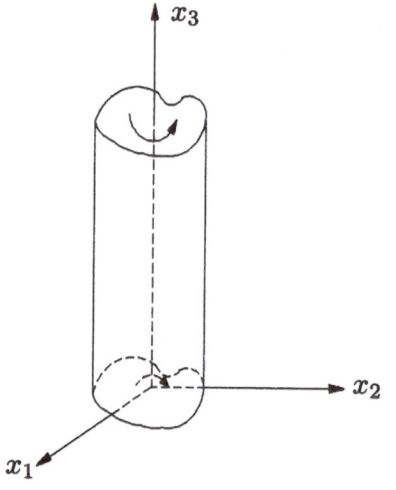

Fig. 34

$$u_1 = -\omega_3(x_3)x_2, \quad u_2 = \omega_3(x_3)x_1^{1)},$$

$$\omega_3 = \omega_{12} = \frac{1}{2}\left(\frac{\partial u_2}{\partial x_1} - \frac{\partial u_1}{\partial x_2}\right). \tag{3.1}$$

For the circular rod, the axial displacement is $u_3 = 0$. That is, each meridian circle of the circular rod remains a meridian circle after the twist,

---
1) See (2.43), §2, Chapter 1.

but rotates through an angle $\omega_3(x_3)$. For general cylinders, however, this conclusion no longer holds and must be changed to

$$u_3 = u_3(x_1, x_2). \tag{3.2}$$

That is, each cross section is not only rotated but also warped and is no longer a plane.

We know from (3.1) and (3.2) that,

$$\varepsilon_{11} = \varepsilon_{22} = \varepsilon_{33} = \varepsilon_{12} = 0. \tag{3.3}$$

This means that torsion is a pure shear deformation without volume change. According to Hooke's law,

$$\sigma_{11} = \sigma_{22} = \sigma_{33} = \sigma_{12} = 0. \tag{3.4}$$

Hence only two shear strains ε_{31}, ε_{32} and two shearing stresses σ_{31}, σ_{32} do not vanish:

$$\varepsilon_{31} = \frac{1}{2}\left(\frac{\partial u_3}{\partial x_1} - \dot{\omega}_3 x_2\right), \varepsilon_{32} = \frac{1}{2}\left(\frac{\partial u_3}{\partial x_2} + \dot{\omega}_3 x_1\right), \tag{3.5}$$

$$\sigma_{31} = 2G\varepsilon_{31}, \quad \sigma_{32} = 2G\varepsilon_{32}, \tag{3.6}$$

where

$$\dot{\omega}_3 = \frac{d\omega_3}{dx_3}. \tag{3.7}$$

Let

$$u_3 = \dot{\omega}_3 \psi(x_1, x_2), \tag{3.8}$$

$\psi(x_1, x_2)$ is called the torsion function. Then

$$\varepsilon_{31} = \frac{1}{2}\dot{\omega}_3\left(\frac{\partial \psi}{\partial x_1} - x_2\right), \quad \varepsilon_{32} = \frac{1}{2}\dot{\omega}_3\left(\frac{\partial \psi}{\partial x_2} + x_1\right), \tag{3.9}$$

$$\sigma_{31} = G\dot{\omega}_3\left(\frac{\partial \psi}{\partial x_1} - x_2\right), \quad \sigma_{32} = G\dot{\omega}_3\left(\frac{\partial \psi}{\partial x_2} + x_1\right). \tag{3.10}$$

The volume density of the strain energy is

$$W = \frac{1}{2}(\sigma_{31}\varepsilon_{31} + \sigma_{13}\varepsilon_{13} + \sigma_{32}\varepsilon_{32} + \sigma_{23}\varepsilon_{23}) = 2G(\varepsilon_{31}^2 + \varepsilon_{32}^2)$$

$$= \frac{1}{2}G\dot{\omega}_3^2\left[\left(\frac{\partial \psi}{\partial x_1} - x_2\right)^2 + \left(\frac{\partial \psi}{\partial x_2} + x_1\right)^2\right].$$

The strain energy per unit length of the cylinder is

$$\bar{W} = \iint_\Omega W \, dx_1 dx_2$$

$$= \frac{1}{2} G \dot{\omega}_3^2 \iint_\Omega \left[\left(\frac{\partial \psi}{\partial x_1} - x_2 \right)^2 + \left(\frac{\partial \psi}{\partial x_2} - x_1 \right)^2 \right] dx_1 dx_2. \qquad (3.11)$$

3.2   Torsion function

Since there are no body or surface loads and the outward normal to the lateral face of the cylinder is $n = (n_1, n_2, 0)^T$, by (5.13) and (5.14) in §5 of Chapter 2, the equilibrium equations are the cross section Ω and its boundary $\partial\Omega$. The outward normal to the lateral face of the cylinder is $n = (n_1, n_2, 0)^T$, where $(n_1, n_2)^T$ is the outward normal to the plane curve $\partial\Omega$. By (5.13) and (5.14) in §5 of Chapter 2, the equilibrium equations on the cross section Ω and its boundary are

$$\begin{cases} \Omega : -\left(\dfrac{\partial\sigma_{31}}{\partial x_1} + \dfrac{\partial\sigma_{32}}{\partial x_2} \right) = 0, & (3.12) \\[3mm] \partial\Omega : \sigma_{31} n_1 + \sigma_{32} n_2 = 0. & (3.13) \end{cases}$$

Substituting (3.10) into the above two equations, we immediately obtain the equilibrium equation and the boundary condition, expressed in terms of ϕ.

$$\begin{cases} \Omega : \Delta\psi = 0, & (3.14) \\[3mm] \partial\Omega : \dfrac{\partial\psi}{\partial n} = x_2 n_1 - x_1 n_2. & (3.15) \end{cases}$$

Thus, ψ is a plane harmonic function and satisfying a boundary condition of the second kind. Moreover, the following compatibility condition holds:

$$\oint_{\partial\Omega} (x_2 n_1 - x_1 n_2) dl = \iint_\Omega \left(\frac{\partial x_2}{\partial x_1} - \frac{\partial x_1}{\partial x_2} \right) dx_1 dx_2 = 0. \qquad (3.16)$$

Hence the boundary value problem (3.14) and (3.15) has a solution unique up to a constant.

If we prescribe $\psi = 0$ on the axis of the cylinder, i.e, $\psi(0,0) = 0$, ψ will be determined uniquely.

When the cylinder is a circular rod, $\partial\Omega$ is a circle. Suppose its radius is R. Then the components of the outward normal to the circle are

$$n_1 = \frac{x_1}{R}, \quad n_2 = \frac{x_2}{R}.$$

Thus the boundary condition (3.15) becomes

$$\partial\Omega : \frac{\partial\psi}{\partial n} = 0,$$

and the solution of (3.14) is $\psi \equiv 0$. That is to say, the cross section of the circular rod does not warp during the twist, which coincides precisely with the conclusion reached B in §5 of Chapter 1.

For general cylinders, the torsion function ψ is independent of the shearing modulus G and of the rate of twise $\dot{w}_3$, depending only on the geometry and size of the cross section Ω. It can be evaluated a priori from (3.14) and (3.15). Then the shear strains and the shearing stresses can be found using (3.9) and (3.10).

The torsion function ψ can also be evaluated using the variational principle. Let

$$D(\psi,\varphi) = \iint_\Omega \left(\frac{\partial\psi}{\partial x_1}\frac{\partial\varphi}{\partial x_1} + \frac{\partial\psi}{\partial x_2}\frac{\partial\varphi}{\partial x_2}\right)dx_1dx_2,$$

$$F(\varphi) = \int_{\partial\Omega} (x_2n_1 - x_1n_2)\varphi dl,$$

then the boundary value problem (3.14) and (3.15) for determining the torsion function ψ is equivalent to the following two variational problems:

1. $\dfrac{1}{2}D(\psi,\psi) - F(\psi) = \min,$

                                             for any φ.

2. $D(\psi,\varphi) - F(\varphi) = 0,$

Clearly, the quadratic functional $D(\varphi,\varphi)$ is degenerate and the strainless state is

$$D(v,v) = 0 \Longleftrightarrow v^{(1)} \equiv \text{const.}$$

Hence solutions of the variational problems can differ by an arbitrary constant. A necessary and sufficient condition for the existence of solutions is $F(v^{(1)}) = 0$, which is just the compatibility condition (3.16).

When the twist angle w_3 of the cross section along the cylinder is given, the stress distribution on each cross section is determined. Having done this, we can go back and discuss the overall problem of the cylinder acting as a one dimensional rod in torsion, i.e., how to determine its twist angle $w_3 = w_3(x_3)$ according to the torque load so as to completely solve the torsion problem for the cylinder.

From (3.11), the strain energy per unit length of, cylinder is

$$\overline{W} = \frac{1}{2}GJ\dot{w}_3^2,$$

where

$$J = \iint_\Omega \Big[\Big(\frac{\partial \psi}{\partial x_1} - x_2\Big)^2 + \Big(\frac{\partial \psi}{\partial x_2} + x_1\Big)^2\Big] dx_1 dx_2. \qquad (3.17)$$

Since ψ is a solution of the boundary value problem (3.14) and (3.15), by Green's formula we get

$$\iint_\Omega \Big[\Big(\frac{\partial \psi}{\partial x_1}\Big)^2 + \Big(\frac{\partial \psi}{\partial x_2}\Big)^2\Big] dx_1 dx_2$$

$$= \oint_{\partial\Omega} \frac{\partial \psi}{\partial n} \psi dl - \iint_\Omega \Big(\frac{\partial^2 \psi}{\partial x_1^2} + \frac{\partial^2 \psi}{\partial x_2^2}\Big) \psi dx_1 dx_2$$

$$= \oint_{\partial\Omega} (x_2 n_1 - x_1 n_2) \psi dl$$

$$= \iint_\Omega \Big(x_2 \frac{\partial \psi}{\partial x_1} - x_1 \frac{\partial \psi}{\partial x_2}\Big) dx_1 dx_2. \qquad (3.18)$$

Thus, J can be rewritten as

$$J = \iint_\Omega \Big[\Big(\frac{\partial \gamma}{\partial x_1}\Big)^2 + \Big(\frac{\partial \psi}{\partial x_2}\Big)^2\Big] dx_1 dx_2 - 2 \iint_\Omega \Big(x_2 \frac{\partial \psi}{\partial x_1}$$

$$- x_1 \frac{\partial \psi}{\partial x_2}\Big) dx_1 dz_2 + \iint_\Omega (x_1^2 + x_2^2) dx_1 dx_2$$

$$= \iint_\Omega \Big[x_1 \Big(\frac{\partial \psi}{\partial x_2} + x_1\Big) - x_2 \Big(\frac{\partial \psi}{\partial x_1} - x_2\Big)\Big] dx_1 dx_2. \qquad (3.19)$$

Let

$$M_3 = \frac{\partial \bar{W}}{\partial \dot\omega_3} = GJ\dot\omega_3$$

$$= \iint_\Omega \Big[x_1 G\dot\omega_3 \Big(\frac{\partial \psi}{\partial x_2} + x_1\Big) - x_2 G\dot\omega_3 \Big(\frac{\partial \psi}{\partial x_1} - x_2\Big)\Big] dx_1 dx_2$$

$$= \iint_\Omega (x_1 \sigma_{32} - x_2 \sigma_{31}) dx_1 dx_2.$$

This means that, M_3 is the resultant moment (i.e. the torque) of the shear stresses on the cross section Ω about the x_3-axis the torque i.e.,. Hence the relation

$$M_3 = GJ\dot\omega_3 = GJ\frac{d\omega_3}{dx_3} \qquad (3.20)$$

can be thought of as the form of Hooke's law for the torsion of cylinders. This is similar to formula (6.3) for torsion of circular rods in §6 of Chapter 1. The factor GJ is called the torsional rigidity of the cross section, while J is called the geometric torsional rigidity. J depends only on the geometry and size of the cross section and is independent of the material. By (3.18) and (3.19), the geometric torsional rigidity J can also be expressed as

$$J = \iint_\Omega (x_1^2 + x_2^2)dx_1dx_2 - \iint_\Omega \left[\left(\frac{\partial\psi}{\partial x_1}\right)^2 + \left(\frac{\partial\psi}{\partial x_2}\right)^2\right]dx_1dx_2. \qquad (3.21)$$

3.3   Stress function

In the preceding subsection, we determined the shearing stress and the torsional rigidity of the cross section using the so-called torsion function. It is also possible to reach the same goal another way, i.e., by the use of the so-called stress function.

Let us first consider the case in which the cross section Ω is a simply-connected domain. Let

$$\sigma_{31} = 2G\dot\omega_3\frac{\partial\varphi}{\partial x_2}, \qquad \sigma_{32} = -2G\dot\omega_3\frac{\partial\varphi}{\partial x_1}, \qquad (3.22)$$

where $\varphi = \varphi(x_1, x_2)$ is called the torsional stress function. Clearly, the function φ determined by the above expression satisfies the equilibrium equation (3.12). On the other hand, by (3.10),

$$\sigma_{31} = G\dot\omega\left(\frac{\partial\psi}{\partial x_1} - x_2\right) = 2G\dot\omega\frac{\partial\varphi}{\partial x_2},$$

$$\sigma_{32} = G\dot\omega_3\left(\frac{\partial\psi}{\partial x_2} + x_1\right) = -2G\dot\omega\frac{\partial\varphi}{\partial x_1},$$

hence

$$\frac{\partial\varphi}{\partial x_1} = 2\frac{\partial\varphi}{\partial x_2} + x_2, \qquad \frac{\partial\psi}{\partial x_2} = -\left(2\frac{\partial\varphi}{\partial x_1} + x_1\right). \qquad (3.23)$$

Differentiating the first expression with respect to x_2 and the second expression with respect to x_1 and subtracting, we obtain

$$\Delta\varphi = -1,$$

so that the stress function φ satisfies Poisson's equation. Furthermore, substituting (3.22) into the boundary condition (3.13), we obtain

$$\sigma_{31}n_1 + \sigma_{32}n_2 = 2G\dot\omega_3\left(\frac{\partial\varphi}{\partial x_2}n_1 - \frac{\partial\varphi}{\partial x_1}n_2\right)$$

$$= 2G\dot\omega_3\left(\frac{\partial\varphi}{\partial x_2}t_2 + \frac{\partial\varphi}{\partial x_1}t_1\right)$$

$$= 2G\dot\omega_3\frac{d\varphi}{dl} = 0, \qquad (3.24)$$

where t_1 and t_2 stand for the direction cosines of the tangent, and l is the arc length (see (2.22) in §2). Therefore, on the boundary $\partial\Omega$, $\varphi \equiv$ const., and there is no loss in taking this constant to be zero. Thus, the stress function φ is a solution of Poisson's equation: with the first kind of boundary condition

$$\begin{cases} \Omega : \Delta\varphi = -1, \\ \partial\Omega : \varphi = 0. \end{cases} \tag{3.25}$$

This problem is simpler than the second kind of boundary value problem (3.14) and (3.15) for the torsion function ψ. However, such as the force method, the stress function method is more complex when Ω is a multiply-connected domain.

Let us now discuss the case of multiply-connected domains. Suppose that the connectivity number of Ω is p. Denoted the external boundary by L_0 and the internal boundaries

by L_m, $m = 1, \cdots, p$, and let $\partial\Omega = L_0 + L_1 + \cdots + L_p$, (Fig. 35). By (3.24), $\dfrac{d\varphi}{dl} = 0$ on the boundary $\partial\Omega$. Hence, φ is constant on each of the external and internal boundaries $\varphi_m(m =$

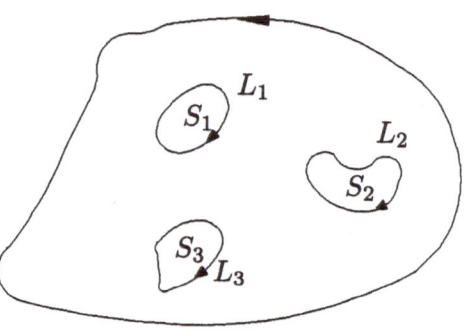

$$L_m : \varphi \equiv \varphi_m \quad (\text{const.}),$$

$$m = 0, 1, \cdots, p.$$

There is no loss in taking the constant $\varphi_0 = 0$ on the external boundary L_0, while the constants

Fig. 35

$1, \cdots, p)$ on the internal boundaries L_m are to be determined. Thus there is a total of p undetermined constants. On the other hand, since the torsion function ψ is single-valued,

$$\oint_{L_m} d\psi = \oint_{L_m} \frac{\partial\psi}{\partial x_1}dx_1 + \frac{\partial\psi}{\partial x_2}dx_2 = 0, \quad m = 1, \cdots, p,$$

substituting (3.23) into this formula, we obtain

$$\oint_{L_m} \left(2\frac{\partial\varphi}{\partial x_2} + x_2 \right)dx_1 - \left(2\frac{\partial\varphi}{\partial x_1} + x_1 \right)dx_2$$

$$= 2\oint_{L_m} \left(\frac{\partial\varphi}{\partial x_2}n_2 + \frac{\partial\varphi}{\partial x_1}n_1 \right)dl + \oint_{L_m} (x_2n_2 + x_1n_1)dl$$

$$= 2\oint_{L_m} \frac{\partial\varphi}{\partial n}dl - 2S_m = 0,$$

where the loop integral about L_m is performed clockwise, and $S_m > 0$ stands for the area bounded by the internal boundary L_m. Therefore

$$\oint_{L_m} \frac{\partial \varphi}{\partial n} dl = S_m, \quad m = 1, 2, \cdots, p. \qquad (3.26)$$

The p constants φ_m, $m = 1, \cdots, p$ are determined by these p constraint conditions. Thus, for multiply-connected domains, the equation and the boundary conditions for determining the stress function φ are

$$\begin{cases} \Omega: \ \Delta\varphi = -1, \\[2mm] L_0: \ \varphi = 0, \\[2mm] L_m: \ \varphi = \varphi_m, \quad \oint_{L_m} \frac{\partial \varphi}{\partial n} dl = S_m, \quad m = 1, \cdots, p. \end{cases} \qquad (3.27)$$

This boundary value problem has a unique solution. The corresponding variational problem is

1. $$\min_{\varphi \in K} \left[\frac{1}{2} D(\varphi, \varphi) - F(\varphi) \right],$$

2. $$\begin{cases} D(\varphi, \varphi) - F(\varphi) = 0, \quad \text{for any } \varphi \in K, \\[2mm] \varphi \in K, \end{cases} \qquad (3.28)$$

where

$$D(\varphi, \psi) = \iint_{\Omega} \left(\frac{\partial \varphi}{\partial x_1} \frac{\partial \psi}{\partial x_1} + \frac{\partial \varphi}{\partial x_2} \frac{\partial \varphi}{\partial x_2} \right) dx_1 dx_2,$$

$$F(\varphi) = \iint_{\Omega} \varphi \, dx_1 dx_2 + \sum_{m=1}^{p} S_m \varphi_m, \quad \varphi|_{L_m} = \varphi_m = \text{const},$$

$$K = \left\{ \varphi(x_1, x_2) \Big| \varphi_{L_0} = 0, \ \frac{\partial \varphi}{\partial l} \Big|_{L_m} = 0, \ m = 1, \cdots, p \right\}.$$

Since the strainless virtual displacement is

$$v \in K: \quad D(v, v) = 0 \Longleftrightarrow v \equiv 0,$$

the quadratic functional $D(\varphi, \varphi)$, $\varphi \in K$ is positive definite, and the variational problem has a unique solution. The equivalence of the variational problem with the boundary value problem (3.27) can be derived using

Green's formula as follows:

$$\varphi \in K : D(\varphi, \psi) - F(\psi) = \iint_\Omega \left(\frac{\partial \varphi}{\partial x_1} \frac{\partial \psi}{\partial x_1} + \frac{\partial \varphi}{\partial x_2} \frac{\partial \psi}{\partial x_2} \right) dx_1 dx_2$$

$$- \iint_\Omega \psi dx_1 dx_2 - \sum_{m=1}^{p} S_m \psi_m$$

$$= \sum_{m=1}^{p} \int_{L_m} \frac{\partial \varphi}{\partial n} \psi dl - \iint_\Omega (\Delta \varphi + 1) \varphi dx_1 dx_2 - \sum_{m=1}^{p} S_m \varphi_m$$

$$= \sum_{m=1}^{p} \left(\int_{L_m} \frac{\partial \varphi}{\partial n} dl - S_m \right) \psi_m - \iint_\Omega (\Delta \varphi + 1) \psi dx_1 dx_2 = 0.$$

Since ψ is arbitrary in the interior of Ω and ψ_m is arbitrary on L_m, we obtain

$$\Delta \varphi = -1, \quad \int_{L_m} \frac{\partial \varphi}{\partial n} dl = S_m, \quad m = 1, \cdots, p.$$

In addition, $\varphi \in K$ hence

$$\varphi|_{L_0} = 0, \quad \varphi|_{L_m} = \varphi_m. \quad \text{(const.)},$$

This is just the boundary value problem (3.27).

Taking $\psi = \varphi$ in (3.28), we have

$$D(\varphi, \varphi) = F(\varphi),$$

i.e., the stress function φ satisfies the integral identity

$$\iint_\Omega \left[\left(\frac{\partial \varphi}{\partial x_1} \right)^2 + \left(\frac{\partial \varphi}{\partial x_2} \right)^2 \right] dx_1 dx_2 = \iint_\Omega \varphi dx_1 dx_2 + \sum_{m=1}^{p} S_m \varphi_m.$$

By substituting (3.23) into the torsional rigidity J (3.27), we obtain the torsional rigidity expressed in terms of the stress function φ:

$$J = 4 \iint_\Omega \left[\left(\frac{\partial \varphi}{\partial x_1} \right)^2 + \left(\frac{\partial \varphi}{\varphi x_2} \right)^2 \right] dx_1 dx_2$$

$$= 4 \left[\iint_\Omega \varphi dx_1 dx_2 + \sum_{m=1}^{p} S_m \varphi_m \right]. \tag{3.29}$$

3.4   Torsion formulas of several specified cross sections

In this section, we give formulas for the torsion function, the stress functions, and the torsional rigidities of several specified cross sections.

1. Circle of radius R.

Torsion function $\psi(x_2, x_1) = 0$.

Stress funtion $\varphi(x_1, x_2) = \dfrac{1}{4}(R^2 - x_1^2 - x_2^2)$.

Geometric torsional rigidity $J = \dfrac{\pi R^4}{2}$.

2. Ellipse (with major axis is $2a$, minor axis $2b$, and coordinate axes coincident with the major and minor axes).

$$\psi(x_1, x_2) = \frac{b^2 - a^2}{a^2 + b^2} x_1 x_2,$$

$$\varphi(x_1, x_2) = \frac{1}{2}\frac{a^2 b^2}{a^2 + b^2}\left(1 - \frac{x_1^2}{a^2} - \frac{x_2^2}{b^2}\right),$$

$$J = \pi \frac{a^3 b^3}{a^2 + b^2}.$$

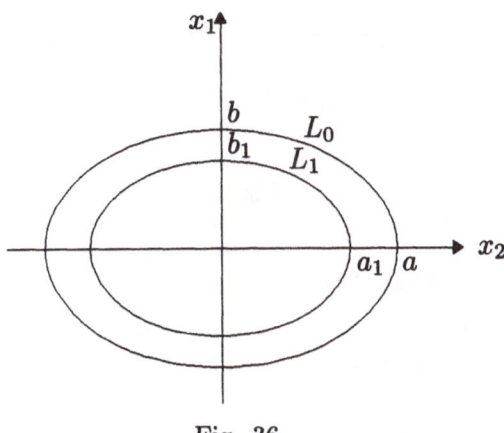

Fig. 36

3. Hollow tube with elliptic cross section (Fig. 36). Equation of the outer ellipse is

$$L_0 : \frac{x_1^2}{a^2} + \frac{x_2^2}{b^2} = 1.$$

Equation of the inner ellipse is

$$L_1 : \frac{x_1^2}{k^2 a^2} + \frac{x_2^2}{k^2 b^2} = 1,$$

$$a_1 = ka, \quad b_1 = kb, \quad k > 1,$$

$$\psi(x_1, x_2) = \frac{b^2 - a^2}{a^2 + b^2} x_1 x_2,$$

$$\varphi(x_1, x_2) = \frac{1}{2}\frac{a^2 b^2}{a^2 + b^2}\left(1 - \frac{x_1^2}{a^2} - \frac{x_2^2}{b^2}\right),$$

$$\varphi\Big|_{L_1} = \frac{1}{2}\frac{a^2 b^2}{a^2 + b^2}(1 - k^2) = \text{const.}$$

$$\int_{L_1} \frac{\partial \varphi}{\partial n} dl = \pi a_1 b_1 = \pi k^2 ab,$$

$$J = (1 - k^4)\frac{\pi a^3 b^3}{a^2 + b^2}.$$

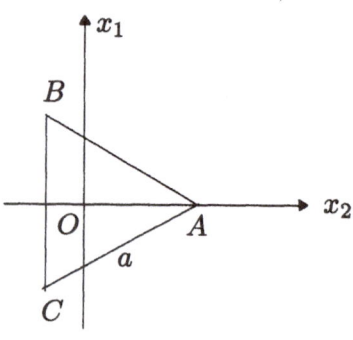

Fig. 37

4. Equilateral triangle (Fig. 37).

Assume the length of a side is a, and that the three altitudes intersect at the origin. The equations of the three sides are

$$AB: \quad x_1 + \sqrt{3}x_2 - \frac{1}{\sqrt{3}}a = 0,$$

$$BC: \quad x_1 + \frac{1}{2\sqrt{3}}a = 0,$$

$$CA: \quad x_1 - \sqrt{3}x_2 - \frac{1}{\sqrt{3}}a = 0.$$

Thus we have

$$\psi(x_1, x_2) = \frac{1}{\sqrt{3}a}x_2(x_2^2 - 3x_1^2),$$

$$J = \frac{\sqrt{3}}{80}a^4,$$

$$\varphi(x_1, x_2) = -\frac{1}{4}(x_1^2 + x_2^2) + \frac{1}{2\sqrt{3}a}(x_1^3 - 3x_1 x_2^2) + \frac{1}{36}a^2.$$

§4 Bending of Thin Plates

4.1 Deformation modes

The elastic deformation of thin plates has two typical modes. The first is deformation by longitudinal stretching and compression under the action of loads longitudinal (in the plane of the plate). The second is the transverse deformation, i.e., the bending under the action of transverse (normal to the plane of the plate) loads.

Deformation by longitudinal stretching and compression of thin plates is just a problem of plane stress and has already been discussed in §1 of the present chapter. Its main features are

(1) $\sigma_{31} = \sigma_{32} = \sigma_{33} = 0$.

(2) The deformation depends absolutely on the longitudinal displacements u_1, u_2 and the corresponding strain components ε_{11}, ε_{22}, ε_{12}. Hence, u_1 and u_2 are the fundamental variables and depending only on the longitudinal coordinates x_1 and x_2, from that the transverse strains ε_{33} is induced. So ε_{33} cannot fully satisfy the spatial geometric compatibility, which is not taken into account in fact.

In the flexural deformation of thin plates, we assume a very thin plate, whose thickness is far smaller than the geometrical sizes in the other two directions. In order to obtain a flexural deformation, it is only necessary to apply a bit of loads on the plane of the plate, which are at least far smaller than the stresses of longitudinal stretching and compression produced by themselves in the interior. Hence, in the boundary equilibrium equations ((5.14) in Chapter 2) for a three dimensional elastic body, the loads g_i can be neglected and we get

$$\sum_{j=1}^{3} \sigma_{ij} n_j = g_i \approx 0,$$

where $\boldsymbol{n} = (n_1, n_2, n_3)^T$ is the outward normal to the boundary surface. Since what under consideration is of small deformation, it can be reputed that, the outward normal direction $\boldsymbol{n}$ of the plate is parallel to the x_3-axis after bending: $\boldsymbol{n} \approx (0, 0, \pm 1)^T$. Thus, on the plane of the plate, we have

$$\sum_{j=1}^{3} \sigma_{ij} n_j \approx \pm \sigma_{i3} \approx 0.$$

Moreover, since the plate is very thin, the above relation also holds in the interior of the plate. Thus we can assume that, within the plate

$$(1')\qquad\qquad \sigma_{31} = \sigma_{32} = \sigma_{33} = 0, \qquad\qquad (4.1)$$

where σ_{3j} at least, are of minor quantities with respect to the other stress components. This point is similar to (1) of the deformation of longitudinal stretching and compression.

On the other hand, the internal longitudinal fibres are stretched or compressed during the flexural deformation of the thin plate. There is stretching on the convex face and compression on the concave face of the plate. Since the transformation from stretching to compression is continuous, there exists a neutral surface without longitudinal stretching or compression. The deformations on the two sides of the neutral surface are reverse in sign. Obviously, the neutral surface is symmetric with respect to the upper and lower surfaces of the plate, i.e., lies in the middle of the thickness of the plate.

We take the neutral surface before deformation to be the (x_1, x_2) plane, i.e., $x_3 = 0$. The displacements in the three directions on the neutral surface are, respectively,

$$u_1^{(0)} = u_2^{(0)} = 0, \quad u_3^{(0)} = u_3^{(0)}(x_1, x_2).$$

Since the plate is very thin, we can assume that, the deflection u_3 is uniform along the thickness of the plate, i.e,

$$u_3(x_1, x_2, x_3) \approx u_3^{(0)}(x_1, x_2).$$

From (4.1), the strains $\varepsilon_{3i} = 0$, $i = 1, 2$, so that

$$\frac{\partial u_i}{\partial x_3} = -\frac{\partial u_3}{\partial x_i} \approx -\frac{\partial u_3^{(0)}}{\partial x_i}, \quad i = 1, 2.$$

After integration, we obtain

$$u_i(x_1, x_2, x_3) = u_i(x_1, x_2, 0) - x_3 \frac{\partial u_3^{(0)}}{\partial x_i}, \quad i = 1, 2.$$

Since the small deformation is small,

$$u_i(x_1, x_2, 0) \approx u_i^{(0)} = 0, \quad i = 1, 2,$$

hence

$$u_i(x_1, x_2, x_3) \approx -x_3 \frac{\partial u_3^{(0)}}{\partial x_i}, \quad i = 1, 2.$$

For the sake of convenience, introduce the abbreviation

$$u_3^{(0)} = w.$$

From now on w denotes the transverse displacement of the neutral surface. Then

$$u_i(x_1, x_2, x_3) = -x_3 \frac{\partial w}{\partial x_i}, \quad i = 1, 2, \tag{4.2}$$

$$u_3(x_1, x_2) = w. \tag{4.3}$$

From this, the expressions of the strains ε_{ij} and the infinitesimal rotation angles ω_{ij} can be obtained in terms of the transverse displacement w as

$$\varepsilon_{ij} = \frac{1}{2}\left(\frac{\partial u_j}{\partial x_i} + \frac{\partial u_i}{\partial x_i}\right) = -x_3 \frac{\partial^2 w}{\partial x_i \partial x_i}, \quad i, j = 1, 2, \tag{4.4}$$

$$\begin{cases} \omega_1 = \omega_{23} = \dfrac{1}{2}\left(\dfrac{\partial u_3}{\partial x_2} - \dfrac{\partial u_2}{\partial x_3}\right) = \dfrac{1}{2}\left(\dfrac{\partial w}{\partial x_2} + \dfrac{\partial w}{\partial x_2}\right) = \dfrac{\partial w}{\partial x_2}, \\[2mm] \omega_2 = \omega_{31} = \dfrac{1}{2}\left(\dfrac{\partial u_1}{\partial x_3} - \dfrac{\partial u_3}{\partial x_1}\right) = \dfrac{1}{2}\left(-\dfrac{\partial w}{\partial x_1} - \dfrac{\partial w}{\partial x_1}\right) = -\dfrac{\partial w}{\partial x_1}, \\[2mm] \omega_3 = \omega_{12} = \dfrac{1}{2}\left(\dfrac{\partial u_2}{\partial x_1} - \dfrac{\partial u_1}{\partial x_2}\right) = \dfrac{1}{2}\left(x_3 \dfrac{\partial^2 w}{\partial x_1 \partial x_2} - x_3 \dfrac{\partial^2 w}{\partial x_2 \partial x_1}\right) = 0. \end{cases} \tag{4.5}$$

Let

$$K_{ij} = -\frac{\partial^2 w}{\partial x_i \partial x_j}, \quad i, j = 1, 2, \tag{4.6}$$

which is just an approximation of the 1st order of the curvature tensor of the neutral surface after the bending. Then

$$\varepsilon_{ij} = x_3 K_{ij}, \quad K_{ij} = K_{ji}, \quad i, j = 1, 2, \tag{4.7}$$

$$\frac{\partial \omega_1}{\partial x_1} = -K_{12}, \quad \frac{\partial \omega_1}{\partial x_2} = -K_{22},$$

$$\frac{\partial \omega_2}{\partial x_1} = K_{11}, \quad \frac{\partial \omega_2}{\partial x_2} = K_{21}. \tag{4.8}$$

Thus, we obtain the second characteristic of the flexural deformation of thin plates:

(2′) The flexural deformation depends entirely on the transverse displacement w, i.e., the so-called deflection. Which depends only on the longitudinal coordinates x_1 and x_2, while the longitudinal displacements u_1, u_2 and the strains ε_{11}, ε_{12}, ε_{22} are determined by the deflection w

through (4.2) and (4.4). This characteristic is entirely different from that of longitudinal stretching and compression.

Since (1) and (1′) are identical, from (1.15) and (1.16), we obtain the stress-strain relation of the flexural deformation as

$$
\begin{cases}
\sigma_{11} = \dfrac{E}{1-\nu^2}(\varepsilon_{11} + \nu\varepsilon_{22}) - \dfrac{E\alpha}{1-\nu}\tau, \\[2mm]
\sigma_{22} = \dfrac{E}{1-\nu^2}(\nu\varepsilon_{11} + \varepsilon_{22}) - \dfrac{E\alpha}{1-\nu}\tau, \\[2mm]
\sigma_{12} = \sigma_{21} = \dfrac{E}{1+\nu}\varepsilon_{12}.
\end{cases}
$$

Or

$$
\begin{cases}
\sigma_{ij} = \sigma'_{ij} - \dfrac{E\alpha}{1-\nu}\tau\delta_{ij}, \\[3mm]
\sigma'_{ij} = \dfrac{E}{1-\nu^2}\left[(1-\nu)\varepsilon_{ij} + \nu\left(\displaystyle\sum_{k=1}^{2}\varepsilon_{kk}\right)\delta_{ij}\right].
\end{cases}
\tag{4.9}
$$

Moreover, from (1.17) and (1.18), we obtain the volume density of strain energy

$$
W = W' - \frac{E\alpha}{1-\nu}\tau\sum_{k=1}^{2}\varepsilon_{kk},
$$

$$
W' = \frac{1}{2}\frac{E}{1-\nu^2}\left[(1-\nu)\sum_{i,j=1}^{2}\varepsilon_{ij}^2 + \nu\left(\sum_{k=1}^{2}\varepsilon_{kk}\right)^2\right].
\tag{4.10}
$$

Therefore, in the modes of bending and stetching and compression of thin plates, Hooke's law and strain energy have identical forms. However, they are essentially different difference between the second characteristics (2′) and (2) of these two kinds of modes.

4.2   Variational principles

At first, we will not consider thermal effects. Then the Hooke's law (4.9) and the strain energy (4.10) can be expressed using relation (4.7) in terms of the curvatures K_{ij} as

$$
\sigma_{ij} = \frac{Ex_3}{1-\nu^2}\left[(1-\nu)K_{ij} + \nu\left(\sum_{k=1}^{2}K_{kk}\right)\delta_{ij}\right], \quad i,j = 1,2,
\tag{4.11}
$$

$$
W = \frac{1}{2}\frac{Ex_3^2}{1-\nu^2}\left[(1-\nu)\sum_{i,j=1}^{2}K_{ij}^2 + \nu\left(\sum_{k=1}^{2}K_{kk}\right)^2\right].
\tag{4.12}
$$

Since the neutral surface is symmetric with respect to the upper and lower surfaces of the plate, if we assume the thickness of the plate is h, let

$$M_{ij} = \int_{-h/2}^{h/2} x_3 \sigma_{ij} dx_3, \quad i, j = 1, 2, \tag{4.13}$$

substitute (4.12) into (4.13) and integrate the result, we obtain

$$M_{ij} = D\left[(1 - \nu)K_{ij} + \nu\left(\sum_{k=1}^{2} K_{kk}\right)\delta_{ij}\right],$$

$$D = \frac{Eh^3}{12(1 - \nu^2)},$$

or

$$M_{11} = D(K_{11} + \nu K_{22}), \quad M_{22} = D(\nu K_{22} + K_{11}),$$

$$M_{12} = M_{21} = D(1 - \nu)K_{12}. \tag{4.14}$$

The above expression is the Hooke's law for flexural deformation constitute of thin plates, which is similar in form to that for stretching and compression deformation. Here, however, what represents the 'strain" is the curvature K_{ij} and what represents the '"stress" is M_{ij} (its mechanical meaning will be shown in Section 4.3). The constant ratio $D = \dfrac{Eh^3}{12(1 - nu^2)}$ is called the flexural rigidity. Compared with the tensile rigidity $D' = \dfrac{Eh}{1 - \nu^2}$ the stretching and compression deformation, the flexural rigidity is a minor quantity of higher order when the thickness h of the plate is quite small.

Similarly, the areal density of the strain energy can be evaluated as

$$\bar{W} = \int_{-h/2}^{h/2} W dx_3 = \frac{1}{2}D\left[(1 - \nu)\sum_{i,j=1}^{2} K_{ij}^2 + \nu\left(\sum_{k=1}^{2} K_{kk}\right)^2\right]. \tag{4.15}$$

We have

$$M_{ij} = \frac{\partial \bar{W}}{\partial K_{ij}}, \quad \bar{W} = \frac{1}{2}\sum_{i,j=1}^{2} M_{ij}K_{ij}. \tag{4.16}$$

The whole plate Ω has the flexural strain energy $P(w)$ and the corresponding functional of virtual work $D(w, v)$:

$$P(w) = \iint_{\Omega} \bar{W}(w) dx_1 dx_2 = \frac{1}{2}\iint_{\Omega}\sum M_{ij}(w)K_{ij}(w) dx_1 dx_2$$

$$= \frac{1}{2}D(w, w), \tag{4.17}$$

$$D(w, v) = \iint_\Omega \sum M_{ij}(w) K_{ij}(v) dx_1 dx_2. \tag{4.18}$$

It is clearly seen from (4.15) that $D(w, w)$ is a quadratic functional of w and is nonnegative, while the functional of virtual work $D(w, v)$ is a bilinear symmetric functional of w and v. Using (4.15) and (4.16), we can write $D(w, w)$ in terms of the derivatives of w as

$$D(w, w) = \iint_\Omega D\Big[\Big(\frac{\partial^2 w}{\partial x_1^2}\Big)^2 + \Big|\frac{\partial^2 w}{\partial x_2^2}\Big|$$

$$+ 2\nu \frac{\partial^2 w}{\partial x_1^2} \frac{\partial^2 w}{\partial x_2^2} + 2(1 - \nu)\Big(\frac{\partial^2 w}{\partial x_1 \partial x_2}\Big)^2\Big] dx dy$$

$$= \iint_\Omega D\Big[\Big(\frac{\partial^2 w}{\partial x_1^2} + \frac{\partial^2 w}{\partial x_2^2}\Big)^2$$

$$- 2(1 - \nu)\Big(\frac{\partial^2 w}{\partial x_1^2} \frac{\partial^2 w}{\partial x_2^2} - \frac{\partial^2 w}{\partial x_1 \partial x_2}\Big)^2\Big] dx dy.$$

For the potential energy of external work, let us first consider the case where the boundary is not subjected to any geometric constraint and there is no elastic support. Then the transverse load P_3 on the plane of the plate Ω and the transverse load q_3 and the moment load m_i on the boundary $\partial\Omega$ yield the potential energy of external work (Fig. 38):

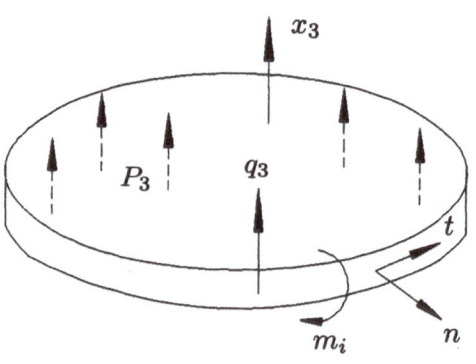

Fig. 38

$$-F(w) = -\Big\{\iint_\Omega P_3 w dx_1 dx_2 + \oint_{\partial\Omega} q_3 w dl + \oint_{\partial\Omega} m_i \omega_i(w) dl\Big\}. \tag{4.19}$$

Let $n = (n_1, n_2, 0)^T$ denote the outward normal direction to the boundary $\partial\Omega$ and $e_3 = (0, 0, 1)^T$ denote a unit vector on the positive x_3-axis, and define a positive tangential direction $t = (t_1, t_2, 0)$ on $\partial\Omega$ to be that makes $\{n, t, e_3\}$ a right-handed system (Fig. 38). $\omega_1(w)$ in the potential energy of external work (4.19) denotes a tangential rotation angle on the boundary $\partial\Omega$ during flexural deformation. The plate after the bending has an infinitesimal of rotation angle with respect to the plate before the bending on $\partial\Omega$. The components of w about the x_1-and x_2-axes are

$$\omega_1 = \frac{\partial w}{\partial x_2}, \quad \omega_2 = -\frac{\partial w}{\partial x_1},$$

respectively, and the tangential component of ω called the tangential rotation angle, is

$$\omega_t(w) = t_1\omega_1 + t_2\omega_2 = -n_2\frac{\partial w}{\partial x_2} + n_1\left(-\frac{\partial w}{\partial x_1}\right) = -\frac{\partial w}{\partial n}. \tag{4.20}$$

This is illustrated in Fig. 39.

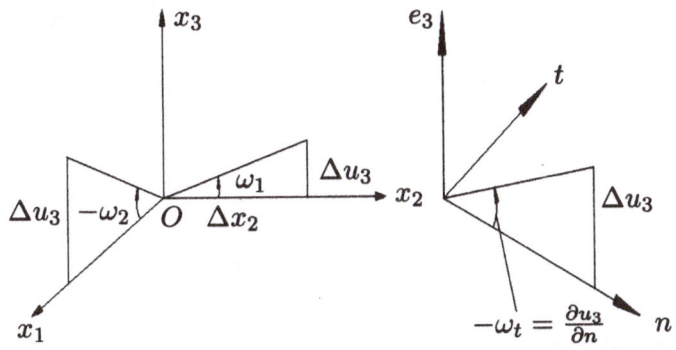

Fig. 39

Substituting (4.20) into $F(w)$, we obtain

$$F(w) = \iint_\Omega P_3 w\, dx_1 dx_2 + \oint_{\partial\Omega} q_3 w\, dl - \oint_{\partial\Omega} m_i \frac{\partial w}{\partial n} dl. \tag{4.21}$$

Then, according to the variational principle, the deflection w of the constrained equilibrium configuration makes the total potential energy a unconditional minimum,

$$J(w) = \frac{1}{2}D(w, w) - F(w) = \text{Min.} \tag{4.22}$$

This is equivalent to the principle of virtual work

$$D(w, v) - F(v) = 0, \quad \text{for any } v. \tag{4.23}$$

4.3   Equilibrium equations

In order to derive the equilibrium equation from the variational principle, we must, as a rule establish Green's formula, that is, we must transform $D(w, v)$ into an expression containing only v itself but not its derivatives by the use of Gauss's integral formula on the domain Ω. Here, since $D(w, v)$ contains the second derivatives of v, it is necessary to use Gauss's integral formula twice. This point is different from the foregoing three dimensional or plane elastic problem but similar to the flexural problem of beams in §7 of Chapter 1.

Let us first consider a plate with uniform thickness and material, i.e., E, ν, and h are all constants in the interior of Ω. Then the corresponding equilibrium solution w is smooth enough to justify the use of Gauss's integral formula. By Gauss's formula, we obtain

$$
\iint_\Omega \sum_{i,j=1}^2 M_{ij}(w) K_{ij}(v) dx_1 dx_2 = - \iint_\Omega \sum_{i,j=1}^2 M_{ij}(w) \frac{\partial^2 v}{\partial x_i \partial x_j} dx_1 dx_2
$$

$$
= \iint_\Omega \sum_{i,j=1}^2 \frac{\partial M_{ij}(w)}{\partial x_j} \frac{\partial v}{\partial x_i} dx_1 dx_2 - \oint_{\partial \Omega} \sum_{i,j=1}^2 M_{ij}(w) n_j \frac{\partial v}{\partial x_i} dl.
$$

$$(4.24)$$

Since on the boundary $\partial \Omega$ we have

$$
\frac{\partial v}{\partial x_i} = n_i \frac{\partial v}{\partial n} + t_i \frac{\partial v}{\partial l},
$$

the line integral term in (4.24) is

$$
\oint_{\partial \Omega} \sum_{i,j=1}^2 M_{ij}(w) n_j \frac{\partial v}{\partial x_i} dl = \oint_{\partial \Omega} \sum_{i,j=1}^2 M_{ij}(w) n_i n_j \frac{\partial v}{\partial n} dl
$$

$$
+ \oint_{\partial \Omega} \sum_{i,j=1}^2 M_{ij}(w) t_i n_j \frac{\partial v}{\partial l} dl.
$$

Let

$$
\Omega : Q_{3i} = \sum_{j=1}^2 \frac{\partial M_{ij}}{\partial x_j}, \quad i = 1, 2, \tag{4.25}
$$

$$
\partial \Omega : Q_{3n} = \sum_{i=1}^2 Q_{3i} n_i, \tag{4.26}
$$

$$
\partial \Omega : M_{nn} = \sum_{i,j=1}^2 M_{ij} n_i n_j, \tag{4.27}
$$

$$
\partial \Omega : M_{tn} = \sum_{i,j=1}^2 M_{ij} t_i n_j, \tag{4.28}
$$

then

$$D(w,v) = \iint_\Omega \sum_{i=1}^{2} Q_{3i}(w)\frac{\partial v}{\partial x_i}dx_1 dx_2 - \oint_{\partial\Omega} M_{nn}(w)\frac{\partial v}{\partial n}dl$$

$$- \oint_{\partial\Omega} M_{tn}(w)\frac{\partial v}{\partial l}dl. \tag{4.29}$$

Using Gauss's formula once more for the first term on the right-hand side of the above equation, we obtain

$$\iint_\Omega \sum_{i=1}^{2} Q_{3i}(w)\frac{\partial v}{\partial x_i}dx_1 dx_2$$

$$= -\iint_\Omega \sum_{i=1}^{2} \frac{\partial Q_{3i}(w)}{\partial x_i}vdx_1 dx_2 + \oint_{\partial\Omega} Q_{3n}(w)vdl.$$

Since $\dfrac{\partial v}{\partial l}$ on the boundary $\partial\Omega$ is determined by v on this boundary, $\dfrac{\partial v}{\partial l}$ can be eliminated using integration by parts for the last term of (4.29). Suppose are all the corner points (i.e., points at which the tangent has a jump) that on the boundary $\partial\Omega$ (Fig. 40). Then after piecewise integration by parts, we obtain

$$-\oint_{\partial\Omega} M_{in}(w)\frac{\partial v}{\partial l}dl = \oint_{\partial\Omega} \frac{\partial M_{tn}(w)}{\partial l}vdl + \sum_{i=1}^{m}[M_{tn}(w)]_{P_i^-}^{P_i^+}\cdots v(P_i).$$

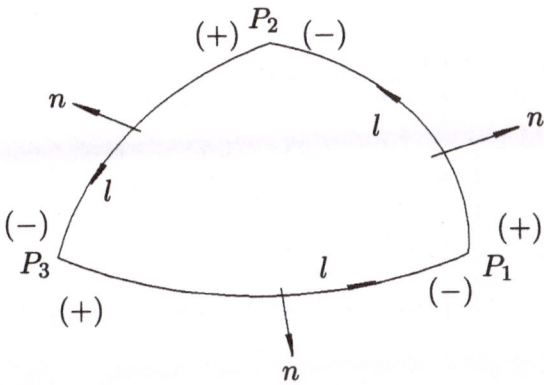

Fig. 40

Thus, we obtain a Green's formulas:

$$
D(w, v) = \iint_\Omega \sum_{i,j=1}^2 M_{ij}(w)K_{ij}(v)dx_1dx_2
$$

$$
= -\iint_\Omega \sum_{i=1}^2 \frac{\partial Q_{3i}(w)}{\partial x_i}vdx_1dx_2 + \oint_{\partial\Omega}\left(Q_{3n}(w) + \frac{\partial M_{tn}(w)}{\partial l}\right)vdl
$$

$$
- \oint_{\partial\Omega} M_{nn}(w)\frac{\partial v}{\partial n}dl + \sum_{i=1}^m [M_{tn}(w)]_{P_i^-}^{P_i^+} v(P_i).
$$

$$(4.30)$$

It follows that

$$
D(w, v) - F(v) = -\iint_\Omega \left(\sum_{i=1}^2 \frac{\partial Q_{3i}(w)}{\partial x_i} + P_3\right)vdx_1dx_2
$$

$$
+ \oint_{\partial\Omega}\left(Q_{3n}(w) + \frac{\partial M_{tn}(w)}{\partial l} - q_3\right)vdl
$$

$$
+ \oint_{\partial\Omega}(-M_{nn}(w) + m_t)\frac{\partial v}{\partial n}dl
$$

$$
+ \sum_{i=1}^m [M_{tn}(w)]_{P_i^-}^{P_i^+} v(P_i)
$$

$$
= 0. \tag{4.31}
$$

Because v is arbitrary in the interior of Ω, on the boundary $\partial\Omega$, and at the point P_i, (4.31) immediately yields the equilibrium equation and the boundary conditions:

$$
\Omega: -\sum_{i=1}^2 \frac{\partial Q_{3i}(w)}{\partial x_i} = P_3, \tag{4.32}
$$

$$
\partial\Omega: \begin{cases} Q_{3n}(w) + \dfrac{\partial M_{tn}(w)}{\partial l} = q_3, & (4.33) \\[2ex] M_{nn}(w) = m_t, & (4.34) \\[2ex] [M_{tn}(w)]_{P_i^-}^{P_i^+} = 0, \quad i = 1, 2, \cdots, m. & (4.35) \end{cases}
$$

Substituting (4.25), (4.6) and Hooke's law (4.14) into (4.32), we obtain the flexural equation of thin plates expressed in terms of the deflection w:

$$\Omega : \frac{\partial^2}{\partial x_1^2} D\Big(\frac{\partial^2 w}{\partial x_1^2} + \nu \frac{\partial^2 w}{\partial x_2^2}\Big) + 2\frac{\partial^2}{\partial x_1 \partial x_2} D(1-\nu)\frac{\partial^2 w}{\partial x_1 \partial x_2}$$

$$+ \frac{\partial^2}{\partial x_2^2} D\Big(\nu \frac{\partial^2 w}{\partial x_1^2} + \frac{\partial^2 w}{\partial x_2^2}\Big) = P_3. \tag{4.36}$$

This is a fourth order elliptic partial differential equation in the deflection w. For thin plates with uniform material and thickness, D and ν are all constants. Then the above equation is reduced to a biharmonic equation

$$\Omega : D\Delta^2 w = P_3. \tag{4.37}$$

We shall now illustrate the meaning of equation (4.32) and the boundary conditions (4.33)–(4.35) in terms of the principle of equilibrium in mechanics. To this end, to give the mechanical meaning of the quantities M_{ij}, M_{tn}, M_{nn}, Q_{3i}, Q_{3n}, etc, introduced in the preceding section.

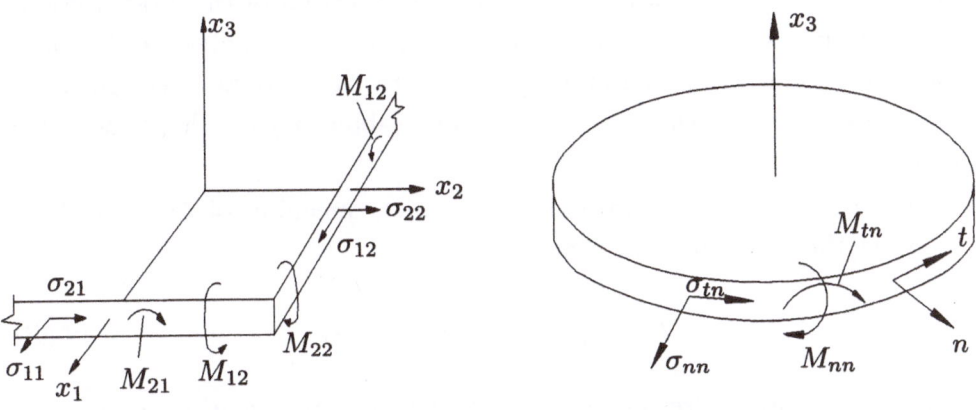

Fig. 41   Fig. 42

(1) Since

$$M_{ij} = \int_{-h/2}^{h/2} x_3 \sigma_{ij} dx_3, \quad i,j = 1,2,$$

from the meaning of the stress components σ_{ij}, and Fig. 41, it is not difficult to see that: M_{11} denotes the bending moment on the x_1-directional cross section (about $+x_2$-axis), M_{21} denotes the twisting moment on the x_1-directional cross section (about $-x_1$-axis), M_{22} denotes the bending moment on the x_2-directional cross section (about $-x_1$-axis), and M_{12} denotes the twisting moment on the x_2-directional cross section (about $+x_2$-axis).

(2) We have

$$M_{nn} = \sum_{i,j=1}^{2} M_{ij} n_i n_j = \sum_{i,j=1}^{2} \left(\int_{-h/2}^{h/2} x_3 \sigma_{ij} dx_3 \right) n_i n_j$$

$$= \int_{-h/2}^{h/2} x_3 \left(\sum_{i,j=1}^{2} \sigma_{ij} n_i n_j \right) dx_3 = \int_{-h/2}^{h/2} x_3 \sigma_{nn} dx_3, \qquad (4.38)$$

$$M_{tn} = \sum_{i,j=1}^{2} M_{ij} l_i n_j = \sum_{i,j=1}^{2} \left(\int_{-h/2}^{h/2} x_3 \sigma_{ij} dx_3 \right) l_i n_j$$

$$= \int_{-h/2}^{h/2} x_3 \left(\sum_{i,j=1}^{2} \sigma_{ij} l_i n_j \right) dx_3 = \int_{-h/2}^{h/2} x_3 \sigma_{tn} dx_3, \qquad (4.39)$$

where σ_{nn} and σ_{tn} are the component of the stress in n and t, respectively, and n and t are the normal the tangential direction of the cross section, respectively. We can see from Fig. 42 that M_{nn} denotes the bending moment (about t) on the cross section with the normal direction of n, and $-M_{tn}$ denotes the twisting moment (about n) on the same cross section.

(3) In order to illustrate the mechanical meaning of Q_{3i} and Q_{3n}, consider the transverse shearing forces

$$\bar{Q}_{3i} = \int_{-h/2}^{h/2} \sigma_{3i} dx_3, \quad i = 1, 2. \qquad (4.40)$$

$\bar{Q}_{31}$ and $\bar{Q}_{32}$ are the resultant force of the transverse shearing stresses σ_{31} and σ_{32} acting on the cross sections with normal directions x_1 and x_2, respectively, and the dimension of the resultant forces is force/length.

Take an arbitrary area element s on the neutral surface, and from it, construct a cylinder V in the plate, whose top and base are denoted $S^+(x_3 = h/2)$ and $S^-(x_3 = -h/2)$ and whose lateral surface is denoted by B. Suppose the transverse body load density is f_3 and the surface load density acting on the top and on the base are g_3^+ and g_3^-, respectively. To have an equilibrium of the x_3-directional forces on this cylinder V, we must have

$$\iint_B \sigma_{3n} dB + \iint_{S^+} g_3^+ dS^+ + \iint_{S^-} g_3^- dS^- + \iiint_V f_3 dV = 0. \qquad (4.41)$$

Since the outward normal direction on the lateral surface B is $\boldsymbol{n} = (n_1, n_2, 0)^T$,

$$\iint_B \sigma_{3n} dB = \iint_B \sum_{l=1}^3 \sigma_{3i} n_i dB = \iint_B \sum_{i=1}^2 \sigma_{3i} n_i dB$$

$$= \oint_{\partial S} \left(\int_{-h/2}^{h/2} \sum_{i=1}^2 \sigma_{3i} n_i dx_3 \right) dl = \oint_{\partial S} \sum_{i=1}^2 \left(\int_{-h/2}^{h/2} \sigma_{3i} dx_3 \right) n_i dl$$

$$= \oint_{\partial S} \sum_{i=1}^2 \bar{Q}_{3i} n_i dl = \iint_S \sum_{i=1}^2 \frac{\partial \bar{Q}_{3i}}{\partial x_i} dx_1 dx_2.$$

The sum of the last three terms in the equilibrium equation (4.41) is

$$\iint_{S^+} g_3^+ dS^+ + \iint_{S^-} g_3^- dS^- + \iiint_V f_3 dV$$

$$= \iint_S \left(g_3^+ + g_3^- + \int_{-h/2}^{h/2} f_3 dx_3 \right) dx_1 dx_2.$$

Let

$$P_3 = g_3^+ + g_3^- + \int_{-h/2}^{h/2} f_3 dx_3. \tag{4.42}$$

This is the transverse surface load acting on the element of area S, whose dimension is force/area. Then the equilibrium equation (4.41) becomes

$$\iint_S \left(\sum_{i=1}^2 \frac{\partial \bar{Q}_{3i}}{\partial x_i} + P_3 \right) dx_1 dx_2 = 0 \quad \text{for any } S \subset \Omega.$$

From this, we obtain a differential form of the equilibrium equation for forces in the x_3-direction:

$$\Omega: \quad -\sum \frac{\partial \bar{Q}_{3i}}{\partial x_i} = P_3. \tag{4.43}$$

Next, consider the equilibrium of moments about the x_1-axis of the above-mentioned cylinder V:

$$\iint_S (x_2 \sigma_{3n} - x_3 \sigma_{2n}) dB + \iint_{S^+} (x_2 g_3^+ - x_3 g_2^+) dS$$

$$+ \iint_{S^-} (x_2 g_3^- - x_3 g_2^-) ds + \iiint_V (x_2 f_3 - x_3 f_2) dV = 0.$$

Because all of the longitudinal loads g_1, g_2, f_1, and f_2 in the flexural problem of plates are zero, and by noting the formulas for the bending moments, i.e.,

$$M_{ij} = \int_{-h/2}^{h/2} x_3 \sigma_{ij} dx_3, \quad i, j = 1, 2,$$

the above-mentioned equation of the moment equilibrium can be reduced
to

$$\iint_B \Big(x_2 \sum_{j=1}^{2} \sigma_{3j} n_j - x_3 \sum_{j=1}^{2} \sigma_{2j} n_j\Big) dB + \iint_{S+} x_2 g_3^+ dS$$

$$+ \iint_{S-} x_2 g_3^- ds + \iiint_V x_2 f_3 dV$$

$$= \oint_{\partial S} \Big(x_2 \sum_{j=1}^{2} \bar{Q}_{3j} n_j - \sum_{j=1}^{2} M_{2j} n_j\Big) dl + \iint_S x_2 P_3 dx_1 dx_2$$

$$= \iint_S \Big[\sum_{j=1}^{2} \frac{\partial}{\partial x_j}(x_2 \bar{Q}_{3j} - M_{2j}) + x_2 P_3\Big] dx_1 dx_2$$

$$= \iint_S \Big[x_2\Big(\sum_{j=1}^{2} \frac{\partial \bar{Q}_{3j}}{\partial x_j} + P_3\Big) + \Big(\bar{Q}_{32} - \sum_{j=1}^{2} \frac{\partial M_{2j}}{\partial x_j}\Big)\Big] dx_1 dx_2$$

$$= \iint_S \Big(\bar{Q}_{32} - \sum_{j=1}^{2} \frac{\partial M_{2j}}{\partial x_j}\Big) dx_1 dx_2$$

$$= 0, \quad \text{for any } S \subset \Omega,$$

Here, the equilibrium equation of the x_3-directional forces (4.43) has been
used. It follows that

$$\Omega : \bar{Q}_{32} = \sum_{j=1}^{2} \frac{\partial M_{2j}}{\partial x_j}.$$

Similarly, considering the equilibrium of the moments about the x_2-axis
of the cylinder V, we get

$$\Omega : \bar{Q}_{31} = \sum_{j=1}^{2} \frac{\partial M_{1j}}{\partial x_j}.$$

To sum up, from the equilibrium of the moments about the x_1-and
x_2-axes, we obtain the relations among the transverse shearing forces $\bar{Q}_{3i}$
and the bending moments M_{ij}

$$\Omega : \quad \bar{Q}_{3i} = \sum_{j=1}^{2} \frac{\partial M_{ij}}{\partial x_j}, \quad j = 1, 2. \tag{4.44}$$

Comparing (4.44) with (4.25), we see that the quantities Q_{3i} introduced in the variational principle are just the transverse shearing forces
$\bar{Q}_{3i}$ on the x_i-directional cross section:

$$Q_{3i} = \bar{Q}_{3i}, \quad i = 1, 2.$$

Hence the equation(4.32) derived from the variational principle is also the equilibrium equation (4.43).

Furthermore, from (4.26), $Q_{3n} = \sum_{i=1}^{2} Q_{3i} n_i$. Thus Q_{3n} is just the transverse shearing force on the cross section with normal direction $\boldsymbol{n}$.

4.4   Boundary conditions and interface conditions

The equilibrium equation (4.36) or (4.37) is the fourth order elliptic partial differential equation of w. In general, two boundary conditions must be prescribed on the boundary to determine a unique solutions.

1. The first kind of boundary condition consists of prescribed geometric constraints. It can be further classified into two cases.

1.1 A prescribed transverse displacement.

$$\Gamma_1: \quad w = \bar{w} \quad \text{is given.} \tag{4.45}$$

1.2 A prescribed tangential angle of rotation.

$$\Gamma_1' : \omega_t(w) = \bar{\omega}_t \quad \text{is given, or}$$

$$\frac{\partial w}{\partial n} = -\bar{\omega}_t, \quad \text{is given.} \tag{4.46}$$

For the above two cases of geometric constraints, the virtual displacement v in the variational problem must satisfy the corresponding annihilating constraint conditions

$$\Gamma_1: \quad v = 0, \quad \Gamma_1': \quad \frac{\partial v}{\partial n} = 0.$$

With these constraints, the functional of the strain energy remains unchanged, while the potential energy of the external work is transformed into

$$-F(v) = -\left\{ \int_{\partial\Omega - \Gamma_1} q_3 v \, dl + \int_{\partial\Omega - \Gamma_1'} m_t \frac{\partial v}{\partial n} \, dl \right\}.$$

Then, by the use of Green's formula, the equilibrium equation (4.32) can be obtained from the variational principle, while the boundary conditions are transformed into

$$\partial\Omega - \Gamma_1: \quad Q_{3n} + \frac{\partial M_{tn}(w)}{\partial l} = q_3,$$

$$\partial\Omega - \Gamma_1': \quad M_{nn}(w) = m_t,$$

which just complement the boundary conditions (4.45) and (4.46). We can see from the foregoing that, if the transverse displacement w has been prescribed on a certain segment of the boundary, then any local transverse load q_3 is out of action. For the same reason, if the tangential rotation angle ω_t has been prescribed, then the local tangential bending moment m_t is also out of action.

The geometric boundary conditions which must be listed out as constraint conditions in variational problems are essential boundary conditions. They contain w and the first derivatives of w, and are comparatively simple in form.

2. The second kind of boundary condition consists of prescribed loads. There are mechanical boundary conditions, which can also be classified into two cases.

2.1 A prescribed transverse load q_3 on Γ_2.

From (4.33), the mathematical statement of the boundary condition is

$$\Gamma_2: \quad Q_{3n}(w) + \frac{\partial M_{tn}(w)}{\partial l} = q_3,$$

which denotes the equilibrium of the transverse shearing forces on the boundary and contains the third derivatives of $\boldsymbol{w}$. It might seem intuitively that only $Q_{3n} = q_3$ is possible, but in fact there is a second term $\dfrac{\partial M_{tn}}{\partial l}$ which can be thought of as an equivalent transverse shearing force $\dfrac{\partial M_{tn}}{\partial l}$ generaled by the drop of the twisting moment on the boundary of the plate. This term, together with Q_{3n}, balances the external load q_3.

2.2 A prescribed moment load m_t on Γ_2'.

From (4.34), the mathematical statement of the boundary condition is

$$\Gamma_2' : M_{nn}(w) = m_t,$$

which denotes the equilibrium of bending moments on the boundary.

Furthermore, we see from (4.35) that, if the corner point P_i of the boundary $\partial\Omega$ is not subjected to any load, then the twisting moment M_{tn} is continuous at that point. If there is a point load r_i at P_i, then a "point term" $-r_i\nu(P_i)$ must be added to the potential energy of external work $-F(\nu)$. Then the equilibrium equation at point P_i will be

$$P_i : [M_{tn}(w)]_{P_i^-}^{P_i^+} = r_i. \tag{4.47}$$

This says that the twisting moment must have a jump at P_i that produces an effective transverse point force to balance the point load r_i (Fig. 44).

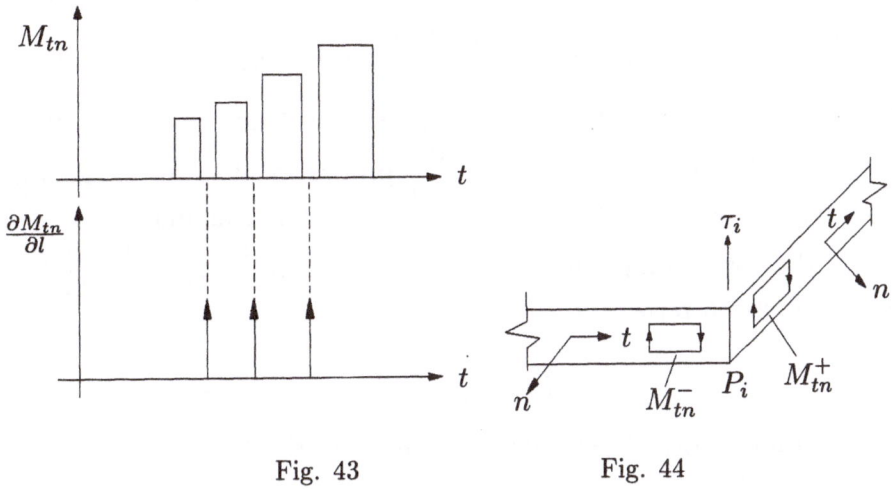

Fig. 43   Fig. 44

The mechanical boundary conditions are natural boundary conditions in variational problems. As for to the internal equilibrium equation, they will be satisfied automatically when the potential energy achieves a minimum. They are, in fact just the equilibrium equations on the boundary. The natural boundary conditions contain the second and third derivatives of w, and their analytic form is very complicated. The advantages of the variational principle approach seem more evident in the flexural problem of thin plates. Historically, the exact form of the mechanical boundary condition for the bending of thin plates was first derived by Kirchhoff just using the variational principle.

3. The third kind of boundary condition consists of elastic supports, appearing as elastic couplings of the plate with the external region on the boundary or on the plane of the plate. These can be classified into three cases.

3.1 On Γ_3, in addition to the transverse load q_3, there is a transverse elastic reaction force $-c_0 w$ proportional to the deflection, where $c_0 > 0$ is the constant of elastic coupling.

In this case, on Γ_3 there will be a per unit length elastic energy $\frac{1}{2}c_0 w^2$, which contributes a term

$$\frac{1}{2}\int_{\Gamma_3} c_2 w^2 dl$$

to the potential energy of external work, and contributes a term

$$\int_{\Gamma_3} c_0 w v dl$$

to the functional of virtual work. The equilibrium equation on Γ_3 will be

$$\Gamma_3: \quad Q_{3n}(w) + \frac{\partial M_{tn}(w)}{\partial l} + c_0 w = q_3. \qquad (4.48)$$

3.2 On Γ_3', in addition to the bending moment load m_i, there will be an elastic reaction moment $-c_1 \omega_t = c_1 \dfrac{\partial w}{\partial n}$ proportional to the tangential rotation angle, where $c_1 > 0$ is the constant of elastic coupling.

On Γ_3', there will be a per unit length elastic energy $\dfrac{1}{2} c_1 \omega_t^2 = \dfrac{1}{2} c_1 \left(\dfrac{\partial w}{\partial n}\right)^2$, which contributes a term

$$\frac{1}{2} \int_{r_3'} c_1 \left(\frac{\partial w}{\partial n}\right)^2 dl$$

to the potential energy of external work, and contributes a term

$$\int_{\Gamma_3'} c_1 \frac{\partial w}{\partial n} \frac{\partial v}{\partial n} dl$$

to the functional of virtual work. The equilibrium equation on Γ_3' will be

$$\Gamma_3' : M_{nn}(w) - c_1 \frac{\partial w}{\partial n} = m_t. \qquad (4.49)$$

The elastic support boundary condition is another kind of mechanical boundary conditions. It is only necessary to absorb the elastic coupling energies into the total potential energy but needless to list them out as the conditions for determining the solution. As for to the second kind of boundary condition, a prescribed load, if the transverse displacement has been prescribed, then the local elastic reaction force is. If the tangential rotation angle has been prescribed, then the local elastic reaction moment is also.

3.3 There exists an elastic coupling between the external region and a plane with the plate, i.e., we consider the plate of elastic foundation.

Suppose the plate is subjected to an elastic reaction force $-cw$ proportional to the deflection on a subdomain Ω' of Ω, where $c > 0$ is the constant of elastic coupling. Then on the plane of the plate Ω', there will be a perr unit area elastic energy $\dfrac{1}{2} cw^2$, which contributes a term

$$\frac{1}{2} \iint_{\Omega'} cw^2 dx_1 dx_2$$

to the total potential energy, and contributes a term

$$\iint_{\Omega'} cwv \, dx_1 dx_2$$

to the functional of virtual work. The equilibrium equations in the interior
of the plate Ω will be

$$
\begin{cases}
\Omega - \Omega' : -\sum_{i=1}^{2} \dfrac{\partial Q_{3i}}{\partial x_i} = P_3, \\[3mm]
\Omega' : -\sum_{i=1}^{2} \dfrac{\partial Q_{3i}}{\partial x_i} + cw = P_3.
\end{cases}
\tag{4.50}
$$

On each segment of the boundary, we may impose two boundary un-
ditions of any of the three kinds discussed above, two geometric or one
geometric, two mechanical or one mechanical, and we may impose different
kinds on the different segments. This can give rise to quite a complicated
combination. One should note that the boundary conditions 1.1 and 2.1
or 3.1 as well as 1.2 and 2.2 or 3.2 repel each other, i.e., are not compatible. In practice, the following three kinds of boundary conditions are
frequently encountered.

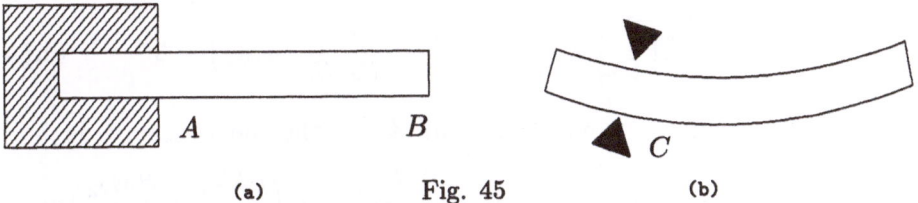

(a)   Fig. 45   (b)

(1)   A fixed edge (Fig. 45a) :

   Displacement $w = 0$ (essential),

   angle of rotation $\dfrac{\partial w}{\partial n} = 0$ (essiential).   (4.51)

(2)   A pinned edge (Fig. 45b) :

   Displacement $w = 0$ (essential),

   bending moment $M_{nn} = 0$ (natural).   (4.52)

(3)   A free edge:

   Bending moment $M_{nn} = 0$ (natural),

   shearing force $Q_{3n} + \dfrac{\partial M_{ln}}{\partial l} = 0$ (natural).   (4.53)

In the following, the two natural boundary conditions of free edge are expressed in terms of the partial derivatives of the displacement w. Since the bending moment is

$$M_{nn} = M_{11}n_1^2 + M_{22}n^2 + 2M_{12}n_1n_2$$

$$= -D\left\{\left(\frac{\partial^2 w}{\partial x_1^2} + \nu\frac{\partial^2 w}{\partial x_2^2}\right)n_1^2 + \left(\nu\frac{\partial^2 w}{\partial x_1^2} + \frac{\partial^2 w}{\partial x_2^2}\right)n_2^2\right.$$

$$\left. + 2(1-\nu)\frac{\partial^2 w}{\partial x_1\partial x_2}n_1n_2\right\}$$

$$= -D\left\{\nu\Delta w + (1-\nu)\left(\frac{\partial^2 w}{\partial x_1^2}n_1^2 + \frac{\partial^2 w}{\partial x_2^2}n_2^2\right.\right.$$

$$\left.\left. + 2\frac{\partial^2 w}{\partial x_1\partial x_2}n_1n_2\right)\right\},$$

the natural boundary condition $M_{nn} = 0$ can be written as

$$\nu\Delta w + (1-\nu)\left(\frac{\partial^2 w}{\partial x_1^2}n_1^2 + \frac{\partial^2 w}{\partial x_2^2}n_2^2 + 2\frac{\partial^2 w}{\partial x_1\partial x_2}n_1n_2\right) = 0. \qquad (4.54)$$

On the other hand, from (4.25) and (4.26), the shearing force is

$$Q_{3n} = Q_{31}n_1 + Q_{32}n_2 = \left(\frac{\partial M_{11}}{\partial x_1} + \frac{\partial M_{12}}{\partial x_2}\right)n_1 + \left(\frac{\partial M_{21}}{\partial x_1} + \frac{\partial M_{22}}{\partial x_2}\right)n_2$$

$$= -D\left\{\left[\frac{\partial}{\partial x_1}\left(\frac{\partial^2 w}{\partial x_1^2} + \nu\frac{\partial^2 w}{\partial x_2^2}\right) + (1-\nu)\frac{\partial}{\partial x_2}\left(\frac{\partial^2 w}{\partial x_1\partial x_2}\right)\right]n_1\right.$$

$$\left. + \left[(1-\nu)\frac{\partial}{\partial x_1}\left(\frac{\partial^2 w}{\partial x_1\partial x_2}\right) + \frac{\partial}{\partial x_2}\left(\nu\frac{\partial^2 w}{\partial x_1^2} + \frac{\partial^2 w}{\partial x_2^2}\right)\right]n_2\right\}$$

$$= -D\left\{\frac{\partial(\Delta w)}{\partial x_1}n_1 + \frac{\partial(\Delta w)}{\partial x_2}n_2\right\} = -D\frac{\partial(\Delta w)}{\partial n}. \qquad (4.55)$$

From (4.28), the twisting moment is

$$M_{tn} = M_{11}t_1n_1 + M_{22}t_2n_2 + M_{12}t_1n_2 + M_{21}t_2n_1.$$

Moreover, since (see (2.22) in §2 of the present chapter)

$$t_1 = -n_2 = \frac{dx_1}{dl}, \quad t_2 = n_1 = \frac{dx_2}{dl},$$

it follows that

$$M_{tn} = -D\left\{ -\left(\frac{\partial^2 w}{\partial x_1^2} + \nu\frac{\partial^2 w}{\partial x_2^2}\right)n_1 n_2 + \left(\nu\frac{\partial^2 w}{\partial x_1^2} + \frac{\partial^2 w}{\partial x_2^2}\right)\right.$$

$$\times n_1 n_2 + (1-\nu)\frac{\partial^2 w}{\partial x_1 \partial x_2}(n_1^2 - n_2^2)\Big\}$$

$$= D(1-\nu)\left\{\left(\frac{\partial^2 w}{\partial x_1^2} - \frac{\partial^2 w}{\partial x_2^2}\right)n_1 n_2 + \frac{\partial^2 w}{\partial x_1 \partial x_2}(n_2^2 - n_1^2)\right\}.$$

Hence, the natural boundary condition $Q_{3n} + \dfrac{\partial M_{tn}}{\partial l} = 0$ can be written as

$$\frac{\partial(\Delta w)}{\partial n} - (1-\nu)\frac{\partial}{\partial l}\left\{\left(\frac{\partial^2 w}{\partial x_1^2} - \frac{\partial^2 w}{\partial x_2^2}\right)n_1 n_2\right.$$

$$\left. + \frac{\partial^2 w}{\partial x_1 \partial x_2}(n_2 - n_1^2)\right\} = 0. \tag{4.56}$$

For the fixed edge, since

$$w = 0,$$

can be the natural boundary condition $M_{nn} = 0$, i.e., (4.54), further simplified. To this end, transform the partial derivatives with respect to x_1 and x_2 in (4.54) into partial derivatives with respect to the tangent t and the normal n. Now

$$\frac{\partial w}{\partial n} = \frac{\partial w}{\partial x_1}n_1 + \frac{\partial w}{\partial x_2}n_2,$$

$$\frac{\partial^2 w}{\partial n^2} = \frac{\partial^2 w}{\partial x_1^2}n_1^2 + \frac{\partial^2 w}{\partial x_2^2}n_2^2 + 2\frac{\partial^2 w}{\partial x_1 \partial x_2}n_1 n_2, \tag{4.57}$$

$$\frac{\partial w}{\partial l} = \frac{\partial w}{\partial x_1}t_1 + \frac{\partial w_2}{\partial x_2}t_2,$$

$$\frac{\partial^2 w}{\partial l^2} = \frac{\partial^2 w}{\partial x_1^2}t_1^2 + \frac{\partial^2 w}{\partial x_2^2}t_2^2 + 2\frac{\partial^2 w}{\partial x_1 \partial x_2}t_1 t_2$$

$$+ \frac{\partial w}{\partial x_1}\frac{d^2 x_1}{dl^2} + \frac{\partial w}{\partial x_2}\frac{d^2 x_2}{dl^2} \tag{4.58}$$

$$= \frac{\partial^2 w}{\partial x_1^2}n_2^2 + \frac{\partial^2 w}{\partial x_2^2}n_1^2 - 2\frac{\partial^2 w}{\partial x_1 \partial x_2}n_1 n_2 - \frac{d\theta}{dl}\frac{\partial w}{\partial n},$$

where θ is the angle included between the tangential direction t and the x_1-axis. Adding up (4.57) and (4.58), we obtain

$$\Delta w = \frac{\partial^2 w}{\partial n^2} + \frac{\partial^2 w}{\partial l^2} + \frac{d\theta}{dl}\frac{\partial w}{\partial n},$$

so that (4.54) can be rewritten as

$$\nu\Delta w + (1-\nu)\frac{\partial^2 w}{\partial n^2} = \frac{\partial^2 w}{\partial n^2} + \nu\left(\frac{\partial^2 w}{\partial l^2} + \frac{d\theta}{dl}\frac{\partial w}{\partial n}\right) = 0. \qquad (4.59)$$

Since $w = 0$ on the fixed edge, $\dfrac{\partial w}{\partial l} = \dfrac{\partial^2 w}{\partial l^2} = 0$. Letting $\dfrac{\partial^2 w}{\partial l^2} = 0$ in (4.59), then we obtain the following two boundary conditions on the fixed edge:

$$\begin{cases} w = 0, \\[2mm] \dfrac{\partial^2 w}{\partial n^2} + \nu\dfrac{d\theta}{dl}\dfrac{\partial w}{\partial n} = 0. \end{cases} \qquad (4.60)$$

If the boundary is a segment of a straight line, then $\dfrac{d\theta}{dl} = 0$. In that case, we obtain the fixed-edge boundary conditions

$$\begin{cases} w = 0, \\[2mm] \dfrac{\partial^2 w}{\partial n^2} = 0. \end{cases} \qquad (4.61)$$

This holds only. Note that, for general curvilinear boundaries, the simply-supported conditions should be (4.60), not (4.61).

Now we discuss interface condition.

During the derivation of equilibrium equations by the variational principle, if the flexural rigidity D of the plate has a discontinuity, for example, that which is introduced by a jump in the thickness along a certain line L in the domain Ω, we can use Green's formula piecewise to obtain the interface conditions,

$$L: \quad \left[Q_{3n} + \frac{\partial M_{tn}(w)}{\partial l}\right]_{-}^{+} = 0, \qquad (4.62)$$

$$L: \quad [M_{nn}(w)]_{-}^{+} = 0. \qquad (4.63)$$

Here, n is the normal direction taken arbitrarily on L. The two equations denote the continuities of the shearing force and the bending moment, respectively.

As for the mechanical boundary conditions (4.33)–(4.35), the interface conditions (4.62) and (4.63) are satisfied automatically when the potential energy achieves a minimum, and are also of a kind of natural boundary conditions that do not need to be listed as conditions for determining a unique solution to the variational problem.

4.5   Strainless states

From (4.17),

$$D(v,v) = \iint_\Omega \sum_{i,j=1}^{2} M_{ij}(v)K_{ij}(v)dx_1dx_2$$

$$= \iint_\Omega D\left[(1-\nu)\sum_{i,j=1}^{2} K_{ij}^2(v) + \nu\left(\sum_{i=1}^{2} K_{ii}(v)\right)^2\right]dx_1dx_2$$

$$\geq 0, \qquad \text{for any } v.$$

So that the strainless condition

$$v: D(v,v) = 0 \Longleftrightarrow K_{ij}(v) = -\frac{\partial^2 v}{\partial x_i \partial x_j} \equiv 0$$

$$\Longleftrightarrow v = v_3 = a_3 + b_1 x_2 - b_2 x_1,$$

$$\omega_1(v) = \frac{\partial v}{\partial x_2} = b_1,$$

$$\omega_2(v) = -\frac{\partial v}{\partial x_1} = b_2.$$

Hence, $D(v,v)$ is degenerat, when the strainless state has three degrees of freedom: one is the transverse rigid translation a_3 and another two degrees of freedom are the two components b_1 and b_2 of the rigid angle of rotation. The strainless v can be expressed as

$$v = a_3 v^{(1)} + b_1 v^{(2)} + b_2 v^{(3)}, \tag{4.64}$$

where

$$v^{(1)} = 1, \omega_1^{(1)} = 0, \omega_2^{(1)} = 0,$$

$$v^{(2)} = x_2, \omega_1^{(2)} = 1, \omega_2^{(2)} = 0,$$

$$v^{(3)} = -x_1, \omega_1^{(3)} = 0, \omega_2^{(3)} = 1.$$

Thus, the necessary and sufficient conditions for the existence of solutions of the equilibrium problem without constraints on the displacement or the angle of rotation are

$$F(v^{(i)}) = 0, \quad i = 1, 2, 3,$$

i.e.,

$$\iint_\Omega P_3 dx_1 dx_2 + \int_{\partial\Omega} q_3 dl = 0, \tag{4.65}$$

$$\iint_\Omega x_2 P_3 dx_1 dx_2 + \int_{\partial\Omega} x_2 q_3 dl + \int_{\partial\Omega} m_t t_1 dl = 0, \tag{4.66}$$

$$\iint_\Omega -x P_3 dx_1 dx_2 + \int_{\partial\Omega} -x_1 q_3 dl + \int_{\partial\Omega} m_t t_2 dl = 0. \tag{4.67}$$

(4.65) denotes the equilibrium of force of the transverse loads, while (4.66) and (4.67) denote the equilibrium of the moments of loads about the x_1- and, x_2-axes respectively. These are the prerequisites for achieving the equilibrium.

As the common practice, the introducing of geometric constraints will always enlarge the subspace on which the quadratic functional of strain energy $D(v,v)$ is positive definite. For example, if the geometric constraints (4.45) and (4.46)

$$\Gamma_1 : w = \bar{w}, \quad \Gamma_1 : \frac{\partial w}{\partial n} = -\bar{\omega}_t$$

have been introduced, then the strainless virtual displacement v must satisfy the corresponding annihilating constraint conditions

$$\Gamma_1 : \quad v = a_3 + b_1 x_2 - b_2 x_1 = 0, \tag{4.68}$$

$$\Gamma_1' : \quad \frac{\partial v}{\partial n} = \frac{\partial v}{\partial x_1} n_1 + \frac{\partial v}{\partial x_2} n_2 = -b_2 n_1 + b_1 n_2 = 0. \tag{4.69}$$

Consequently, the degrees of freedom of the strainless state will be decreased such as to make one among the three parameters or all of them vanish. In the latter last case, for any virtual displacement v, $D(v,v) = 0 \iff v = 0$, that is just the positive definiteness of $D(v,v)$ which guarantees the equilibrium problem has a unique solution. We can see from (4.68) that, one can ensure the positive definiteness of $D(v,v)$ so long as the boundary segment Γ_1 of the prescribed displacement contains three points not situated on the same straight line, e.g., there exists a segment of bent pinned boundary on Γ_1 (Fig. 46a). Moreover, the positive definiteness of $D(v,v)$ can also be ensured if Γ_1 and Γ_1' hold on the same segment, i.e., there exists a segment of fixed boundary (Fig. 46b).

As for the geometric constraint condition, whenever the elastic support condition is added, the number of degrees of freedom of the strainless state will always be decreased, too. The reason is that

$$D(v,v) = \iint_\Omega D\left[(1-\nu)\sum_{i,j=1}^{2} K_{ij}^2(v) + \nu\left(\sum_{i=1}^{2} K_{ii}(v)\right)^2\right]dx_1dx_2$$

$$+ \iint_{\Omega'} cv^2 dx_1 dx_2 + \int_{\Gamma_3} c_0v^2 dl$$

$$+ \int_{\gamma_3'} c_1\left(\frac{\partial v}{\partial n}\right)^2 dl, c, c_0, c_1 > 0.$$

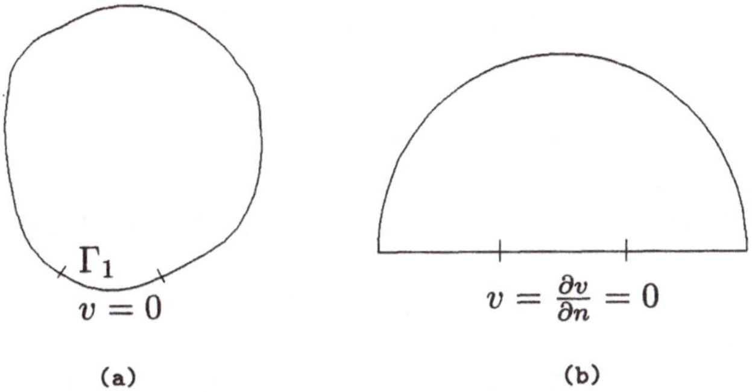

$$\Gamma_1$$
$$v = 0$$

(a)

$$v = \frac{\partial v}{\partial n} = 0$$

(b)

Fig. 46

Hence, for any v, $D(v,v) = 0$ is equivalent to

$$\Omega: \quad K_{ij}(v) \equiv 0, \text{ i. e., } v = a_3 + b_1x_2 - b_2x_1,$$

$$\Omega': \quad v = 0; \ \Gamma_3 : v = 0; \ \Gamma_3' : \frac{\partial v}{\partial n} = 0.$$

Under these additional conditions, the number of degrees of freedom of the strainless state will be reduced to zero, making $D(v,v)$ positive definite. Obviously, $D(v,v)$ is positive definite so long as there exists a constant of elastic support $c > 0$ on a subdomain Ω' of Ω with an area no matter how small but not vanishing. For improvements in the positive definiteness of $D(v,v)$, the results induced by the two kinds of the elastic supports on the boundaries Γ_3' and Γ_3 are the same as the results induced by the two kinds of geometric constraints.

4.6 Thermal effects

Having given the stress-strain relations in (4.9) and the strain energy density formulas (4.10), let us now return to the problem of thermal effects.

Assume that the temperature field τ in the interior of the plate body is linearly in the thickness, i.e., in the x_3-direction:

$$\tau = \tau(x_1, x_2, x_3) = \tau(x_1, x_2, 0) + x_3\tau'(x_1, x_2). \qquad (4.70)$$

Let

$$\bar{\tau} = \bar{\tau}(x_1, x_2) = \frac{1}{h}\int_{-h/2}^{h/2}\tau(x_1, x_2, x_3)dx_3$$

be the mean temperature averaged over the thickness of the plate. Integrating both the sides of (4.70) with respect to the thickness of the plate, we obtain

$$\bar{\tau}(x_1, x_2) = \tau(x_1, x_2, 0),$$

so that

$$\tau = \bar{\tau}(x_1, x_2) + x_3\tau'(x_1, x_2).$$

Substituting this expression into (4.9) and (4.10), we obtain

$$\left\{\begin{array}{l} \sigma_{ij} = \sigma'_{ij} - \dfrac{E\alpha}{1-\nu}(\bar{\tau} + x_3\tau')\delta_{ij}, \\[3mm] \sigma'_{ij} = \dfrac{E}{1-\nu^2}\left[(1-\nu)\varepsilon_{ij} + \nu\Big(\sum_{k=1}^{2}\varepsilon_{kk}\Big)\delta_{ij}\right] \\[3mm] \quad = \dfrac{Ex_3}{1-\nu^2}\left[(1-\nu)K_{ij} + \nu\Big(\sum_{k=1}^{2}K_{kk}\Big)\delta_{ij}\right]. \end{array}\right. \qquad (4.71)$$

$$\left\{\begin{array}{l} W = W' - \dfrac{E\alpha x_3}{1-\nu}(\bar{\tau} + x_3\tau')\sum_{k=1}^{2}K_{kk}, \\[3mm] W' = \dfrac{1}{2}\dfrac{Ex_3^2}{1-\nu^2}\left[(1-\nu)\sum_{i,j=1}^{2}K_{ij}^2 + \nu\Big(\sum_{k=1}^{2}K_{kk}\Big)^2\right]. \end{array}\right. \qquad (4.72)$$

The relations between σ'_{ij} or W' and K_{ij} are identical with those that hold when there is no thermal effect. Here, however, they stand for not the stress or the strain energy but only. Let

$$M_{ij} = \int_{-h/2}^{h/2} x_3\sigma_{ij}dx_3, \quad i, j = 1, 2,$$

$$\bar{W} = \int_{-h/2}^{h/2} W dx_3.$$

After integration, if we note that

$$\int_{-h/2}^{h/2} x_3 dx_3 = 0, \quad \int_{-h/2}^{h/2} x_3^2 dx_3 = \frac{h^3}{12},$$

we obtain

$$\begin{cases} M_{ij} = M'_{ij} - \bar{D}\tau'\delta_{ij}, \\ M'_{ij} = D\left[(1-\nu)K_{ij} + \nu\sum_{k=1}^{2} K_{kk}\right], \end{cases} \qquad (4.73)$$

$$\begin{cases} \bar{W} = \bar{W}' - \bar{D}\tau'\sum_{k=1}^{2} K_{kk}, \\ \bar{W}' = \frac{1}{2}D\left[(1-\nu)\sum_{i,j=1}^{2} K_{ij}^2 + \nu\left(\sum_{k=1}^{2} K_{kk}\right)^2\right] = \frac{1}{2}\sum M'_{ij}K_{ij}, \end{cases} \qquad (4.74)$$

where

$$D = \frac{Eh^3}{12(1-\nu^2)}, \qquad \bar{D} = \frac{Eh^3\alpha}{12(1-\nu)} = D(1+\nu)\alpha.$$

We also have,

$$M_{ij} = \frac{\partial \bar{W}}{\partial K_{ij}}, \quad i,j = 1,2.$$

(4.73) is just Hooke's law with thermal effects. When $\tau' \equiv 0$, $\sigma_{ij} = \sigma'_{ij}$, $M_{ij} = M'_{ij}$, and $\bar{W} = \bar{W}'$, they reduce to the case without thermal effects. Note that, under the assumption that the temperature is linear in the thickness, for flexural deformation of plates, the thermal effect only acts through the transverse temperature difference τ', while the transverse average temperature $\bar{\tau}$ is out of action. But for the longitudinal deformation, the thermal effect acts through the transverse average temperature, which is just contrary to the case of the flexural deformation.

For the deflection field w, the total strain energy of the plate is

$$\begin{aligned} P(w) &= \iint_\Omega \bar{W}(w)dx_1dx_2 \\ &= \frac{1}{2}\iint_\Omega \sum_{i,j=1}^{2} M'_{ij}(w)K_{ij}(w)dx_1dx_2 - \iint_\Omega \bar{D}\tau'\sum_{k=1}^{2} K_{kk}(w)dx_1dx_2. \end{aligned} \qquad (4.75)$$

Compared to the case without thermal effect, (4.75) has an extra term that depends on the temperature difference:

$$-F_\tau(w) = -\iint_\Omega \overline{D}\tau'\sum_{k=1}^{2} K_{kk}(w)dx_1dx_2.$$

This term is a linear functional of w. It can be considered to be the potential energy of external work induced by the temperature load and can be included in the potential energy of external work $-F(v)$. Thus, there is no loss in letting the strain energy be a second order functional independent of temperature

$$P(w) = \frac{1}{2} \iint_\Omega \sum_{i,j=1}^2 M'_{ij}(w) K_{ij}(w) dx_1 dx_2 \tag{4.76}$$

$$= \frac{1}{2} D(w, w).$$

The corresponding bilinear functional of virtual work is

$$D(w, v) = \iint_\Omega \sum_{i,j=1}^2 M_{ij}(w) K_{ij}(v) dx_1 dx_2.$$

These formulas have the same form as (4.17) and (4.18).

For the load condition (4.19), we have

$$F(v) = \iint_\Omega P_3 v dx_1 dx_2 + \oint_{\partial\Omega} q_3 v dl - \oint_{\partial\Omega} m_t \frac{\partial v}{\partial n} dl + F_\tau(v).$$

The total potential energy is

$$J(w) = \frac{1}{2} D(w, w) - F(w).$$

Since

$$\sum_{k=1}^2 K_{kk}(v) = -\left(\frac{\partial^2 v}{\partial x_1^2} + \frac{\partial^2 v}{\partial x_2^2}\right) = -\Delta v,$$

using Green's formula, it is easy to obtain

$$-F_\tau(v) = \iint_\Omega \bar{D}\tau' \Delta v dx_1 dx_2$$

$$= \iint_\Omega \Delta(\bar{D}\tau') v dx_1 dx_2 - \oint_{\partial\Omega} \frac{\partial(\bar{D}\tau')}{\partial n} v dl + \oint_{\partial\Omega} \bar{D}\tau' \frac{\partial v}{\partial n} dl.$$

So that

$$F(v) = \iint_\Omega [P_3 - \Delta(\bar{D}\tau')] v dx_1 dx_2 + \oint_{\partial\Omega} \left[g_3 + \frac{\partial(\bar{D}\tau')}{\partial n}\right] v dl$$

$$- \oint_{\partial\Omega} (m_t + \bar{D}\tau') \frac{\partial v}{\partial n} dl.$$

Then, substituting (4.73) into (4.76), from the variational principle

$$D(w, v) - F(v) = 0, \quad \text{for any } v,$$

we can derive the equilibrium equation, the mechanical boundary conditions, and the prerequisites for equilibrium when the geometric constraints are absent, etc. These are formally identical with (4.32)–(4.35) provided that the load terms P_3, q_3, in (4.32)–(4.35) are increased by a temperature load, transforming them into

$$P_3 - \Delta(\bar{D}\tau'), \quad q_3 + \frac{\partial(\bar{D}\tau')}{\partial n}, \quad m_t + \bar{D}\tau',$$

respectively.

§5   Bending of Spatial Beams

5.1   Deformation modes

Several kinds of modes of elastic deformation of rods were discussed in Chapter 1: the first was longitudinal stretching and compression under longitudinal (in the axial direction) loads, and the second was shear deformation including longitudinal simple shear and torsion. Another typical deformation is transverse deformation, i.e., bending under transverse (normal to the axial direction) loads. Since it is to a plane, we have already given a preliminary discussion of this kind of bending. Comparing with the one dimensional stress mode in §1 and the bending mode in §4, the general case of bending of spatial beams will be further discussed in the present chapter on the basis of the three-dimensional elasticity.

Deformation by longitudinal stretching and compression (free in the two transverse directions) of slender rods is called the one dimensional stress mode. Its main features are

(1) $\sigma_{12} = \sigma_{22} = \sigma_{32} = 0$, $\sigma_{13} = \sigma_{23} = \sigma_{33} = 0$.

(2) The deformation depends only on the longitudinal displacement u_1 and the corresponding strain ε_{11}. Hence, u_1 is taken as the fundamental variable depending only on the longitudinal coordinate x_1.

Thus the bending of slender rods can be considered as the further simplification of the bending of thin plates. In the following, the discussion will be performed in the form of §4 but being reduced to one dimension from two dimensions.

Since the rod is very thin (the size of the cross section is far smaller than the length), in order to obtain a flexural deformation, it is only necessary to apply some not too large loads g_2 and g_3 on the surface of the rod which are far smaller than the stresses of longitudinal stretching and compression that they produce in the interior. Hence, on the surface of the rod, we have

$$\sum_{j=1}^{3} \sigma_{ij} n_j \approx 0, \quad i = 1, 2, 3,$$

where $\boldsymbol{n} = (n_1, n_2, n_3)^T$ indicates the outward normal direction to the surface of the rod. Since we are considering only a, the outward normal direction to the surface of the rod is $\boldsymbol{n} \approx (0, n_2, n_3)^T$. Hence

$$\sum_{j=1}^{3} \sigma_{ij} n_j = \sigma_{i2} n_2 + \sigma_{i3} n_3 \approx 0 \quad \text{for any } n_2, n_3.$$

Consequently, on the surface of the rod,

$$\sigma_{i2} \approx 0, \quad \sigma_{i3} \approx 0, \quad i = 1, 2, 3.$$

Moreover, since the area of the cross section is very small, we can consider that, in the interior of the rod, we also have

(1') $\sigma_{12} = \sigma_{22} = \sigma_{32} = 0$, $\sigma_{13} = \sigma_{23} = \sigma_{33} = 0$, or at least, σ_{i2} and σ_{i3} are small compared with the remaining stress component σ_{11}.

On the other hand, in taking that there exists a neutral axis without longitudinal stretching or compression during the bending of the slender rod. There is no loss to take neutral axis before deformation as the x_1-axis, i.e., $x_2 = x_3 = 0$. The there components of the displacement of the neutral axis are denoted in turn by

$$u_1^{(0)} = 0, \ u_2^{(0)} = u_2^{(0)}(x_1), \quad u_3^{(0)} = u_3^{(0)}(x_1). \tag{5.1}$$

Since the area of the cross section is very small, the transverse displacements, i.e, the deflections u_2 and u_3, can be considered as consistent over the cross section, i.e., approximately equal to the deflections on the neutral axis

$$u_2(x_1, x_2, x_3) \approx u_2^{(0)}(x_1), \quad u_3(x_1, x_2, x_3) \approx u_3^{(0)}(x_1).$$

Since $\sigma_{12} = \sigma_{13} = 0$, then $\varepsilon_{12} = \varepsilon_{13} = 0$, and we have

$$\frac{\partial u_1}{\partial x_2} = -\frac{\partial u_2}{\partial x_1} \approx -\frac{du_2^{(0)}}{dx_1}, \tag{5.2}$$

$$\frac{\partial u_1}{\partial x_3} = -\frac{\partial u_3}{\partial x_1} \approx -\frac{du_3^{(0)}}{dx_1}. \tag{5.3}$$

Integrating (5.2) with respect to x_2, we obtain

$$u_1(x_1, x_2, x_3) = -x_2 \frac{du_2^{(0)}}{dx_1} + f(x_1, x_3).$$

Differentiating the above expression with respect to x_3, and using (5.3), we obtain

$$\frac{\partial f}{\partial x_3} = -\frac{du_3^{(0)}}{dx_1}.$$

Integrating again with respect to x_3, we have

$$f(x_1, x_3) = -x_3 \frac{du_3^{(0)}}{dx_1} + g(x_1).$$

Thus

$$u_1(x_1, x_2, x_3) = -x_2 \frac{du_2^{(0)}}{dx_1} - x_3 \frac{du_3^{(0)}}{dx_1} + g(x_1).$$

Letting $x_2 = x_3 = 0$, we have

$$u_1(x_1, 0, 0) = g(x_1) \approx u_1^{(0)}(x_1) = 0,$$

so that

$$u_1(x_1, x_2, x_3) = -x_2 \frac{du_2^{(0)}}{dx_1} - x_3 \frac{du_3^{(0)}}{dx_1}.$$

For the sake of convenience in what follows, which dispense with the superscripts in the deflections of the neutral axis,

$$u_2^{(0)} = u_2(x_1), u_3^{(0)} = u_3(x_1).$$

Now the transverse displacements u_2 and u_3 stand for the deflections of the neutral axis.

Because

$$u_1(x_1, x_2, x_3) = -x_2 u_2' - x_3 u_3', \tag{5.4}$$

the longitudinal displacement u_1 can be expressed in terms of the derivatives of the deflections u_2 and u_3. Besides, u_1 still depends on x_2 and x_3 linearly. Accordingly, we immediately obtain the strain ε_{11} and the infinitesimal angle of rotation ω_{ij} expressed in terms of the deflections u_2 and u_3 obtain as follows:

$$\varepsilon_{11} = \frac{\partial u_1}{\partial x_1} = -(x_2 u_2'' + x_3 u_3''), \tag{5.5}$$

$$\omega_1 = \omega_{23} = \frac{1}{2}\left(\frac{\partial u_3}{\partial x_2} - \frac{\partial u_2}{\partial x_3}\right) = 0,$$

$$\omega_2 = \omega_{31} = \frac{1}{2}\left(\frac{\partial u_1}{\partial x_3} - \frac{\partial u_3}{\partial x_1}\right) = \frac{1}{2}\left(-\frac{du_3}{dx_1} - \frac{du_3}{dx_1}\right) = -u_3', \tag{5.6}$$

$$\omega_3 = \omega_{12} = \frac{1}{2}\left(\frac{\partial u_2}{\partial x_1} - \frac{\partial u_1}{\partial x_2}\right) = \frac{1}{2}\left(\frac{du_2}{dx_1} + \frac{du_2}{dx_1}\right) = u_2'. \tag{5.7}$$

Let

$$K_2 = \frac{d\omega_2}{dx_1} = -u_3'', \quad K_3 = \frac{d\omega_3}{dx_1} = u_2'', \tag{5.8}$$

these are the first order approximations of the curvature vector of the spatial curve formed by the neutral axis after bending. Then

$$\varepsilon_{11} = x_3 K_2 - x_2 K_3. \tag{5.9}$$

Thus, we obtain the second characteristic of the flexural deformation of slender rods:

(2′) The flexural deformation depends only on the deflections u_2 and u_3, which depend only on the longitudinal coordinate x_1. The longitudinal displacement u_1 and the strain ε_{11} are determined by the deflections u_2 and u_3.

Since the first characteristics of the deformation of stretching and compression (1) and the bending deformation of slender rods (1′) are identical, according to the one dimensional stress mode in Section 1.4 of the present chapter, we obtain

$$\varepsilon_{22} = \varepsilon_{33} = -\nu\varepsilon_{11} + (1 + \nu)\alpha\tau,$$

$$\varepsilon_{12} = \varepsilon_{21} = \varepsilon_{13} = \varepsilon_{31} = \varepsilon_{23} = \varepsilon_{23} = 0.$$

Hooke's law takes the form

$$\sigma_{11} = E\varepsilon_{11} - E\alpha\tau, \tag{5.10}$$

and the volume density of the strain energy

$$W = \frac{1}{2}\sigma_{11}\varepsilon_{11} = \frac{1}{2}E\varepsilon_{11}^2 - E\alpha\tau\varepsilon_{11}. \tag{5.11}$$

Hence, the forms of the Hooke's law and the strain energy of rods in bending are identical with those of rods in tension. However, in these two kinds of deformation modes, the schemes (2) and (2') for determining the longitudinal displacement u_1 differ essentially.

5.2   Variational principles

At first, we do not consider thermal effect. Since there are no longitudinal loads in the flexural problem of rods, the resultant force of the longitudinal stresses on the individual cross sections must vanish, i.e.,

$$\iint_s \sigma_{11} dx_2 dx_3 = \iint_s E(x_3 K_2 - x_2 K_3) dx_2 dx_3$$

$$= EK_2 \iint_s x_3 dx_2 dx_3 - EK_3 \iint_s x_2 dx_2 dx_3$$

$$= 0 \quad \text{for any } K_2, K_3.$$

Hence

$$\iint_s x_2 dx_2 dx_3 = 0, \quad \iint_s x_3 dx_2 dx_3 = 0.$$

This implies that, the centroid of the cross section lies on the x_1-axis. Hence the moment of inertia of cross section S on the $x_2 - x_3$-plane is expressed as

$$I_{ij} = \iint_S [(x_2^2 + x_3^2)\delta_{ij} - x_i x_j] dx_2 dx_3, \tag{5.12}$$

i.e.,

$$I_{22} = \iint_S x_3^2 dS \text{—the moment of inertia with respect to the } x_2\text{-axis,}$$

$$I_{33} = \iint_S x_2^2 dS \text{—the moment of inertia with respect to the } x_3\text{-axis,}$$

$$I_{23} = I_{22} = - \iint_s x_2 x_3 dS \text{—the product of inertia with respect to the}$$

$$x_2\text{-and } x_3\text{-axes.}$$

Hence, the line density of strain energy of the rod is

$$\bar{W} = \iint_s W dx_2 dx_3 = \frac{1}{2} \iint_s E \varepsilon_{11}^2 dx_2 dx_3$$

$$= \frac{1}{2} \iint_s E(x_3 K_2 - x_2 K_3)^2 dx_2 dx_3$$

$$= \frac{1}{2} E(I_{22} K_2^2 + 2I_{23} K_2 K_3 + I_{33} K_3^2)$$

$$= \frac{1}{2} \sum_{i,j=2}^{3} E I_{ij} K_i K_j. \tag{5.13}$$

Define

$$M_i = \frac{\partial \bar{W}}{\partial K_i} = \sum_{j=2}^{3} E I_{ij} K_j, \quad i = 2, 3,$$

or

$$\begin{cases} M_2 = E(I_{22} K_2 + I_{23} K_3), \\ M_3 = E(I_{32} K_2 + I_{33} K_3), \end{cases} \tag{5.14}$$

then we have

$$\bar{W} = \frac{1}{2}(M_2 K_2 + M_3 K_3). \tag{5.15}$$

In mechanics M_i represent, the resultant moment of the stresses, i.e., the bending moment, about the $x_i(i = 2, 3)$-axes on the cross section. The bending moment about the x_2-axis is

$$\iint_s (x_3\sigma_{11} - x_1\sigma_{31})dS = \iint_s x_3\sigma_{11}dS = \iint_s (Ex_3^2 K_2 - Ex_2 x_3 K_3)dS$$

$$= EI_{22}K_2 + EI_{23}K_3 = M_2. \tag{5.16}$$

The bending moment about he x_3-axis is

$$\iint_s (x_1\sigma_{21} - x_2\sigma_{11})dS = -\iint_s x_2\sigma_{11}dS$$

$$= \iint_s (-Ex_2 x_3 K_2 + Ex_2^2 K_3)dS$$

$$= EI_{32}K_2 + EI_{33}K_3 = M_3. \tag{5.17}$$

This is Hooke's law for the bending of rods. Here the strain is described by the curvatures K_2 and K_3, and the stress is described by the bending moments M_2 and M_3. The constant EI_{ij} is called the flexural rigidity of the rod, where I_{ij} is the moment of inertia of the cross section depending only on the geometric size and shape but independing of the material.

The bending strain energy of the whole rod is

$$P(u) = \int_a^b \bar{W}dx_1 = \int_a^b \frac{1}{2}(M_2(u)K_2(u) + M_3(u)K_3(u))dx_1$$

$$= \frac{1}{2}\int_a^b (M_3 u_2'' - M_2 u_3'')dx_1 = \frac{1}{2}D(\boldsymbol{u}, \boldsymbol{u}), \tag{5.18}$$

where $\boldsymbol{u} = (u_2, u_3)^T$. The bilinear functional of the virtual work is

$$D(\boldsymbol{u}, \boldsymbol{v}) = \int_a^b (M_3(\boldsymbol{u})v_2'' - M_2(\boldsymbol{u})v_3'')dx_1. \tag{5.19}$$

Suppose there are transverse linear loads q_2 and q_3 acting along the length of the rod, the potential energy of the external work is

$$-\int_a^b (q_2 u_2 + q_3 u_3)dx_1.$$

Furthermore, suppose that at the two end points $x_1 = a, b$ there are transverse point loads of force r_2 and r_3 and point the loads of moment

m_2 and m_3 about the x_2-and x_3-axes, respectively. Then the potential energy of external work will be

$$-\sum_{p=a,b}(r_2u_2 + r_3u_3 + m_2\omega_2 + m_3\omega_3)_p$$

$$= -\sum_{p=a,b}(r_2u_2 + r_3u_3 + m_3u_2' - m_2u_3')_p.$$

Hence the total potential energy of external work is

$$-F(\boldsymbol{u}) = -\Big\{\int_a^b (q_2u_2 + q_3u_3)dx_1$$

$$+ \sum_{p=a,b}(r_2u_2 + r_3u_3 + m_3u_2' - m_2u_3')_p\Big\}, \qquad (5.20)$$

and the total potential energy is

$$J(\boldsymbol{u}) = \frac{1}{2}D(\boldsymbol{u}, \boldsymbol{u}) - F(\boldsymbol{u}).$$

Suppose that no constraint conditions of displacement or angle of rotation are imposed at the two end points of the rod. Then according to the variational principle, the deflection of the equilibrium configuration $\boldsymbol{u} = (u_2, u_3)$ makes the total potential energy $J(\boldsymbol{u})$ a unconditional minimum

$$J(\boldsymbol{u}) = \frac{1}{2}D(\boldsymbol{u}, \boldsymbol{u}) - F(\boldsymbol{u}) = \text{Min}, \qquad (5.21)$$

or

$$D(\boldsymbol{u}, \boldsymbol{v}) - F(\boldsymbol{v}) = 0, \quad \text{for any } \boldsymbol{v} = (v_2, v_3)^T. \qquad (5.22)$$

5.3   Equilibrium equations

Integrating by parts twice, we obtain

$$D(\boldsymbol{u}, \boldsymbol{v}) = \int_a^b (M_3v_2'' - M_2v_3'')dx_1$$

$$= \int_a^b (M_3''v_2 - M_2''v_3)dx_1$$

$$+ \sum_{p=a,b}(\varepsilon M_3v_2' - \varepsilon M_3'v_2 - \varepsilon M_2v_3' + \varepsilon M_2'v_3)_p,$$

where, $\varepsilon(b) = 1$, $\varepsilon(a) = -1$. Hence, by (5.20)

$$
\begin{aligned}
D(\boldsymbol{u}, \boldsymbol{v}) - F(\boldsymbol{v}) = &\int_a^b [(M_3'' - q_2)v_2 + (-M_2'' - q_3)v_3] dx_2 \\
&+ \sum_{p=a,b} [(-\varepsilon M_3' - r_2)v_2 + (\varepsilon M_2' - r_3)v_3 + (\varepsilon M_3 - m_3)v_2'] \\
&+ (-\varepsilon M_2 + m_2)v_3']_p \\
= &\,0, \quad \text{for any } v_2, v_3.
\end{aligned}
$$

This expression is equivalent to the equilibrium equations in the interior and at the two end points of the rod

$$
\begin{cases}
M_3'' = q_2, -M_2'' = q_3; \quad a < x_1 < b, & (5.23) \\[2mm]
-\varepsilon M_3' = r_2, \ \varepsilon M_2' = r_3, \quad x_1 = a, b, & (5.24) \\[2mm]
\varepsilon M_2 = m_2, \varepsilon M_3 = m_3. & (5.25)
\end{cases}
$$

The boundary conditions can be written as

$$
\begin{cases}
M_3'(a) = r_2(a), \quad\ -M_2'(a) = r_3(a), \\
-M_2(a) = m_2(a), \ \ -M_3(a) = m_3(a);
\end{cases} \quad x_1 = a \qquad (5.26)
$$

$$
\begin{cases}
-M_3'(b) = r_2(b), \ \ M_2'(b) = r_3(b), \\
M_2(b) = m_2(b), \quad\ M_3(b) = m_3(b);
\end{cases} \quad x_1 = b \qquad (5.27)
$$

respectively. Now we have

$$
M_2 = EI_{22}K_2 + EI_{23}K_3 = EI_{23}u_2'' - EI_{22}u_3'',
$$

$$
M_3 = EI_{32}K_2 + EI_{33}K_3 = EI_{33}u_2'' - EI_{32}u_3''.
$$

Substituting these two expressions into (5.23)–(5.25), the bending equations for a spatial beam become

$$
\begin{cases}
(EI_{33}u_2'')'' - (EI_{32}u_3'')'' = q_2, \\
-(EI_{23}u_2'') + (EI_{22}u_3'')'' = q_3,
\end{cases} \quad a < x_1 < b. \qquad (5.28)
$$

The boundary conditions at the two end points can now be expressed in terms of the deflections u_2 and u_3 by

$$\begin{cases} -\varepsilon(EI_{33}u_2'' - EI_{32}u_3'')' = r_2, \\ \varepsilon(EI_{23}u_2'' - EI_{22}u_3'')' = r_3, \\ \varepsilon(EI_{23}u_2'' - EI_{22}u_3'') = m_2, \\ \varepsilon(EI_{33}u_2'' - EI_{32}u_3'') = m_3, \end{cases} \qquad x_1 = a, b. \qquad (5.29)$$

In what follows, let us consider a special case, i.e., the products of inertia of the cross section s are zero: $I_{23} = I_{32} = 0$. This implies that, the x_2 and x_3 are the principal axes of inertia on the cross section S. For example, if the cross section S is symmetric with respect to the x_2-or x_3-axis, then the x_2-and x_3-axes are just the principal axes of inertia. In general, the x_2-and x_3-axes can be made principal axes of inertia by rotating the coordinates. If exist the principal axes of inertia, the equations (5.28) for the bidirectional bending of spatial beams and the boundary conditions (5.29) can be simplified to two separate equations of single directional bending and the corresponding boundary conditions, which have already been discussed in §7 of Chapter 1.

The equation and the boundary conditions for the deflection u_2 are

$$\begin{cases} (EI_{33}u_2)'' = q_2, \quad a < x_1 < b, \\ -\varepsilon(EI_{33}u_2'')' = r_2, \quad \varepsilon EI_{33}u_2'' = m_3; \quad x_1 = a, b. \end{cases}$$

And the equation and the boundary conditions for the deflection u_3 are

$$\begin{cases} (EI_2u_3'')'' = q_3, \quad a < x_1 < b, \\ -\varepsilon(EI_{22}u_3'')' = r_3, \quad -\varepsilon EI_{22}u_3'' = m_2; \quad x_1 = a, b. \end{cases}$$

Let us now illustrate the mechanical meaning of the equations (5.23) and the boundary conditions (5.24) and (5.25). Consider an arbitrary segment (x_1', x_1'') of the rod, which is acted on by body loads of density f_2 and f_3 and surface loads of density g_2 and g_3. The resultant forces of the transverse shearing stresses acting on a cross sections of the rod are the shearing forces Q_2 and Q_3:

$$Q_i = \iint_s \sigma_{i1} dS, \quad i = 2, 3. \qquad (5.30)$$

The resultant moment about the x_i-axis of the stresses on the cross section is just the bending moment M_i.

Equilibrium of the x_i-directional forces on line element (x_1', x_1'') the rod is given by the equation

$$Q_i(x_1'') - Q_i(x_1') + \int_{x_1'}^{x_1''} dx_1 \oint_{\partial s} g_i dl + \int_{x_1'}^{x_1''} dx_1 \iint_s f_i dS = 0. \qquad (5.31)$$

Let

$$q_i = \oint_{\partial s} g_i dl + \iint_s f_i ds, \quad i = 2, 3,$$

which is the line density of a transverse load prescribed in the flexural problem of slender rods. Its dimension is force/length. Thus, (5.31) becomes

$$Q_i(x_1'') - Q_i(x_1') + \int_{x_1'}^{x_1''} q_i dx_1 = \int_{x-1'}^{x_1''} \left(\frac{dQ_i}{dx_1} + q_i \right) dx_1 = 0.$$

The above expression holds for any x_1', $x_1'' \in [a, b]$. Hence we have the equilibrium equations of shearing forces

$$-\frac{dQ_i}{dx_1} = q_i, \quad i = 2, 3, \ a < x_1 < b. \qquad (5.32)$$

Next, determine the equilibrium of moment about the x_2-axis of the same line element of the rod

$$\iint_{s(x_1'')} (x_3 \sigma_{11} - x_1 \sigma_{31}) dS - \iint_{s(x_1'')} (x_3 \sigma_{11} - x_1 \sigma_{31}) dS$$

$$+ \int_{x_1'}^{x_1''} dx_1 \oint_{\partial s} (x_3 g_1 - x_1 g_3) dl + \int_{x_1'}^{x_1''} dx_1 \iint_S (x_3 f_1 - x_1 f_3) dS = 0.$$

Since $g_1 = f_1 = 0$, we obtain

$$(M_2(x_1'') - x_1'' Q_3(c_1'')) - (M_2(x_1') - x_1' Q_3(x_1')) - \int_{x_1'}^{x_1''} x_1 q_3 dx_1$$

$$= \int_{x_1'}^{x_1''} \left[\frac{dM_2}{dx_1} - \frac{d}{dx_1}(x_1 Q_3) - x_1 q_3 \right] dx_1 = 0.$$

The above expression holds for any x_1', $x_1'' \in [a, b]$. Hence we have

$$\frac{dM_2}{dx_1} - \frac{d}{dx_1}(x_1 Q_3) - x_1 q_3 = 0.$$

Moreover, using (5.32), we obtain the equilibrium equation of moments

$$\frac{dM_2}{dx_1} = Q_3. \qquad (5.33)$$

For the same reason, examining the equilibrium of moments about the x_3-axis of the element, we obtain

$$-\frac{dM_3}{dx_1} = Q_2. \qquad (5.34)$$

By substituting (5.33) and (5.34) into (5.32), the equilibrium equations can be expressed in terms of the bending moments M_i:

$$a < x_1 < b: \quad M_3'' = q_2, \quad -M_2'' = q_3.$$

These are just the equations (5.23) derived from the variational principle. Furthermore, we can see from (5.33) and (5.34) that the boundary conditions (5.24) stand for the equilibrium of shearing forces at the end points, while the boundary conditions (5.25) stand for the equilibrium of moments at the end points.

5.4   Boundary conditions and interface conditions

The equilibrium equations (5.28) for the bending of spatial beams are two fourth order ordinary differential equations in two unknown functions. In order to determine a unique solution, it is necessary to give four boundary conditions at each end. Appropriately, (5.29) provides the right number of boundary conditions.

The boundary conditions (5.29), which are derived from the variational principle, are mechanical boundary conditions, i.e., the natural boundary conditions which denote the equilibrium of the shearing forces and of the bending moments at the end points. Another kind of boundary conditions are geometric boundary conditions including fixed displacement and fixed angle of rotation, which must be listed out as constraint conditions in the variational problem and are essential boundary conditions. When a displacement boundary condition is introduced, an equilibrium condition of shearing forces in the same direction will be replaced locally by the displacement boundary condition, and when an angle of rotation boundary condition is introduced, an equilibrium condition of bending moments in the same direction will be replaced locally by the angfle of rotation boundary condition. These two kinds of boundary conditions are mutually complementary but incompatible. The corresponding boundary condition of shearing forces or bending moments can be derived from the variational principle automatically at places where the displacement or the angle of rotation is unconstrained. The end point may have an elastic support as well, which is also a kind of natural boundary condition provided that the elastic energy of the coupling has been absorbed into the total energy. For detailed discussion of boundary condition, see §7 of Chapter 1.

Interface conditions. When the flexural rigidity EI_{ij} of the rod has a discontinuity, e.g., due to the cross section having a jump some point $x = p$, the interface conditions

$$[M_3']_{p-0}^{p+0} = 0, \quad [M_2']_{p-0}^{p+0} = 0, \tag{5.35}$$

$$[M_2]_{p-0}^{p+0} = 0, \quad [M_3]_{p-0}^{p+0} = 0. \tag{5.36}$$

can be derived from the variation principle by piecewise integration by parts over $a \le x \le p$ and $p \le x \le b$. (5.35) denotes the continuity of the shearing forces Q_2 and Q_3 at the interface, while (5.36) denotes the continuity of the bending moments M_2 and M_3 at the interface. Suppose there are transverse point loads $r_2(p)$ and $r_3(p)$ at the point $x = p$. Then the potential energy of external work will be increased by

$$-r_2(p)u_2(p) - r_3((p)u_3(p).$$

Using the same method, we can verify that

$$[M_3']_{p-0}^{p+0} = r_2(p), \quad -[M_2']_{p-0}^{p+0} = r_3(p), \tag{5.37}$$

$$[M_2]_{p-0}^{p+0} = 0, \quad [M_3]_{p-0}^{p+0} = 0. \tag{5.38}$$

The equations (5.37) denote a stet of the shearing forces at the interface, namely jumps that balance the point loads. The equations (5.35) are particular cases. (5.38) says that, the bending moments are still continuous even if there are point loads at the interface. These two interface conditions belong to the natural boundary conditions, and will be satisfied automatically when the potential energy achieves a minimum.

5.5   Strainless states

Because footnote connfused with superscript

$$\iint_s x_2^2 ds \cdot \iint_s x_3^2 ds > \left(\iint_s x_2 x_3 ds \right)^2,$$

it follows that

$$I_{22}I_{33} > (I_{23})^2. \tag{5.39}$$

Thus

$$EI_{22}K_2^2 + 2EI_{23}K_2K_3 + EI_{33}K_3^2$$

---

[1] The proof of this inequality is as follows: Take arbitrarily a real number t. Then $\iint_s (tx_2 + x_2)^2 ds > 0$. Hence

$$\iint_s (cx_2 + x_3)^2 ds = t^2 \iint_s x_2^2 ds + 2t \iint_s x_2 x_3 ds + \iint_s x_3^2 ds > 0.$$

is a positive definite quadratic form of K_2 and K_3. Now

$$D(\boldsymbol{u}, \boldsymbol{u}) = \int_a^b (EI_{22}K^2 + 2EI_{23}K_2K_3 + EI_{33}K_3^2)dx_1.$$

Hence, a state ν is strainless iff

$$D(\boldsymbol{v}, \boldsymbol{v}) = 0 \Longleftrightarrow K_2 = K_3 = 0 \Longleftrightarrow v_2'' = v_3'' = 0$$

$$\Longleftrightarrow v_2 = a_2 + b_3 x_1, \quad v_3 = a_3 - b_2 x_1.$$

We can write this in the form

$$\boldsymbol{v} = \begin{bmatrix} v_2 \\ v_3 \end{bmatrix} = a_2 \begin{bmatrix} 1 \\ 0 \end{bmatrix} + a_3 \begin{bmatrix} 0 \\ 1 \end{bmatrix} + b_3 \begin{bmatrix} x_1 \\ 0 \end{bmatrix} + b_2 \begin{bmatrix} 0 \\ -x_1 \end{bmatrix}$$

$$= a_2 \boldsymbol{v}^{(1)} + a_3 \boldsymbol{v}^{(2)} + b_3 \boldsymbol{v}^{(3)} + b_2 \boldsymbol{v}^{(4)}. \tag{5.40}$$

These are exactly the projections of the following infinitesimal rigid displacement on the planes of x_2 and x_3:

$$\boldsymbol{a} + \boldsymbol{b} \wedge \boldsymbol{x} = \begin{bmatrix} a_1 \\ a_2 \\ a_3 \end{bmatrix} + \begin{bmatrix} b_2 x_3 - b_3 x_2 \\ b_3 x_1 - b_1 x_3 \\ b_1 x_2 - b_2 x_1 \end{bmatrix}_{x_2 = x_3 = 0} \tag{5.41}$$

Each transverse displacement v_2 or v_3 has two degrees of freedom. The degrees of freedom of v_2 are the x_2-directional rigid translation a_2 and the infinitesimal rigid rotation b_3 about the x_3-axis. The degrees of freedom of v_3 are the x_3-directional rigid translation a_3 and the infinitesimal rigid rotation b_2 about the x_2-axis. Necessary and sufficient conditions for the existence of solutions of the variational problem (5.21) or (5.22) are

$$F(\boldsymbol{v}^{(i)}) = 0, \quad i = 1, 2, 3, 4.$$

Of these, $F(\boldsymbol{v}^{(1)}) = F(\boldsymbol{v}^{(3)}) = 0$ denote the equilibrium of the x_2-directional forces and the equilibrium of the moments about the x_2-axis, and $F(\boldsymbol{v}^{(2)}) = F(\boldsymbol{v}^{(4)}) = 0$ denote the equilibrium of the directional forces and the equilibrium of the moments about x_3-axis.

---

for any real number t. Since the coefficient of t^2 satisfies $\iint_s x_2^2 ds > 0$, the discriminant

$$\left(\iint_s x_2 x_3 ds \right)^2 - \iint_s x_2^2 ds \cdot \iint_s x_3^2 ds < 0.$$

Q.E.D.

As for the relation between geometrical constraints, elastic support and the degrees of freedom of the strainless state, see §7 of Chapter 1 for details.

5.6   Thermal effects

Suppose the distribution of the temperature field τ on the cross section S i.e.

$$S : \tau(x_1, x_2, x_3) = \bar{\tau}(x_1) + x_2\tau_2(x_1) + x_3\tau_3(x_1), \qquad (5.42)$$

where $\bar{\tau}(x_1)$ is the average temperature on the cross section. By (5.11), the strain energy density with thermal effects is

$$W = \frac{1}{2}E\varepsilon_{11}^2 - E\alpha\tau\varepsilon_{11}$$

$$= \frac{1}{2}E(x_3K_2 - x_2K_3)^2$$

$$- E\alpha(\bar{\tau} + x_2\tau_2 + x_3\tau_3)(x_3K_2 - x_2K_3). \qquad (5.43)$$

Since the centroid of the cross section passes through the x_1-axis,

$$\iint_S x_2 dS = \iint_S x_3 dS = 0.$$

Hence the total strain energy is

$$P(\boldsymbol{u}) = \int_a^b dx_1 \iint_s W dS$$

$$= \frac{1}{2}\int_a^b (EI_{22}K_2^2 + 2EI_{23}K_2K_3 + EI_{33}K_3^2)dx_1$$

$$- \int_a^b E\alpha[(I_{23}\tau - I_{33}\tau_2) - (I_{23}\tau_2 - I_{22}\tau_3)K_2]dx_1. \qquad (5.44)$$

Let

$$F_\tau(\boldsymbol{u}) = \int_a^b E\alpha[(I_{23}\tau_3 - I_{33}\tau_2)K_3 - (I_{23}\tau_2 - I_{22}\tau_3)K_2]dx_1$$

$$= \int_a^b E\alpha[(I_{23}\tau_3 I_{33}\tau_2)u_2'' + (I_{23}\tau_2 - I_{22}\tau_3)u_3'']dx_1$$

$$= \int_a^b (E_2 u_2'' + E_3 u_3'')dx_1, \qquad (5.45)$$

$$E_2 = E\alpha(I_{23}\tau_3 - I_{33}\tau_2), \quad E_3 = E\alpha(I_{23}\tau_2 - I_{22}\tau_3). \qquad (5.46)$$

This is a linear functional of $\boldsymbol{u} = (u_2, u_3)^T$. Compared to the case without thermal effect there is an extra linear term $-F_\tau(\boldsymbol{u})$ depending on the temperature differences of τ_2 and τ_3,

$$P(\boldsymbol{u}) = \frac{1}{2}D(\boldsymbol{u}, \boldsymbol{u}) - F_\tau(\boldsymbol{u}).$$

Merging this term into the potential energy of external work $-F(\boldsymbol{u})$, from (5.20), we obtain

$$F(\boldsymbol{u}) = \int_a^b (q_2 u_2 + q_3 u_3)dx_1$$

$$+ \sum_{P=a,b} [r_2 u_2 + r_3 u_3 m_3 u_2' - m_2 u_3']_P + F_\tau(\boldsymbol{u}). \qquad (5.47)$$

Then the total potential energy is

$$J(\boldsymbol{u}) = \frac{1}{2}D(\boldsymbol{u}, \boldsymbol{u}) - F(\boldsymbol{u}).$$

Integrating $F_\tau(\boldsymbol{u})$ by parts twice, we have

$$F_\tau(\boldsymbol{u}) = \int_a^b (E_2 u_2'' + E_3 u_3'')dx_1$$

$$= \int_a^b (E_2'' u_2 + E_3'' u_3)dx_1$$

$$+ \sum_{P=a,b} [-\varepsilon E_2' u_2 - \varepsilon E_3' u_3 + \varepsilon E_2 u_2' + \varepsilon E_3 u_3']_P, \qquad (5.48)$$

where the meaning of the notation ε is the same as before: $\varepsilon(b) = 1$, $\varepsilon(a) = -1$. The equilibrium equations and the load boundary conditions can be derived from the variational principle as follows:

$$M_3'' = q_2 + E_2'', \qquad -M_2'' = q_3 + E_3'', \qquad a < x_1 < b, \qquad (5.49)$$

$$-\varepsilon M_3' = r_2 - \varepsilon E_2', \quad \varepsilon M_2' = r_3 - \varepsilon E_3', \qquad x_1 = a, b. \qquad (5.50)$$

$$\varepsilon M_2 = m_2 - \varepsilon E_3, \quad \varepsilon M_3 = m_3 + \varepsilon E_2. \qquad (5.51)$$

(5.50) represents the equilibrium of shearing forces, (5.51) the equilibrium of bending moments.

These forms are identical with (5.23)–(5.25) in the case without thermal effect, except that the right hand side contains an additional temperature load term.

Chapter 4

Composite Elastic Structures

§1   Introduction

Deformation problems for various kinds of individual elastic bodies,such as rods, plates and three-dimensional elastic bodies, have been discussed in the preceding chapters. What we often encounter in engineering practice, however, is often not the individual rod, plate or body but a certain combination of them, i.e., there contain various kinds of members coupled in one way or another in the system. A system consisting of several similar or different kinds of elastic members is called a composite elastic structure.

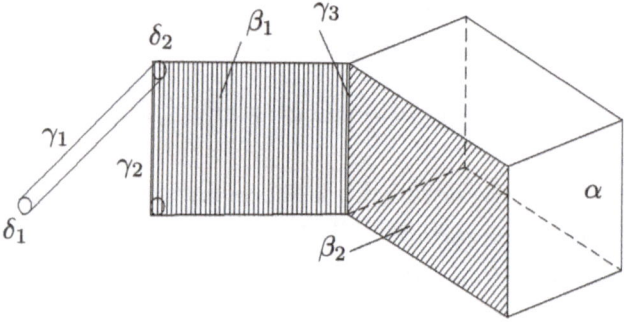

Fig. 47

The structure as shown in Fig. 47 is composed of one body member α, two plate members β_1 and β_2, and three rod members γ_1, γ_2 and γ_3. In a structure, a plate member may be a separate entity, like such as β_1, which is connected with plate member β_2 through rod member γ_3, or may be the boundary surface of a body member like β_2. Similarly, a rod member may be a separate entity, like γ_1, which intersects the rod member γ_2 at the point δ_2, or may be n boundary line of a plate member like γ_2 and γ_3.

In order to solve a large number of problems arising from engineering practice it is necessary to generalize the theory of three dimensional elasticity to the composite structure. Since a composite structure has various kinds of members, the couplings among the members are of great significance. This is the key point distinguishing the composite structure from the individual elastic body. So it cannot be treated individually as an individual member. On the other hand, it is also inconvenient to look on the whole structure as an individual and indistinguishable three dimensional elastic body and to solve it in a uniform way by the theory of three dimensional elasticity. A reasonable method is to take the simplified models of plates and rods in the preceding chapters, i.e., to look on the plate as a two dimensional elastic plane without thickness geometrically, to look on the rod as an one dimensional line without cross section, and in the meantime to consider the elastic couplings among the members of various dimensions. This is in a sense also a simplification of the theory of three-dimensional elasticity in a sence.

In the following model of the composite structures, we consider only three kinds of members, i.e., three dimensional body members, two dimensional plate members and one dimensional rod members. For convenience, we also include null dimensional point members. Members of these four kinds are denoted by α, β, γ and δ, respectively.

In this model there is still a fundamental convention: when two members are coupled, the dimension (begin the null dimension, one dimension, and two dimension, correspondingly) at a joint (being a point, a line or a surface) can only differ from the dimensions of the two adjacent members at most by one dimension. A joint at which the dimension difference is higher than one is called the nonstandard joint such joints will not be considered in the present chapter. All the joints shown in Fig. 48 are nonstandard joints, but they can often be transformed into standard joints by including an additional member or changing some individual member. For example, in Fig.(a) the edge where a plate is connected with a rod can be in Fig.(a), transformed into a rod member in Fig.(c); the surface where a corner point of a body is connected with a rod can be transformed into a plate member and its edge through this corner point transformed into a rod member, in Fig.(d) the two surfaces by the two sides of a joint of two bodies can be transformed into plate members in Fig.(e), the two surfaces at the joint of the bodies and plates can be transformed into plate members; and so on. All joints become standard after these transformation.

Because composite structures can be very complicated both geometrically and mechanically, it is from the view of mathematical treatment it is of great advantage to adopt the displacement method and the variational principle emphasized throughout this book. The mathematical basis of the composite structure model presented in this chapter is rigorous and complete so long as the connections between members follow with the fundamental convention state above.

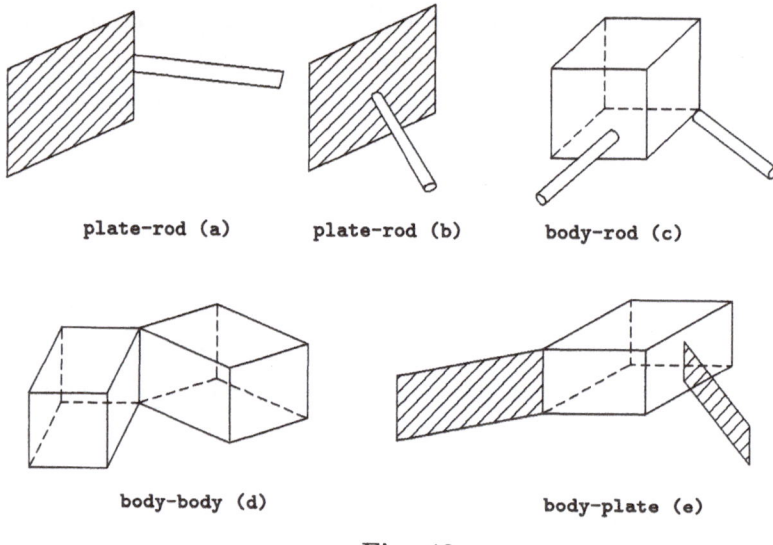

plate-rod (a)   plate-rod (b)   body-rod (c)

body-body (d)   body-plate (e)

Fig. 48

We must emphasize that the equilibrium equation derived from the variational principle is extremely complicated and tedious due to the complexity of the composite structure, so that it is quite difficult to solve practical problems starting from this equation. However, the variational principle for composite structures is not complicated in itself its formulation is very simple. Basically it is similar to the variational principle for individual members. Hence, starting from the variational principle, the application of the finite element method will be a straightforward and fruitful way to solve the equilibrium problem of composite structures.

In the following, this idea will be detailed in two parts. As a special case, plane composite structures will be discussed first. The problem in this case is comparatively simple, but it suffices to explain all concepts and methods. Afterward, we will generalize the problem to that of three-dimensional structures.

§2 Plane Composite Structures

2.1 Geometric description

Plane composite structures form a special case of composite structures. All members lie in the same plane, and the loads are also restricted to this plane, as for a plane plate-rod structure and a plane frame. Plane systems of rods in tension, already discussed in Section 2.4 of Chapter 1, form just such an example. Since the elastic displacement and the bending are all restricted in the same plane, a plate member is just a thin plane plate in stretching when transverse flexural deformation is not considered. A rod member is also a plane member when only stretching and bending in the plane are considered, but bending in other directions and torsion are neglected.

Suppose a composite structure Ω is composed of plate members $\beta_1, \cdots,$ β_N in stretching and rod members $\gamma_1, \cdots, \gamma_N$, in stretching and bending, all lying in the same plane:

$$\Omega = \{\beta_1, \cdots, \beta_N; \gamma_1, \cdots, \gamma_2\}.$$

For example, in Fig. 49 doublelines stand for rod members, while shaded blocks stand for plate members. Here, $\Omega = \{\beta_1, \beta_2; \gamma_1, \gamma_2, \gamma_3, \gamma_4\}$. We denote the one dimensional and the two dimensional parts of the structure, respectively, by

$$\Omega^1 = \{\gamma_1, \cdots, \gamma_{N_1}\}, \quad \Omega^2 = \{\beta_1, \cdots, \beta_{N_2}\}.$$

In Fig. 49, $\Omega^1 = \{\gamma_1, \gamma_2, \gamma_3, \gamma_4\}$, $\Omega^2 = \{\beta_1, \beta_2\}$.

Geometrically, all rod members are straight line segments. The boundaries ∂r of each rod member γ are the two end points. All the boundary points of the rod members are denoted by $\delta_1, \cdots, \delta_{N_0}$, which are the point elements of the structure. Let

$$\Gamma^0 = \{\delta_1, \cdots, \delta_{N_0}\}.$$

Γ^0 can be considered as the null dimensional boundary of the structure Ω. In Fig. 49, $\Gamma^0 = \{\delta_1, \delta_2, \delta_3, \delta_4\}$.

All plate members are polygons, which may as well be assumed to be simply-connected. The boundary $\partial \beta$ of each plate member β is composed of several line segments. Geometrically, a rod member may be a boundary line segment of a plate member ($\gamma_1, \gamma_4 \in \partial \beta_2$ in Fig. 49), but need not be (γ_2 and γ_3 in Fig. 49). Conversely, a boundary line segment of a plate member may be a rod member (γ_1 and γ_4 in Fig. 49), but need not be, i.e., being its own boundary of the plate member (γ_5, γ_6 and γ_7, in Fig. 49). By denoting all such boundary line segments of the plate members their own as $\gamma_{N+1}, \cdots, \gamma_{N'_1}$, and by letting

$$\Gamma^1 = \{\gamma_{N_1+1}, \cdots, \gamma_{N'_1}\}.$$

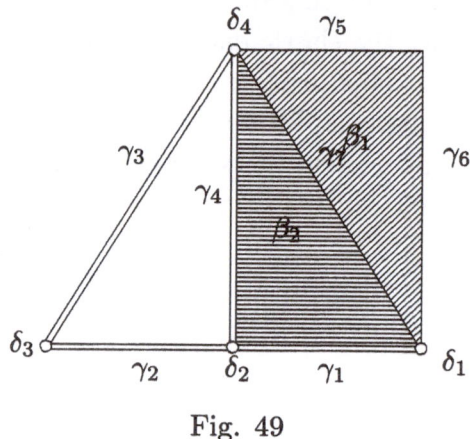

Fig. 49

Γ^1 can be considered as the one dimensional boundary of the structure Ω. In Fig. 49, $\Gamma^1 = \{\gamma_5, \gamma_6, \gamma_7\}$. The rod members Ω^1 and the one dimensional boundaries Γ^1 are merged into

$$\Omega^1 + \Gamma^1 = \{\gamma_1, \cdots, \gamma_{N_1}, \gamma_{N_1+1}, \cdots, \gamma_{N'_1}\}$$

the set of all line elements of the structure. The boundary of the structure Ω can be considered to be composed of $\Gamma = \{\Gamma^1, \Gamma^0\}$.

Inverse boundary

The connection of various members in a structure is mainly described using the boundary relation ∂, where the boundary ∂A of a p-dimensional element A is $(p-1)$-dimensional. We can also recognize the internal connection among the members through an inverse relation ∂^{-1}, i.e., defining a $(p+1)$-dimensional inverse boundary $\partial^{-1}A$ for a p-dimensional element A. For an arbitrary line element. $\gamma \in \Omega^1 + \Gamma^1$, its inverse boundary $\partial^{-1}\gamma$ is defined as a set of all plate members having γ as a segment of the boundary. Similarly, for an arbitrary point element $\delta \in \Gamma^0$, $\partial^{-1}\delta$ is defined as a set of all rod members having δ as a boundary point. For example, in Fig. 49, $\partial^{-1}\gamma_7 = \{\beta_1, \beta_2\}, \partial^{-1}\gamma_1 = \{\beta_2\}, \partial^{-1}\delta_4 = \{\gamma_3, \gamma_4\}$. It is easy to see that,

$$\text{for } \gamma \in \Omega^1 + \Gamma^1, \ \beta \in \Omega^2 : \ \gamma \in \partial\beta \iff \beta \in \partial^{-1}\gamma,$$
$$\text{for } \delta \in \gamma^0, \ \gamma \in \Omega^1 : \ \delta \in \partial\gamma \iff \gamma \in \partial^{-1}\delta.$$

· Global coordinates and local coordinates

In the plane of a structure, take a right hand orthogonal coordinate system (x_1, x_2) as the reference coordinates, called the global coordinates. Further, for each point (δ), line (γ) or area (β) element in Ω, choose a local right hand orthogonal coordinate system $(x_1^\delta, x_2^\delta), (x_1^\gamma, x_2^\gamma)$, or (x_1^β, x_2^β), respectively. In the plane, however, there is no loss in using the global coordinates as the local coordinates of point elements and of area elements:

$$(x_1, x_2) = (x_1^\delta, x_2^\delta) = (x_1^\beta, x_2^\beta).$$

For a line element γ, we adopt the convention that x_1^γ is its longitudinal direction and x_2^γ is its transverse direction, so that x_1^γ and x_2^γ form a right hand system and an end point δ of γ is the origin of the local coordinates. Suppose the global coordinates of δ are (ξ_1, ξ_2), and the angle measured from the x_1-axis to x_1^γ-axis in the anticlockwise direction is θ^γ. Then the transformation relations between the global coordinates and the local coordinates are (See Fig. 50),

$$\begin{cases} x_1 = \xi_1 + x_1^\gamma \cos\theta^\gamma - x_2^\gamma \sin\theta^\gamma, \\ x_2 = \xi_2 + x_1^\gamma \sin\theta^r + x_2^\gamma \cos\theta^\gamma. \end{cases} \tag{2.1}$$

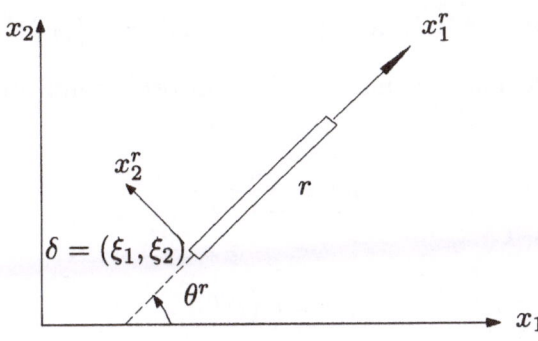

Fig. 50

Let

$$A^\gamma = \begin{bmatrix} a_{11}^\gamma & a_{12}^\gamma \\ a_{21}^\gamma & a_{22}^\gamma \end{bmatrix} = \begin{bmatrix} \cos\theta^\gamma & -\sin\theta^\gamma \\ \sin\theta^\gamma & \cos\theta^r \end{bmatrix}. \tag{2.2}$$

The coordinate unit vectors e_1^γ and e_2^γ of the local coordinate system have global components

$$
\begin{aligned}
e_1^\gamma &= (a_{11}^\gamma, a_{21}^\gamma)^T = (\cos\theta^\gamma, \sin\theta^\gamma)^T, \\
e_2^\gamma &= (a_{12}^\gamma, a_{22}^\gamma)^T = (-\sin^\gamma, \cos\theta^\gamma)^T.
\end{aligned}
\tag{2.3}
$$

A plane displacement vector u can be expressed in terms of the unit vectors $e_1 = (1,0)^T, e_2 = (0,1)^T$ of the system of global coordinates as

$$
u = \sum_{i=1}^{2} u_i e_i,
$$

where u_j are called the global components of the displacement. The same displacement can also be expressed in terms of the unit vectors on the system of local coordinates e_1^γ and e_2^γ as

$$
u = \sum_{i=1}^{2} u_i^\gamma e_i^\gamma,
$$

where u_i^γ are called the local components of the displacement. We can see from the transformation relations (2.1) that the transformation relations between the global and local components of the displacement are

$$
u_i = \sum_{j=1}^{2} a_{ij}^\gamma u_j^\gamma,
\tag{2.4}
$$

i.e.,

$$
u = A^\gamma u^\gamma, \quad u = (u_1, u_2)^T, \quad u^\gamma = (u_1^\gamma, u_2^\gamma)^T.
$$

Since the transformation matrix A^γ is an orthogonal matrix, $(A^\gamma)^{-1} = (A^\gamma)^T$. Hence

$$
u_j^\gamma = \sum_{i=1}^{2} a_{ij}^\gamma u_i,
\tag{2.5}
$$

i.e.,

$$
u^\gamma = (A^\gamma)^T u.
$$

These transformation relations are also applicable to any plane vector such as the force f. The rotation angle ω_3 in the plane is essentially a rotation about x_3-axis perpendicular to the $(x_1 - x_2)$-plane, and as a rotation vector, it points to the x_3-axis. Hence its magnitude remains unchanged after the rotation of the coordinates in the plane, i.e.,

$$\omega_3 = \omega_3^\gamma.$$

A similar invariant relation also applies to the moment M_3 in the plane and so on.

Oriented connection between the members

Since we have prescribed an individual right hand system (global or local) for the point, line, or area element, we have also prescribed a "direction" for them. In all of the following discussion, every element will be considered as oriented.

The boundary $\partial\beta$ of each area element β has an outward normal direction n^β, which lies in the plane of the area element. Choose a tangential direction t^β to the boundary $\partial\beta$ so as to construct a right hand system $\{n^\beta, t^\beta\}$. On the other hand, suppose there is ∂ line element $\gamma \in \partial\beta$ the tangential direction t^β on γ is denoted by $[t^\beta]^\gamma$. According to the convention for local coordinates of a line element, the longitudinal unit vector of γ is e_1^γ. We say that β and γ are conforming when e_1^γ has the same direction as $[t^\beta]_\gamma$, and we say that β and γ are inverse when e_1^γ is opposite in direction which is reverse to $[t^\beta]_\gamma$ (Fig. 51). Then, for arbitrary area element β and line element γ, we define the oriented connection relations by

$$\varepsilon(\beta, \gamma) = \begin{cases} 0, & \text{if } \gamma \notin \partial\beta, \\ 1, & \text{if } \gamma \in \partial\beta \quad \text{if } \beta \text{ and } \gamma \text{ are conforming,} \\ -1 & \text{if } \gamma \in \partial\beta \quad \text{if } \beta \text{ and } \gamma \text{ are inverse .} \end{cases}$$

Consequently,

$$\gamma \in \partial\beta \Rightarrow [t^\beta]_\gamma = \varepsilon(\beta, \gamma) e_1^\gamma, \quad [n^\beta]_\gamma = -\varepsilon(\beta, \gamma) e_1^\gamma. \tag{2.6}$$

Similarly, each line element γ has a longitudinal unit vector e_1^γ. For a point element $\delta \in \partial\gamma$, if e_1^γ is directed toward the point δ, then we say that γ and δ are conforming (δ_2 in Fig. 51). If e_1^γ is directed away from the point δ, then we say that γ and δ are inverse (δ_1 in Fig. 51). Then, for arbitrary line element γ and point element δ, the relations of oriented connection can be defined by

$$\varepsilon(\beta, \gamma) = \begin{cases} 0, & \text{if } \delta \notin \partial\gamma, \\ 1, & \text{if } \delta \in \partial\gamma, \quad \text{if } \gamma \text{ and } \delta \text{ are conforming,} \\ -1, & \text{if } \delta \in \partial\gamma, \quad \text{if } \gamma \text{ and } \delta \text{ are inverse .} \end{cases}$$

Using this notation, the ordinary definite integral formula can be written as

$$\int_\gamma \frac{\partial \varphi}{\partial x_1^\gamma} = \int_{\delta_1}^{\delta_2} \frac{\partial \varphi}{\partial x_1^\gamma} d\gamma = \varphi(\delta_2) - \varphi(\delta_1) = \sum_{\delta \in \partial r} \varepsilon(\gamma, \delta) \varphi(\delta).$$

This expression is convenient when the problem becomes complicated.

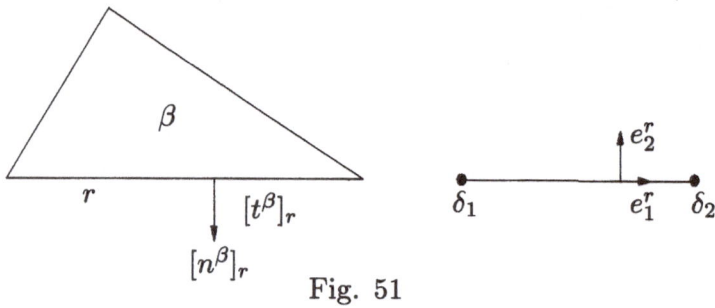

Fig. 51

As stated above, a connection relation $\varepsilon(\beta, \gamma)$ can be determined for each area element β and line element γ. Hence a matrix $\varepsilon(\beta_j, \gamma_k)$ can be formed for all area elements and line elements. Similarly, a matrix $\varepsilon(\gamma_k, \delta_1)$ can be formed for all line elements and point elements. These two matrices are called the connection matrices. They qualitatively describe the connection relationship, i.e., the topological relationship among the various members inside the structure.

2.2   Fundamental members

A plane structure has only two basic kinds of members: the one dimensional plane rod in stretching and bending and the two dimensional thin plate in stretching. Using the notation of the preceding chapters, we now summarize their geometric and mechanical essentials as follows.

1. One dimensional rods.

Local coordinates: $x_i^\gamma, i = 1, 2$. Longitudinal coordinate is x_1^γ, and transverse coordinate is x_2^γ. $x_2^\gamma = 0$ on the rod.

Fundamental variables: displacements $u_i^\gamma(x_1), i = 1, 2$, longitudinal displacement u_1^γ, transverse displacement u_2^γ and rotation angle $\omega_3 = \dfrac{du_2^\gamma}{dx_1^\gamma}$.

Transformation of coordinates:

$$\boldsymbol{u} = A^\gamma \boldsymbol{u}^\gamma, \ u_i = \sum_{j=1}^{2} a_{ij}^\gamma u_j^\gamma, \ i = 1, 2,$$

$$u_j^\gamma = \sum_{i=1}^{2} a_{ij}^\gamma u_i, \ j = 1, 2.$$

(2.7)

All the following quantities are expressed in the local coordinates.
Strains:

bending

tension

$$\varepsilon_u = \varepsilon_{11}^\gamma(\boldsymbol{u}) = \frac{du_1}{dx_1};$$

bending

$$K_3 = K_3^\gamma(\boldsymbol{u}) = \frac{d^2 u_2}{dx_1^2} = \frac{d\omega_3}{dx_1}.$$

Stresses:

tensile force

$$Q_1 = Q_1^\gamma(\boldsymbol{u}) = E A \varepsilon_{11}(\boldsymbol{u});$$

bending moment

$$M_3 = M_3^\gamma(\boldsymbol{u}) = E I_{33} K_3(\boldsymbol{u});$$

shearing force

$$Q_2 = Q_2^r(\boldsymbol{n}) = -\frac{dM_3(\boldsymbol{u})}{dx_1}.$$

Here, A denotes the area of the cross section and I_{33} denotes the moment of inertia of the cross section about the x_3-axis which is normal to the $(x_1 - x_2)$-plane (Fig. 52).
The strain energies:

tension

$$\frac{1}{2} D_1^\gamma(\boldsymbol{u}, \boldsymbol{u}) = \frac{1}{2} \int_\gamma Q_1(\boldsymbol{u}) \varepsilon_{11}(\boldsymbol{u}) d\gamma = \frac{1}{2} \int_\gamma E A \varepsilon_{11}^2(\boldsymbol{u}) d\gamma,$$

which depends only on u_1;

bending

$$\frac{1}{2} D_2^\gamma(\boldsymbol{u}, \boldsymbol{u}) = \frac{1}{2} \int_\gamma M_3(\boldsymbol{u}) K_3(\boldsymbol{u}) d\gamma = \frac{1}{2} \int_\gamma E I_{33} K_3^2(\boldsymbol{u}) d\gamma,$$

which depends only on u_2;

total

$$\frac{1}{2} D^\gamma(\boldsymbol{u}, \boldsymbol{u}) = \frac{1}{2} D_1^\gamma(\boldsymbol{u}, \boldsymbol{u}) + \frac{1}{2} D_2^\gamma(\boldsymbol{u}, \boldsymbol{u}).$$

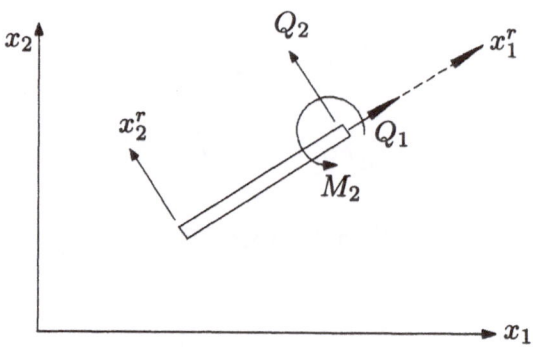

Fig. 52

Functionals of virtual work:

tension

$$D_1^\gamma(\boldsymbol{u}, \boldsymbol{v}) = \int_\gamma Q_1(\boldsymbol{u})\varepsilon_{11}(\boldsymbol{v})d\gamma = \int_\gamma EA\varepsilon_{11}(\boldsymbol{u})\varepsilon_{11}(\boldsymbol{v})d\gamma;$$

bending

$$D_2^\gamma(\boldsymbol{u}, \boldsymbol{v}) = \int_\gamma M_3(\boldsymbol{u})K_3(\boldsymbol{v})d\gamma = \int_\gamma EI_{33}K_3(\boldsymbol{u})K_3(\boldsymbol{v})d\gamma;$$

total

$$D^\gamma(\boldsymbol{u}, \boldsymbol{v}) = D_1^\gamma(\boldsymbol{u}, \boldsymbol{v}) + D_2^\gamma(\boldsymbol{u}, \boldsymbol{v}).$$

Strainless states:

tension

$$D_1^\gamma(\boldsymbol{v}, \boldsymbol{v}) = 0 \iff \varepsilon_{11}(\boldsymbol{v}) \equiv 0 \iff v_1 = a_1;$$

bending

$$D_2^\gamma(\boldsymbol{v}, \boldsymbol{v}) = 0 \iff K_3(\boldsymbol{v}) \equiv 0 \iff v_2 = a_2 + b_3 x_1, \ \omega_3 = b_3;$$

total

$$D^\gamma(\boldsymbol{v}, \boldsymbol{v}) = 0 \iff \varepsilon_{11}(\boldsymbol{v}) \equiv K_3(\boldsymbol{v}) \equiv 0$$

$$\iff v_1 = a_1, \ v_2 = a_2 + b_3 x_1, \ \omega_3 = b_3.$$

i.e.,

$$\begin{bmatrix} v_1 \\ v_2 \end{bmatrix} = \begin{bmatrix} a_1 - b_3 x_2 \\ a_2 + b_3 x_1 \end{bmatrix}_{x_2=0}. \tag{2.8}$$

Loads:

forces

$$f_i = f_i^\gamma(x_1), \ i = 1, 2.$$

Potential energies of external work:

$$\text{tension} \ -F_1^\gamma(\boldsymbol{v}) = - \int_\gamma f_1 v_1 d\gamma,$$

which is depending only on v_1;

$$\text{bending} \ -F_2^\gamma(\boldsymbol{v}) = - \int_\gamma f_2 v_2 d\gamma, \tag{2.9}$$

which is depending only on v_2;

$$\text{total} \ -F^\gamma(\boldsymbol{v}) = -(F_1^\gamma(\boldsymbol{v}) + F_2^\gamma(\boldsymbol{v})).$$

Total potential energy:

$$J^\gamma(\boldsymbol{u}) = \frac{1}{2} D^\gamma(\boldsymbol{u}, \boldsymbol{u}) - F^\gamma(\boldsymbol{u}).$$

Green's formulas:

tension

$$D_1^\gamma(\boldsymbol{u}, \boldsymbol{v}) = \int_\gamma \left(- \frac{dQ_1(\boldsymbol{u})}{dx_1} \right) v_1 d\gamma + \sum_{\delta \in \partial\gamma} \varepsilon(\gamma, \delta) [Q_1(\boldsymbol{u}) v_1]_\delta;$$

bending

$$D_2^\gamma(\boldsymbol{u}, \boldsymbol{v}) = \int_\gamma \left(- \frac{dQ_2(\boldsymbol{u})}{dx_1} \right) v_2 d\gamma$$

$$+ \sum_{\delta \in \partial\gamma} \varepsilon(\gamma, \varepsilon) [Q_2(\boldsymbol{u}) v_2 + M_3(\boldsymbol{u}) \omega_3(\boldsymbol{v})]_\delta;$$

total

$$D^\gamma(\boldsymbol{u}, \boldsymbol{v}) = \int_\gamma \sum_{i=1}^2 \left(- \frac{dQ_i(\boldsymbol{u})}{dx_1} \right) v_i d\gamma + \sum_{\delta \in \partial\gamma} \varepsilon(\gamma, \delta)$$

$$\times \left[\sum_{j=1}^2 Q_j(\boldsymbol{u}) v_j + M_3(\boldsymbol{u}) \omega_3(\boldsymbol{v}) \right]_\delta. \tag{2.10}$$

Elastic support: the elastic energy of coupling should be included in the strain energy when the rod member has elastic couplings with the external region.

Strain energies:

$$\frac{1}{2} D^\gamma(\boldsymbol{u}, \boldsymbol{u}) \ \text{must be increased by} \ \frac{1}{2} \int_\gamma c_1 u_1^2 d\gamma,$$

$$\frac{1}{2} D_2^\gamma(\boldsymbol{u}, \boldsymbol{u}) \ \text{must be increased by} \ \frac{1}{2} \int_\gamma c_2 u_2^2 d\gamma.$$

Functionals of virtual work

$$D_2^\gamma(\boldsymbol{u}, \boldsymbol{u}) \text{ must be increased by } \int_\gamma c u_1 u_1 v_1 d\gamma,$$

$$D_2^\gamma(\boldsymbol{u}, \boldsymbol{v}) \text{ must be increased by } \int_\gamma c_2 u_2 v_2 d\gamma.$$

The first term at the right hand side of the Green's formula for $D^\gamma(\boldsymbol{u}, \boldsymbol{v})$ should be rewritten as

$$\int_\gamma \sum_{i=1}^2 \left(-\frac{dQ_i(u)}{dx_1} + c_i u_i \right) v_i d\gamma,$$

where $c_i \geq 0$ is the constant of coupling.

2. The two-dimensional plate member

Local coordinates = the global coordinates:

$$x_i = x_i^\beta, \quad i = 1, 2.$$

Fundamental variables:

displacements $u_i = u_i^\beta(x_1, x_2), \ i = 1, 2;$

rotation angle $w_3 = \dfrac{1}{2}\left(\dfrac{\partial u_2}{\partial x_1} - \dfrac{\partial u_1}{\partial x_2}\right).$

Transformation of coordinates:

$$u = u^\beta.$$

Strains:

$$\varepsilon_{ij} = \varepsilon_{ij}^\beta(u) = \frac{1}{2}\left(\frac{\partial u_j}{\partial x_i} + \frac{\partial u_i}{\partial x_j}\right), \quad i, j = 1, 2.$$

Stresses:

$$Q_{ij} = Q_{ij}^\beta(\boldsymbol{u}) = h\sigma_{ij} = D'\left[(1 - \nu)\varepsilon_{ij} + \nu\left(\sum_{k=1}^2 \varepsilon_{kk}\right)\delta_{ij}\right], \quad i, j = 1, 2,$$

Tensile rigidity:

$$D' = \frac{Eh}{1 - \nu^2},$$

where h is the thickness (Fig. 53a).

Strain energy:

$$\frac{1}{2}D^\beta(\boldsymbol{u}, \boldsymbol{u}) = \frac{1}{2}\iint_\beta \sum_{i,j=1}^2 Q_{ij}(\boldsymbol{u})\varepsilon_{ij}(\boldsymbol{u})d\beta$$

$$= \frac{1}{2}\iint_\beta D'\left[(1 - \nu)\sum_{i,j=1}^2 \varepsilon_{ij}^2 + \nu\left(\sum_{k=1}^2 \varepsilon_{kk}\right)^2\right]d\beta.$$

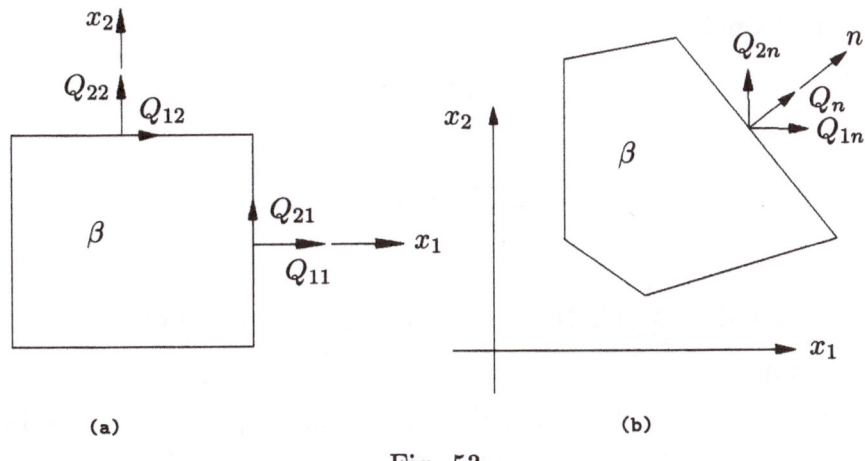

Fig. 53

Functional of virtual work:

$$D^\beta(\boldsymbol{u}, \boldsymbol{v}) = \iint_\beta \sum_{i,j=1}^{2} Q_{ij}(\boldsymbol{u})\varepsilon_{ij}(\boldsymbol{v})d\beta$$

$$= \iint_\beta D'\Big[(1-\nu)\sum_{i,j=1}^{2}\varepsilon_{ij}(\boldsymbol{u})\varepsilon_{ij}(\boldsymbol{v}) + \nu\Big(\sum_{k=1}^{2}\varepsilon_{kk}(\boldsymbol{u})\Big)$$

$$\times\Big(\sum_{k=1}^{2}\varepsilon_{kk}(\boldsymbol{v})\Big)\Big]d\beta.$$

Strainless states:

$$D^\beta(\boldsymbol{v}, \boldsymbol{v}) = 0 \iff v_1 = a_1 - b_3 x_2, \ v_2 = a_2 + b_3 x_1, \ \omega_3 = b_3,$$

i.e.,

$$\begin{bmatrix} v_1 \\ v_2 \end{bmatrix} = \begin{bmatrix} a_1 - b_3 x_2 \\ a_2 + b_3 x_1 \end{bmatrix}. \tag{2.11}$$

Loads:

forces

$$f_i = f_i^\beta(x_1, x_2), \ i = 1, 2.$$

Potential energy of external work:

$$-F^\beta(\boldsymbol{v}) = -\iint_\beta \sum_{i=1}^{2} f_i v_i d\beta. \tag{2.12}$$

Total potential energy:

$$J^\beta(\boldsymbol{u}) = \frac{1}{2}D^\beta(\boldsymbol{u}, \boldsymbol{u}) - F^\beta(\boldsymbol{u}).$$

Green's formula:

$$D^\beta(\boldsymbol{u}, \boldsymbol{v}) = \iint_\beta \sum_{i=1}^{2} \left(-\sum_{j=1}^{2} \frac{\partial Q_{ij}(\boldsymbol{u})}{\partial x_j} \right) v_i d\beta$$

$$+ \sum_{r \in \partial\beta} \int_\gamma \sum_{j=1}^{2} Q_{jn}(\boldsymbol{u}) v_j d\gamma \qquad (2.13)$$

where $Q_{jn}(\boldsymbol{u}) = \sum_{k=1}^{2} Q_{jk}(\boldsymbol{u}) n_k$, $\boldsymbol{n} = (n_1, n_2)^T$ the outward normal to $\partial\beta$ (Fig. 53b).

Elastic support: the potential energy of the coupling must be included in the strain energy when the plate member β has elastic couplings with the external region. Suppose the elastic reaction force per unit area is $-\sum_{j=1}^{2} c_{ij}^\beta u_j$, $i = 1, 2$. Then, strain energy $\frac{1}{2} D^\beta(\boldsymbol{u}, \boldsymbol{u})$ should be increased by $\frac{1}{2} \iint_\beta \sum_{i,j=1}^{2} c_{ij}^\beta c_{ij}^\beta u_i u_j d\beta$, functional of virtual work $D^\beta(\boldsymbol{u}, \boldsymbol{v})$ should be increased by $\iint_\beta \sum_{i,j=1}^{2} c_{ij}^\beta u_i v_j d\beta$.

Green's formula: the first term in the Green's formula (2.13) for $D^\beta(\boldsymbol{u}, \boldsymbol{v})$ should be rewritten as

$$\iint_\beta \sum_{i=1}^{2} \left[\sum_{j=1}^{2} \left(-\frac{\partial Q_{ij}(\boldsymbol{u})}{\partial x_j} + c_{ij}^\beta u_j \right) \right] v_i d\beta,$$

where $[c_{ij}^\beta]_{i,j=1,2}$ is a symmetric positive definite or positive semi-definite matrix of the couplings.

2.3   The rigid connection

For a compound elastic structure, it is still necessary to satisfy specified continuity conditions at the joints of the various members. Let us first discuss the case where the members are rigidly connected. In this case, we have the continuity of displacements and the specified continuity of rotation angles.

For the sake of convenience, the values taken by the global displacement $\boldsymbol{u}$ (correspondingly, the local displacement $\boldsymbol{u}^\gamma$) on the area element β, on the line element γ, and on the point element δ are denoted in terms of $[\boldsymbol{u}]_\beta, [\boldsymbol{u}]_\gamma, [\boldsymbol{u}]_\delta([\boldsymbol{u}^\gamma]_\beta, [\boldsymbol{u}^\gamma]_\gamma, [\boldsymbol{u}^\gamma]_\delta)$, respectively. The value on the boundary line element γ taken by the global displacement $[u]_\beta$ on the area element β is denoted by $[u]_{\beta,\gamma}$, and so on. Sometimes, $[u^\gamma]_\gamma$ will be abbreviated u^γ.

Continuity of displacements (to transfer forces).

1. The continuity of displacements at a joint of members with the same dimension.

(i) The displacement is continuous when two rod members γ and γ' are connected at the point δ:

$$[u]_{\gamma,\delta} = [u]_{\gamma',\delta}.$$

(ii) The displacement is continuous when two plate members β and β' are connected on a line element γ which is not a rod member

$$[u]_{\beta,\gamma} = [u]_{\beta',\gamma}.$$

It follows that, on the boundary elements $\delta \in \Gamma^0$ and $\gamma \in \Gamma^1$ there exist single-valued displacements $[u]_\delta$ and $[u]_\gamma$ satisfying

$$[u]_\delta = [u]_{\gamma,\delta}$$

for any rod member γ connected at the point element δ,

$$[u]_\gamma = [u]_{\beta,\gamma}$$

for any plate member β connected on the line element γ.

2. The displacement is continuous: $[u]_\gamma = [u]_{\beta,\gamma}$, when the members with different dimensions, i.e., the rod member γ and the plate member $\beta(r \in \partial\beta)$, are connected. It follows that, on the rod member $r \in \Omega^1$, there also exists a single-valued displacement $[u]_\gamma$.

The continuity of rotation angles (to transfer moments).

The rotation angle is continuous when two rod members γ and γ' are connected at the point $\delta : [\omega_3]_{\gamma,\delta} = [\omega_3]_{\gamma',\delta}$. It follows that, on every point element $\delta \in \Gamma^0$, there exists a single-valued rotation angle $[\omega_3]_\delta$ satisfying

$$[\omega_3]_\delta = [\omega_3]_{\gamma,\delta}$$

for any rod member γ connected at the point member δ.

Note that, in relation to the continuity of rotation angles of the rigid connection, there is only one requirement for the plane structure, i.e., the different rod members have the same rotation angle ω_3 at the joint point. It is not required that two members have the same rotation angle ω_3, on a joint line γ when a plate member is connected with another plate member on the line element γ or a rod member γ is taken as the boundary of the plate member. This is illustrated by the following two examples.

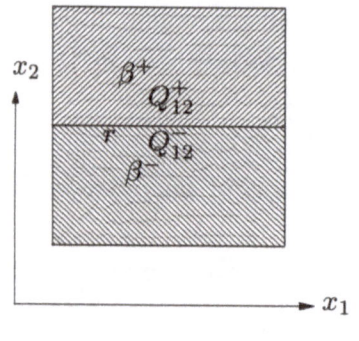

x_2

β^+

Q_{12}^+

γ

Q_{12}^-

β^-

x_1

Fig. 54

Example 1. Suppose two blocks of thin plates β^+ and β^- are rigidly connected on a line segment γ (Fig. 54). For the plate β^-, the rotation angle is

$$\omega_3^+ = \frac{1}{2}\Big(\frac{\partial u_2^-}{\partial x_1} - \frac{\partial u_2^-}{\partial x_2}\Big),$$

For the plate β^+, the rotation angle is

$$\omega_3^- = \frac{1}{2}\Big(\frac{\partial u_2^+}{\partial x_1} - \frac{\partial u_1^+}{\partial u_1^+}\Big).$$

If we require that the rotation angles on the joint line γ are equal to each other: $[\omega_3^+]_\gamma = [\omega_3^-]_\gamma$, then

$$\Big[\frac{\partial u_2^+}{\partial x_1} - \frac{\partial u_1^+}{\partial x_2}\Big]_\gamma = \Big[\frac{\partial u_2^-}{\partial x_1} - \frac{\partial u_1^-}{\partial x_2}\Big]_\gamma.$$

Since β^+ and β^- are rigidly connected on γ, from the continuity of displacement, we have

$$[u_2^+]_\gamma = [u_2^-]_\gamma,$$

so that,

$$\Big[\frac{\partial u_2^+}{\partial x_1}\Big]_\gamma = \Big[\frac{\partial u_2^-}{\partial x_1}\Big]_\gamma,$$

hence

$$\Big[\frac{\partial u_1^+}{\partial x_2}\Big]_\gamma = \Big[\frac{\partial u_1^-}{\partial x_2}\Big]_\gamma.$$

It follows that

$$\frac{1}{2}\Big[\frac{\partial u_2^+}{\partial x_1} + \frac{\partial u_1^+}{\partial x_2}\Big]_\gamma = \frac{1}{2}\Big[\frac{\partial u_2^-}{\partial x_1} + \frac{\partial u_1^-}{\partial x_2}\Big]_\gamma.$$

This implies that, on the joint line γ, the shear strains satisfy $[\varepsilon_{12}^+]_\gamma = [\varepsilon_{12}^-]_\gamma$.

On the other hand, one can see from the interface condition, which must be satisfied on the internal interface in an elastic body (the equilibrium equations on the interface derived as the natural boundary condition of the variational problem) in §5 of Chapter 2, that the shearing force Q_{12} satisfies

$$[Q_{12}^+]_\gamma = [Q_{12}^-]_\gamma, Q_{12} = \frac{Eh}{1+\nu}\varepsilon_{12}.$$

Hence the requirement that the rotation angles are equal on the joint line γ, leads to

$$\Big[\frac{Eh}{1+\nu}\Big]^{+} = \Big[\frac{Eh}{1+\nu}\Big]^{-}.$$

The plate members β^{+} and β^{-}, however, may have different thickness or may be of different materials, hence the constants $\Big[\dfrac{Eh}{1+\nu}\Big]^{+}$ and $\Big[\dfrac{Eh}{1+\nu}\Big]^{-}$ may be different, a contradiction. Thus, the rotation angles ω_3 on the joint line of two plate members will not in general be equal.

Example 2. Suppose a rod γ is taken as the boundary of the plate β and is rigidly connected with the boundary (Fig. 55).

The rotation angle of the plate is

$$[\omega_3]_\beta = \frac{1}{2}\Big(\frac{\partial[u_2]_\beta}{\partial x_1} - \frac{\partial[u_1]_\beta}{\partial x_2}\Big),$$

and the rotation angle of the rod is

$$[\omega_3]_\gamma = \frac{d[u_2]_\gamma}{dx_1}.$$

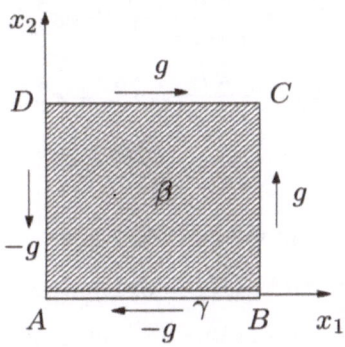

Fig. 55

If the rotation angles on γ are required to be equal to each other, $[\omega_3]_{\beta,\gamma} = [\omega_3]_\gamma$, then

$$\frac{1}{2}\Big[\frac{\partial[u_2]_\beta}{\partial x_1} - \frac{\partial[u_1]_\beta}{\partial x_2}\Big]_\gamma = \frac{d[u_2]_\gamma}{dx_1}.$$

Since γ and β are rigidly connected, from the continuity of displacements, we have,

$$[u_2]_{\beta,\gamma} = [u_2]_\gamma,$$

so that

$$\Big[\frac{\partial[u_2]_\beta}{\partial x_1}\Big]_\gamma = \frac{\partial[u_2]_{\beta,\gamma}}{\partial x_1} = \frac{d[u_2]_\gamma}{dx_1}.$$

Thus

$$\frac{1}{2}\left[\frac{\partial [u_2]_\beta}{\partial x_1} + \frac{\partial [u_1]_\beta}{\partial x_2}\right]_\gamma = 0.$$

This implies that, on γ, the shear strain of the plate satisfies $[\varepsilon_{12}^\beta]_\gamma = 0$, so that the shearing force vanishes,

$$[Q_{12}^\beta]_\gamma = \frac{Eh}{1+\nu}[\varepsilon_{12}^\beta]_\gamma = 0.$$

As illustrated in Fig.55, let us now apply the loads to the three segments of the boundary of plate β, and apply a longitudinal line load to the rod γ. Let $g = \dfrac{Eh}{1+\nu}$. Suppose the surface load is equal to zero. Then the equilibrium equations of this plane composite structures are

$$\text{plate } \beta : \begin{cases} \dfrac{\partial Q_{11}^\beta}{\partial x_1} + \dfrac{\partial Q_{12}^\beta}{\partial x_2} = 0, \\[3mm] \dfrac{\partial Q_{21}^\beta}{\partial x_1} + \dfrac{\partial Q_{12}^\beta}{\partial x_2} = 0, \end{cases}$$

$$Q_{11}^\beta = \frac{Eh}{1-\nu^2}(\varepsilon_{11}^\beta + \nu\varepsilon_{22}^\beta),$$

$$Q_{22}^\beta = \frac{Eh}{1-\nu^2}(\nu\varepsilon_{11}^2 + \varepsilon_{22}^\beta),$$

$$Q_{12}^\beta = Q_{21}^\beta = \frac{Eh}{1+\nu}\varepsilon_{12}^\beta.$$

The boundary conditions for the plate are:

$$\begin{cases} BC : Q_{11}^\beta = 0, Q_{21}^\beta = g, \\ DC : Q_{12}^\beta = g, Q_{12}^\beta = 0, \\ AD : Q_{11}^\beta = 0, Q_{21}^\beta = g. \end{cases}$$

The equilibrium equations for the rod γ are:

$$\begin{cases} \text{tension} : -EA\dfrac{d^2[u_1]_\gamma}{dx_1^2} - Q_{12}^\beta = -g, \\[3mm] \text{bending} : EI_{33}\dfrac{d^4[u_1]_\gamma}{dx_1^4} + Q_{22}^\beta = 0. \end{cases}$$

It is easy to verify that, this equilibrium problem has a (nonunique) displacement solution of the following form:

$$\beta : \begin{cases} [u_1]_\beta = u_1(x_1, x_2) = x_2, \\ [u_2]_\beta = u_2(x_1, x_2) = x_1. \end{cases} \qquad \gamma : \begin{cases} [u_1]_\gamma = 0, \\ [u_2]_\gamma = x_1. \end{cases}$$

Because of the rigid connection, this solution satisfies the continuity conditions of the displacements for the rod γ and the plate β:

$$[u_1]_{\beta,\gamma} = u_1(x_1,0) = 0 = [u_1]_\gamma,$$

$$[u_2]_{\beta,\gamma} = u_2(x_1,0) = x_1 = [u_2]_\gamma.$$

Further, in the plate β, we have

$$Q_{11}^\beta = Q_{22}^\beta = 0, \ \ Q_{12}^\beta = Q_{21}^\beta = \frac{Eh}{1+\nu} = g.$$

So that, the equilibrium equation and the boundary conditions of the plate are satisfied. On the rod γ, since $\dfrac{d^2[u_1]_\gamma}{dx_1^2} = \dfrac{d^4[u_1]_\gamma}{dx_1^4} = 0$, the equilibrium equation of the rod is also satisfied.

Although this displacement solution is not unique and may differ by a plane rigid displacement, the stress is unique. However, we have the shearing force $Q_{12}^\beta = g$ in the interior of the plate β, so that

$$[Q_{12}^\beta]_\gamma = g = \frac{Eh}{1+\nu} \neq 0.$$

This contradicts the conclusion that $[Q_{12}^\beta]_\gamma = 0$, which was derived above from the requirement that on γ the rotation angles are equal. Therefore, when a rod is taken as the boundary of a plate and is rigidly connected with the boundary, it is not in general true that the rod and the plate have the same angle of rotation on the joint line.

2.4 Boundary conditions

The boundary of a plane composite structure is simply divided into two parts: the one dimensional boundary Γ^1 and the null dimensional boundary Γ^0. These will be illustrated as follows.

1. The one dimensional boundary $\Gamma^1, \gamma \in \Gamma^1$.

The plane stress mode for thin plates, as the geometric boundary conditions, the displacement can be prescribed. E.g.

$$\gamma : u_i^\gamma = \bar{u}_i^\gamma, \ i = 1,2.$$

When γ is subjected to line loads $f_i^r, i = 1,2$, the potential energy of external work is

$$-F^\gamma(v) = -\int_\gamma \sum_{i=1}^2 f_i^\gamma v_i^\gamma d\gamma = -\int_\gamma \sum_{i=1}^2 [f_i^\gamma]_\gamma [v_i^\gamma]_\gamma d\gamma. \tag{2.14}$$

γ may also be subjected to the elastic support. Suppose γ is subjected to per unit length elastic reaction forces $-\sum\limits_{j=1}^{2} c_{ij}^{\gamma} u_j^{\gamma}, i = 1, 2$. Then the strain energy is

$$\frac{1}{2} D^{\gamma}(\boldsymbol{u}, \boldsymbol{u}) = \frac{1}{2} \int_{\gamma} \sum_{i,j=1}^{2} c_{ij}^{\gamma} u_i^{\gamma} u_j^{\gamma} d\gamma,$$

and the functional of virtual work is

$$D^{\gamma}(\boldsymbol{u}, \boldsymbol{v}) = \int_{\gamma} \sum_{i,j=1}^{2} c_{ij}^{\gamma} u_i^{\gamma} v_j^{\gamma} d\gamma.$$

2. The null dimensional boundary $\Gamma^0, \delta \in \Gamma^0$.

In the stretching and bending mode for plane beams, the displacements and the rotation angle or only a part of them can be prescribed as geometric boundary conditions

$$\delta : u_i = \bar{u}_i, \ i = 1, 2; \ \omega_3 = \bar{\omega}_3;$$

Suppose the point δ is subjected to point force loads $[f_i]_\delta, \ i = 1, 2$ and a moment load $[m_3]_\delta$. Then

the potential energy of external work is

$$-F_{12}^{\delta}(\boldsymbol{v}) = = \sum_{i=1}^{2} [f_i]_\delta [v_i]_\delta, \quad -F_3^{\delta}(\boldsymbol{v}) = -[m_3]_\delta [\omega_3]_\delta,$$

$$-F^{\delta}(\boldsymbol{v}) = -(F_{12}^{\delta}(\boldsymbol{v}) + F_3^{\delta}(\boldsymbol{v})). \tag{2.15}$$

The point δ may also be elasticly supported. Suppose the elastic reaction forces are $-\sum\limits_{j=1}^{2} c_{ij}^{\delta} [u_j]_\delta, \ i = 1, 2$, and the elastic reaction moment is $-c_3 \omega_{3,\delta}$. Then the strain energies are

$$\frac{1}{2} D_{12}^{\delta}(\boldsymbol{u}, \boldsymbol{u}) = \frac{1}{2} \sum_{i,j=1}^{2} c_{i,j=1}^{\delta} c_{ij}^{\delta} [u_i]_\delta [u_j]_\delta,$$

$$\frac{1}{2} D_3^{\delta}(\boldsymbol{u}, \boldsymbol{u}) = \frac{1}{2} c_3 [\omega_3]_\delta^2,$$

$$\frac{1}{2} D^{\delta}(\boldsymbol{u}, \boldsymbol{u}) = \frac{1}{2} (D_{12}^{\delta}(\boldsymbol{u}, \boldsymbol{u}) + D_3^{\delta}(\boldsymbol{u}, \boldsymbol{u})),$$

and the functionals of virtual work are

$$D_{12}^{\delta}(\boldsymbol{u}, \boldsymbol{v}) = \sum_{i,j=1}^{2} c_{ij}^{\delta}[u_i]_{\delta}[v_j]_{\delta},$$

$$D_3^{\delta}(\boldsymbol{u}, \boldsymbol{v}) = c_3[\omega_3(\boldsymbol{u})]_{\delta}[\omega_3(\boldsymbol{v})]_{\delta},$$

$$D^{\delta}(\boldsymbol{u}, \boldsymbol{v}) = D_{12}^{\delta}(\boldsymbol{u}, \boldsymbol{v}) + D_3^{\delta}(\boldsymbol{u}, \boldsymbol{v}).$$

Note that, the corresponding loads $[f_i]_{\delta}$ (or $[m_3]_{\delta}$) are not in play whenever the geometrical constraint u_i (or ω_3) has been prescribed.

2.5   The pinned connection

We have discussed above the completely rigid-connected plane structure, at the joints of its members, in which the three continuity conditions of the displacements and the rotation angles (see Section 2.3) are satisfied. In practice structure may also have nonrigid connections, i.e., at some specified "joints", the continuity condition may be relaxed or partially relaxed, which brings about an increase in the number of degrees of freedom of the displacement or of the rotation angle.

One kind of nonrigid connection is to relax the continuity of the displacement, i.e., the displacement at a joint may be discontinuous, so that a dislocation can arise. We will not discuss this possibility. Another kind of nonrigid connection encountered more frequently in practice still retains the continuity of displacement but relaxes the continuity of rotation angle, i.e., there is a certain rotational freedom between the members. Such a joint that can transfer the forces but not moments is a loose-leaf hinge. In general, this kind of joint is called a pinned connection. For plane composite structures, the pinned connection only appears at joints between rods. At the pin-connected point, the rotation angles are no longer single-valued. As a result, the loaded mode, the expression of potential energy, and even the form of the equilibrium equations will all be modified.

Suppose $\delta \in \Gamma^0$ is a pin-connected point. Let $\partial^{-1}\delta$ be a set of all rod members $\gamma_1, \cdots, \gamma_q$ intersecting at δ. These rod members are divided into p groups of pinned connection, and denoted by $\partial_1^{-1}\delta, \cdots, \partial_p^{-1}\delta$:

$$\partial^{-1}\delta = |\partial_1^{-1}\delta + \cdots + \partial_p^{-1}\delta| = \{\gamma_1, \cdots, \gamma_q\}.$$

In each group $\partial_j^{-1}\delta$, the individual rod members are mutually rigidly connected, and the rotation angles at the point δ have a common value which is denoted by $[\omega_3]_{\partial_j^{-1}\delta,\delta}$, then

$$[\omega_3]_{\partial_j^{-1}\delta,\delta} = [\omega_3]_{\gamma,\delta}, \text{ for any } \gamma \in \partial_j^{-1}\delta.$$

The various groups are mutually pin-connected. Hence at the point δ there are p independent values of the rotation angle,

$$[\omega_3]_{\partial_j^{-1}\delta,\delta}, \ j = 1, \cdots, p,$$

instead of a single-valued $[\omega_3]_\delta$ for the completely rigid connection.

The case $p = 1$ corresponds to a completely rigid connection, and the case $p = q$ to a completely pinned connection: $\partial_j^{-1}\delta = \gamma_j, j = 1, \cdots, q$. In the latter case, the rotation angles at the joints of various rod members are completely independent, having totally q independent values: $[\omega_3]_{rj,\delta}, \ j = 1, \cdots, q$.

In the group-pinned connection above, each rigidly connected group $\partial_j^{-1}\delta$ can be subjected to moment loads

$$[m_3]_{\partial_j^{-1}\delta,\delta}, \ j = 1, \cdots, p,$$

at the point δ, totalling P moment loads instead of a single $[m_3]_\delta$ for the rigid connection. Consequently, it produces the potential energy of external work

$$-F_3^\delta(v) = -\sum_{j=1}^{p}[m_3]_{\partial_j^{-1}\delta,\delta}[\omega_3]_{\partial_j^{-1}\delta,\delta} \tag{2.16}$$

instead of $-F_3^\delta(v) = -[m_3]_\delta[\omega_3]_\delta$ for the rigid connection. In general, all such moment loads vanish at the pin-connected point, i.e.,

$$F_3^\delta(v) = 0.$$

Principally, the values of the rotation angles $[\omega_3]_{\partial_j^{-1}\delta,\delta}, j = 1, \cdots, p$ can also be prescribed as geometric constraint conditions at the pin-connected point δ.

2.6 Variational principles

Suppose we have a plane structure Ω as defined above. A two dimensional member $\beta \in \Omega^2$ and an one dimensional member $\gamma \in \Omega^1$ of Ω have their own elastic structures and strain energies. All of them can be subjected to loads and produce potential energies of the external work. A one dimensional boundary element $\gamma \in \Gamma^1$ and a null dimensional boundary element $\delta \in \Gamma^0$ of Ω can also be elasticly supported and then have elastic energies of the coupling to be included as the strain energies. They can also be subjected to loads and have the corresponding potential energies of external work. Adding up all such potential energies, we obtain the total strain energy, the corresponding functional of virtual work and the total potential energy of the external work:

$$\frac{1}{2}D(u, u) = \sum_{\beta \in \Omega^2} \frac{1}{2}D^\beta(u, u) + \sum_{\gamma \in \Omega^1 + \Gamma^1} \frac{1}{2}D^\gamma(u, u)$$

$$+ \sum_{\delta \in \Gamma^0} \frac{1}{2}D^\delta(u, u), \tag{2.17}$$

$$D(u, v) = \sum_{\beta \in \Omega^2} D^\beta(u, v) + \sum_{\gamma \in \Omega^1 + \Gamma^1} D^\gamma(u, v)$$

$$+ \sum_{\delta \in \Gamma^0} D^\delta(u, v), \tag{2.18}$$

$$-F(u) = -\left\{ \sum_{\beta \in \Omega^2} F^\beta(u) + \sum_{\gamma \in \Omega^1 + \Gamma^1} F^\gamma(u) \right.$$

$$\left. + \sum_{\delta \in \Gamma^0} F^\delta(u) \right\}. \tag{2.19}$$

The total potential energy of the whole structure Ω is

$$J(u) = \frac{1}{2}D(u, u) - F(u).$$

Suppose the structure is completely rigid-connected. Then u_i, $i = 1, 2$ and ω_3 must satisfy the continuity conditions 1. to 3. in Section 2.3. Besides this, we can impose geometric boundary conditions on the boundary of Ω, i.e., we can fix the displacement on a boundary line $\gamma \in \Gamma^1$ and both the displacement and the rotation angle on a boundary point $\delta \in \Gamma^0$, thus we have prescribed a set of geometric constraints. By the variational principle, the displacement of the equilibrium configuration u under the prescribed constraints makes the total potential energy a minimum:

$$J(u) = \text{Min.} \tag{2.20}$$

This is equivalent to the principle of virtual work,

$$D(u, v) - F(v) = 0, \text{ for any virtual displacement } v. \tag{2.21}$$

From (2.21), the equilibrium equations on the various parts of the structure and on its internal and external boundaries can then be derived by the use of Green's formula.

2.7   Equilibrium equations

For the sake of simplicity, no elastic support will be considered below. Using the Green's formula (2.13) for $D^\beta(\boldsymbol{u}, \boldsymbol{v})$ and transforming the displacements v_j expressed in global coordinates in the line integral term into the local coordinates of γ through the transformation formula (2.7), we obtain

$$
D^\beta(\boldsymbol{u}, \boldsymbol{v}) = \iint_\beta \sum_{i=1}^2 \Big(-\sum_{j=1}^2 \frac{\partial Q_{ij}(\boldsymbol{u})}{\partial x_j} \Big) [v_i]_\beta d\beta
$$

$$
+ \sum_{\gamma \in \partial\beta} \int_\gamma \sum_{j=1}^2 Q_{jn}(\boldsymbol{u}) [v_j]_{\beta,\gamma} d\gamma
$$

$$
= \int_B \int \sum_{i=1}^2 \Big(-\sum_{j=1}^2 \frac{\partial Q_{ij}(\boldsymbol{u})}{\partial x_j} \Big) [v_i]_B d\beta
$$

$$
+ \sum_{\gamma \in \partial\beta} \int_\gamma \sum_{i=1}^2 \sum_{j=1}^2 \Big(\sum Q_{jn}(\boldsymbol{u}) a_{ji}^\gamma \Big) [v_i^\gamma]_{\beta,\gamma} d\gamma. \tag{2.22}
$$

Similarly, using the Green's formula (2.10) for $D^\gamma(\boldsymbol{u}, \boldsymbol{v})$ and transforming the displacements v_f^γ expressed in local coordinates on the boundary points into global coordinates (the coordinates of a point δ are just the global coordinates), we obtain

$$
D^\gamma(\boldsymbol{u}, \boldsymbol{v}) = \int_\gamma \sum_{i=1}^2 \Big(-\frac{dQ_i^\gamma(\boldsymbol{u})}{dx_1^\gamma} \Big) [u_i^\gamma]_\gamma d\gamma
$$

$$
+ \sum_{\delta \in \partial\gamma} \varepsilon(\gamma, \delta) \Big[\sum_{j=1}^2 Q_j^\gamma(\boldsymbol{u}) [v_i^\gamma]_{\gamma,\delta}
$$

$$
+ [M_3(\boldsymbol{u})]_{\gamma,\delta} [\omega_3(\boldsymbol{v})]_{\gamma,\delta} \Big]
$$

$$
= \int_\gamma \sum_{i=1}^2 \Big(-\frac{dQ_i^\gamma(\boldsymbol{u})}{dx_1^\gamma} \Big) [v_i^\gamma]_\gamma d\gamma
$$

$$
+ \sum_{\delta \in \partial\gamma} \varepsilon(\gamma, \delta) \Big[\sum_{i=1}^2 \sum_{j=1}^2 \Big(\sum Q_j^\gamma(\boldsymbol{u}) a_{ij}^\gamma \Big) [v_i]_{\gamma,\delta}
$$

$$
+ [M_3(\boldsymbol{u})]_{\gamma,\delta} [\omega_3(\boldsymbol{v})]_{\gamma,\delta} \Big], \tag{2.23}
$$

Substituting (2.22) and (2.23) into (2.18), we obtain

$$
D(\boldsymbol{u}, \boldsymbol{v}) = \sum_{\beta \in \Omega^2} \iint_{\beta} \sum_{i=1}^{2} \left(-\sum_{j=1}^{2} \frac{\partial Q_{ij}(\boldsymbol{u})}{\partial x_j} \right) [v_i]_\beta d\beta
$$

$$
+ \sum_{\gamma \in \Omega^1} \int_{\gamma} \sum_{i=1}^{2} \left[\left(\frac{-dQ_i^\gamma(\boldsymbol{u})}{dx_1^\gamma} \right) [v_i^\gamma]_\gamma \right.
$$

$$
+ \sum_{\beta \in \partial^{-1}\gamma} \sum_{j=1}^{2} Q_{jn}(\boldsymbol{u}) a_{ij}^\gamma [v_i^\gamma]_{\beta,\gamma} \right] d\gamma
$$

$$
+ \sum_{\gamma \in \Gamma^1} \sum_{\beta \in \partial^{-1}\gamma} \int_{\gamma} \sum_{i=1}^{2} \sum_{j=1}^{2} Q_{jn}(\boldsymbol{u}) a_{jn}^\gamma(\boldsymbol{u}) a_{ji}^\gamma [v_i^\gamma]_{\beta,\gamma} d\gamma
$$

$$
+ \sum_{\delta \in \Gamma^0} \sum_{\gamma \in \partial^{-1}\delta} \varepsilon(\gamma, \delta) \left[\sum_{ji=1}^{2} \left(\sum_{j=1}^{2} Q_j^\gamma(\boldsymbol{u}) a_{ij}^\gamma \right) [v_i]_{\gamma,\delta} \right.
$$

$$
+ [M_3(\boldsymbol{u})]_{\gamma,\delta} [\omega_3(\boldsymbol{v}))]_{\gamma,\delta} \bigg]. \tag{2.24}
$$

From (2.19) and the potential energies of external work on various elements, (2.9), (2.12), (2.14) and (2.15), we obtain

$$
F(\boldsymbol{v}) = \sum_{\beta \in \Omega^1} \iint_{\beta} \sum_{i=1}^{2} [f_i]_\beta [v_i]_\beta d\beta + \sum_{\gamma \in Q^1} \int_{\gamma} \sum_{i=1}^{2} [f_i^\gamma]_\gamma [v_i^\gamma]_\gamma d\gamma
$$

$$
+ \sum_{\gamma \in \Gamma^1} \int_{\gamma} \sum_{i=1}^{2} [f_1^\gamma]_\gamma [v_i^\gamma]_\gamma d\gamma + \sum_{\delta \in \Gamma^0} \left(\sum_{i=1}^{2} [f_i]_\delta [v_i]_\delta + [m_3]_\delta [\omega_3]_\delta \right). \tag{2.25}
$$

(2.24) is just the general Green's formula no matter whether the structure is rigidly connected or not. According to the continuity conditions in Section 2.3, when Ω is completely rigidly connected we have

$$
[v_i^\gamma]_\gamma = [v_i^\gamma]_{\beta,\gamma}, \ i = 1, 2, \ \text{for any } \beta \in \partial^{-1}\gamma,
$$

$$
[v_i]_\delta = [v_i]_{\gamma,\beta}, \ i = 1, 2, \ \text{for any } \gamma \in \partial^{-1}\delta,
$$

$$
[\omega_3]_\delta = [\omega_3]_{\gamma,\delta}, \ \text{for any } \gamma \in \partial^{-1}\delta.
$$

Then, from (2.24) and (2.25), we obtain

$$D(\boldsymbol{u}, \boldsymbol{v}) - F(\boldsymbol{v}) = \sum_{\beta \in \Omega^2} \iint_\beta \sum_{i=1}^2 \Big[-\sum_{j=1}^2 \frac{\partial Q_{ij}(\boldsymbol{u})}{\partial x_j}$$

$$-[f_i]_\beta \Big][v_i]_\beta d\beta + \sum_{r \in \Omega^1} \int_\gamma \sum_{i=1}^2 \Big[\frac{-dQ_i^\gamma(\boldsymbol{u})}{dx_1^\gamma}$$

$$+ \sum_{\beta \in \partial^{-1}\gamma} \sum_{j=1}^2 Q_{jn}(\boldsymbol{u})a_{ji}^\gamma - [f_i^\gamma]_\gamma \Big][v_i^\gamma]_\gamma d\gamma$$

$$+ \sum_{\gamma \in \Gamma^1} \int_\gamma \sum_{i=1}^2 \Big[\sum_{\beta \in \partial^{-1}\gamma} \sum_{j=1}^2 Q_{jn}(\boldsymbol{u})d_{ji}^\gamma - [f_i^\gamma]_\gamma \Big][v_i^\gamma]_\gamma d\gamma$$

$$+ \sum_{\delta \in \Gamma^0} \sum_{i=1}^2 \Big[\sum_{\gamma \in \partial^{-1}\delta} \varepsilon(\gamma, \delta) \sum_{j=1}^2 Q_j^\gamma(\boldsymbol{u})a_{ij}^\gamma - [f_i]_\delta \Big][v_i]_\delta$$

$$+ \sum_{\delta \in \Gamma^0} \Big[\sum_{\gamma \in \partial^{-1}\delta} \varepsilon(\gamma_1 \delta)[M_3(\boldsymbol{u})]_{\gamma,\delta} - [m_3]_\delta \Big][\omega_3]_\delta.$$

$$(2.26)$$

Now suppose that no geometric constraint is imposed on the boundaries Γ^1 and Γ^0. Then the necessary and sufficient condition for the validity of (2.21) is that every bracket in (2.26) vanishes, i.e., $[\cdots] = 0$. From the above, we obtain the following equilibrium equations

$$-\sum_{j=1}^2 \frac{\partial Q_{ij}(\boldsymbol{u})}{\partial x_j} = [f_i]_\beta, \ i = 1, 2, \quad \beta \in \Omega^2, \tag{2.27}$$

$$\frac{-dQ_i^\gamma(\boldsymbol{u})}{dx_1^\gamma} + \sum_{\beta \in \gamma} \sum_{j=1}^2 [Q_{jn}(\boldsymbol{u})]_{\beta,\gamma} a_{ji}^\gamma = [f_i^\gamma]_\gamma, \ i = 1, 2 \quad \gamma \in \Omega^1, \tag{2.28}$$

$$\sum_{\beta \in \partial^{-1}\gamma} \sum_{j=1}^2 [Q_{jn}(\boldsymbol{u})]_{\beta,\gamma} a_{ji}^\gamma = [f_i^\gamma]_\gamma, \ i = 1, 2, \quad \gamma \in \Gamma^1, \tag{2.29}$$

$$\sum_{\gamma \in \partial^{-1}\delta} \varepsilon(\gamma, \varepsilon) \sum_{j=1}^2 [Q_j^\gamma(\boldsymbol{u})]_{\gamma,\delta} a_{ij}^\gamma = [f_i]_\delta, \quad \delta \in \Gamma^0, \tag{2.30}$$

$$\sum_{\gamma \in \partial^{-1}\delta} \varepsilon(\gamma, \varepsilon)[M_3(\boldsymbol{u})]_{\gamma,\delta} = [m_3]_\delta, \quad \delta \in \Gamma^0. \tag{2.31}$$

We can see from the equations above that the equilibrium equations (2.27) on the two-dimensional plate member Ω^2 are just the equilibrium equations for a single plane stress plate without considering the coupling. Compared with the ordinary equilibrium equation of a single rod member, there is one more term, i.e., the second term on the left hand side of the equations (2.28), which arise from the plate couplings. Because of the continuity of displacements, forces are transferred from the plate to the rod. Whenever geometric constraints are prescribed on Γ^1 and Γ^0 so as to fix the displacement component u_i or the rotation angle ω_3 at some point, the equations containing f_i or m_3 are displaced locally, while all the other remain unchanged.

Equilibrium equations of moment at pin-connected points.

Assume that the structure described above is completely rigid-connected. Let us now see how to modify the equilibrium equation for the moment (2.31) at the pin-connected point δ. Consider the simplest and also the most often encountered case, that in which the point δ is completely pin-connected. In that case, the rotation angles $[\omega_3]_{rj,\delta}$ at the point δ of the individual rods γ_j in $\partial^{-1}\delta = \{\gamma_1, \cdots, \gamma_q\}$ are independent of one another, so the rotation angle has q degrees of freedom at δ. Suppose all moment loads for the individual degrees of freedom vanish. Then the potential energy of external work at the point δ is $F_3^\delta(v) = 0$ according to (2.16). Due to the independence of the rotation angles $[\omega_3]_{rj,\delta}$, $j = 1, \cdots, q$, we can see from the principle of virtual work (2.21) that, there are q equilibrium equations for the moment at the point δ,

$$\delta \in \Gamma^0 : [M_3(\boldsymbol{u})]_{\gamma j,\delta} = 0, \; j = 1, \cdots, q, \tag{2.32}$$

instead of the single equation (2.31) for a completely rigid connection. (2.32) says that the bending moment of the individual pin-connected rod members at the pin-connected point vanishes.

Since the displacements v_1 and v_2 at a pin-connected point are continuous, as in the case of a rigid connection, the other two equilibrium equations (2.30) at the point δ remain unchanged.

2.8   Strainless states

Consider a geometrically unrestrained, free elasticaly supported, and completely rigid-connected plane structure. For the strainless states are given by

$$D(\boldsymbol{v}, \boldsymbol{v}) = 0 \iff D^\beta(\boldsymbol{v}, \boldsymbol{v}) = D^\gamma(\boldsymbol{v}, \boldsymbol{v}) = 0, \text{ for any } \beta \text{ and } \gamma.$$

Hence, from (2.8) and (2.11), that any strainless state is a plane rigid body displacement on each plate or rod member. In global coordinates, they are

$$v_1 = a_1^\beta - b_3^\beta x_2, \ v_2 = a_2^\beta + b_3 x_1, \ (x_1, x_2) \in \beta,$$
$$v_1 = a_1^\gamma - b_3^\gamma x_2, \ v_2 = a_2^\gamma + b_3^\gamma x_1, \ (x_1, x_2) \in \gamma,$$

respectively. On each member, there are two translation parameters a_1 and a_2 and one rotation angle parameter b_3. It follows easily from the rigid connection condition that these three parameters are equal on the different members. For example, if the plate members β_1 and β_2 are rigid-connected on the line element γ, then, according to the continuity of displacements,

$$[v_1]_{\beta_1, \gamma} = [v_1]_{\beta_2, \gamma}, \ [v_2]_{\beta_1, \gamma} = [v_2]_{\beta_2, \gamma}.$$

That is, for $(x_1, x_2) \in \gamma$, we must have

$$(a_1^{\beta_1} - a_1^{\beta_2}) - (b_3^{\beta_1} - b_3^{\beta_2})x_2 = 0,$$
$$(a_2^{\beta_1} - a_2^{\beta_2}) + (b_3^{\beta_1} - b_3^{\beta_2})x_1 = 0.$$

It follows that

$$a_1^\beta = a_1^{\beta_2}, \ a_2^{\beta_1} = a_1^{\beta_2}, \ b_3^{\beta+1} = b_3^\beta.$$

Similarly, if a plate member β is rigid-connected with a rod member γ, then, according to the continuity of displacements,

$$[v_1]_\gamma = [v_1]_{\beta, \gamma}, \ [v_2]_\gamma = [v_2]_{\beta, \gamma},$$

That is, for $(x_1, x_2) \in \gamma$, we must have

$$(a_1^\beta - a_1^\gamma) - (b_3^\beta - b_3^\gamma)x_2 = 0,$$
$$(a_2^\beta - a_2^\gamma) + (b_3^\beta - b_3^\gamma)x_1 = 0,$$

so that

$$a_1^\beta = a_1^\gamma, \ a_2^\beta = a_2^\gamma, \ b_3^\beta = b_3^\gamma.$$

Moreover, if the rod members γ_1 and γ_2 are rigid-connected at a point element δ, then, from the continuity of displacements,

$$[v_1]_{\gamma_1, \delta} = [v_1]_{\gamma_2, \delta}, \ [v_2]_{\gamma_1, \delta} = [v_2]_{\gamma_2, \delta}.$$

That is, for the point $(x_1, x_2) = (x^\delta, x_2^\delta)$, we must have

$$(a_1^{\gamma_1} - a_1^{\gamma_1}) - (b_3^{\gamma_1} - b_3^{\gamma_2})x_2^\delta = 0,$$
$$(a_2^{\gamma_1} - a_2^{\gamma_2}) + (b_1^{\gamma_1} - b_3^{\gamma_2})x_1^\delta = 0.$$

Again, from the continuity of rotation angles we have

$$[\omega_3]_{\gamma_1, \delta} = [\omega_3]_{\gamma_2, \delta},$$

i.e.,

$$\left[\frac{\partial v_2}{\partial x_1}\right]_{\gamma_1,\delta} = \left[\frac{\partial v_2}{\partial x_1}\right]_{\gamma_2,\delta}.$$

Hence $b_3^{\gamma_1} = b_3^{\gamma_2}$. Substituting this equation into the expression above, we immediately obtain

$$a_1^{\gamma_1} = a_1^{\gamma_2}, \; a_2^{\gamma_1} = a_2^{\gamma_2}, \; b_3^{\gamma_1} = b_3^{\gamma_2}.$$

This shows that the displacement field of the strainless state of the whole structure is a plane rigid displacement having three degrees of freedom in all. Therefore the strain energy $D(v, v)$ is degenerate under the unrestrained condition. Let

$$v^{(1)} = \begin{bmatrix} 1 \\ 0 \end{bmatrix}, \; v^{(2)} = \begin{bmatrix} 0 \\ 1 \end{bmatrix}, \; v^{(3)} = \begin{bmatrix} -x_2 \\ x_1 \end{bmatrix}, \qquad (2.33)$$

then

$$\omega_3(v^{(1)}) = 0, \; \omega_3(v^{(2)}) = 0, \; \omega_3(v^{(3)}) = 1.$$

Any strainless state v can be expressed as a linear combination

$$v = a_1 v^{(1)} + a_2 v^{(2)} + b_3 v^{(3)}.$$

The prerequisites for the existence of solutions to an unconstrained equilibrium problem are

$$F(v^{(i)}) = 0, \quad i = 1, 2, 3, \qquad (2.34)$$

where

$$F(v) = \sum_{\beta \in \Omega^2} \iint_\beta \sum_{i=1}^{2} [f_i]_\beta [v_i]_\beta d\beta + \sum_{\gamma \in \Omega^1 + \Gamma^1} \int_\gamma \sum_{i=1}^{2} [f_i^\gamma]_\gamma [v_i^\gamma]_\gamma d\gamma$$

$$+ \sum_{\delta \in \Gamma^0} \left(\sum_{i=1}^{2} [f_i]_\delta [v_i]_\delta + [m_3]_\delta [\omega_3]_\delta \right).$$

Transforming the second term in the integrand of the above expression from local coordinates into global coordinates, and letting

$$f^\gamma = \begin{bmatrix} f_1^\gamma \\ f_2^\gamma \end{bmatrix}, \; v^\gamma = \begin{bmatrix} v_1^\gamma \\ v_2^\gamma \end{bmatrix},$$

according to the transformation formula (2.4), in global coordinates the force f and the displacement v are

$$f = A^\gamma f^\gamma, \; v = A^\gamma v^\gamma,$$

where, A^γ is an orthogonal matrix, i.e., $(A^\gamma)^{-1} = (A^\gamma)^T$. Thus, we have

$$\sum_{i=1}^{2} f_i^\gamma v_i^\gamma = (\boldsymbol{f}^\gamma, \boldsymbol{v}^\gamma) = ((A^\gamma)^{-1}\boldsymbol{f}, (A^\gamma)^{-1}\boldsymbol{v})$$

$$= ((A^\gamma)^{-1}\boldsymbol{f}, (A^\gamma)^T\boldsymbol{v}) = (A^\gamma(A^\gamma)^{-1}\boldsymbol{f}, \boldsymbol{v})$$

$$= (\boldsymbol{f}, \boldsymbol{v}) = \sum_{i=1}^{2} f_i v_i.$$

So that

$$F(\boldsymbol{v}) = \sum_{\beta \in \Omega^2} \iint_\beta [f_i]_\beta [v_i]_\beta [v_i]_\beta d\beta + \sum_{\gamma \in \Omega^1 + \Gamma'} \int_\gamma \sum_{i=1}^{2} [f_i]_\gamma [v_i]_\gamma d\gamma$$

$$+ \sum_{\delta \in \Gamma^0} \left(\sum_{i=1}^{2} [f_i]_\delta [v_i]_\delta + [m_3]_\delta [\omega_3]_\delta \right).$$

Substituting (2.33) into this equation, the prerequisites (2.34) are met i.e., the resultant forces in the x_1 and x_2 directions and the resultant moment about the x_3-axis all vanish:

$$F(\boldsymbol{v}^{(i)}) = \sum_{\beta \in \Omega^2} \iint_\beta [f_i]_\beta d\beta + \sum_{\gamma \in \Omega^1 + \Gamma^1} \int_\gamma [f_i]_\gamma d\gamma + \sum_{\delta \in \Gamma^0} [f_i]_\delta$$

$$= 0, \qquad i = 1, 2,$$

$$F(\boldsymbol{v}^{(3)}) = \sum_{\beta \in \Omega^2} \iint_\beta [x_1 f_2 - x_2 f_1]_\beta d\beta + \sum_{\gamma \in \Omega^1 + \Gamma^1} \int_\gamma [x_1 f_2 - x_2 f_1]_\gamma d\gamma$$

$$+ \sum_{\delta \in \Gamma^0} [x_1 f_2 - x_2 f_1]_\delta + \sum_{\delta \in \Gamma^0} [m_3]_\delta$$

$$= 0.$$

The equilibrium problem is solvable when these three prerequisites have been satisfied. However, the displacement solution is only unique up to a plane rigid displacement. The stress solution is still unique. As the number of geometric constraints or of elastic supports is increased, the number of degrees of freedom of the strainless state is reduced until the strain energy $D(\boldsymbol{v}, \boldsymbol{v})$ becomes positive definite and a unique displacement is guaranteed. These conclusions are identical with those in the case of the single elastic body.

§3   Space Composite Structures

3.1   Geometric description

A space composite structure is composed of three kinds of elastic members: three dimensional elastic body members, two dimensional plate members, i.e., thin plates in stretching and in bending, and one dimensional rod members, i.e., slender rods in tension, in bending and in torsion.

For the sake of simplicity, assume that each three dimensional elastic body member α is a geometrically simply-connected polyhedron, and its boundary $\partial\alpha$ is composed of four or more polygons (Fig. 56). Next, assume that the two dimensional plate member β is a geometrically simply-connected polygon, and as its boundary $\partial\beta$ is composed of three or more line segments. A plate member may be connected with a body member its boundary surface (Fig. 57), and also may exist independently. An one dimensional rod member γ is geometrically a straight line segment, and its boundary $\partial\gamma$ is composed of two end points. As for plane structures, a rod member may be part of the boundary of a plate member, i.e., a stiffener of a plate member (Fig. 58), and also may exist independently.

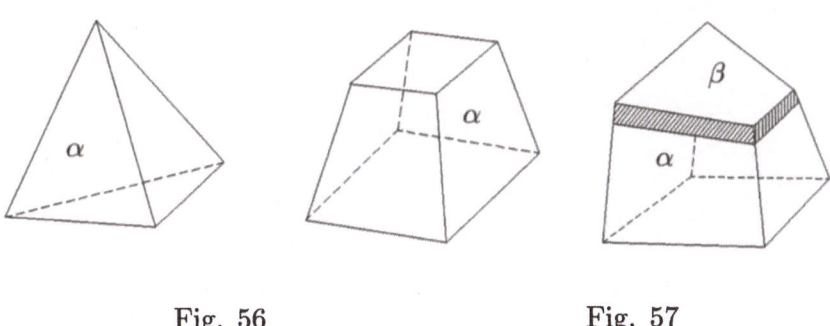

Fig. 56   Fig. 57

Two body members may have a common interface and be connected directly (Fig. 59a), or may be connected indirectly through a plate member as their common interface (Fig. 59b).

Two plate members or more than two plate members, may be connected directly through a common interface line (Fig. 60a), and also may be connected indirectly through a rod member as their common interface (Fig. 60b).

Two rod members or more than two rod members may have a common boundary point (Fig. 61).

Suppose there are N_3 body members $\alpha_1, \cdots, \alpha_N$, N_2 plate members $\beta_1, \cdots, \beta, \beta_{N_1}$, and N_1 rod members $\gamma_1, \cdots, \gamma_{N_1}$.

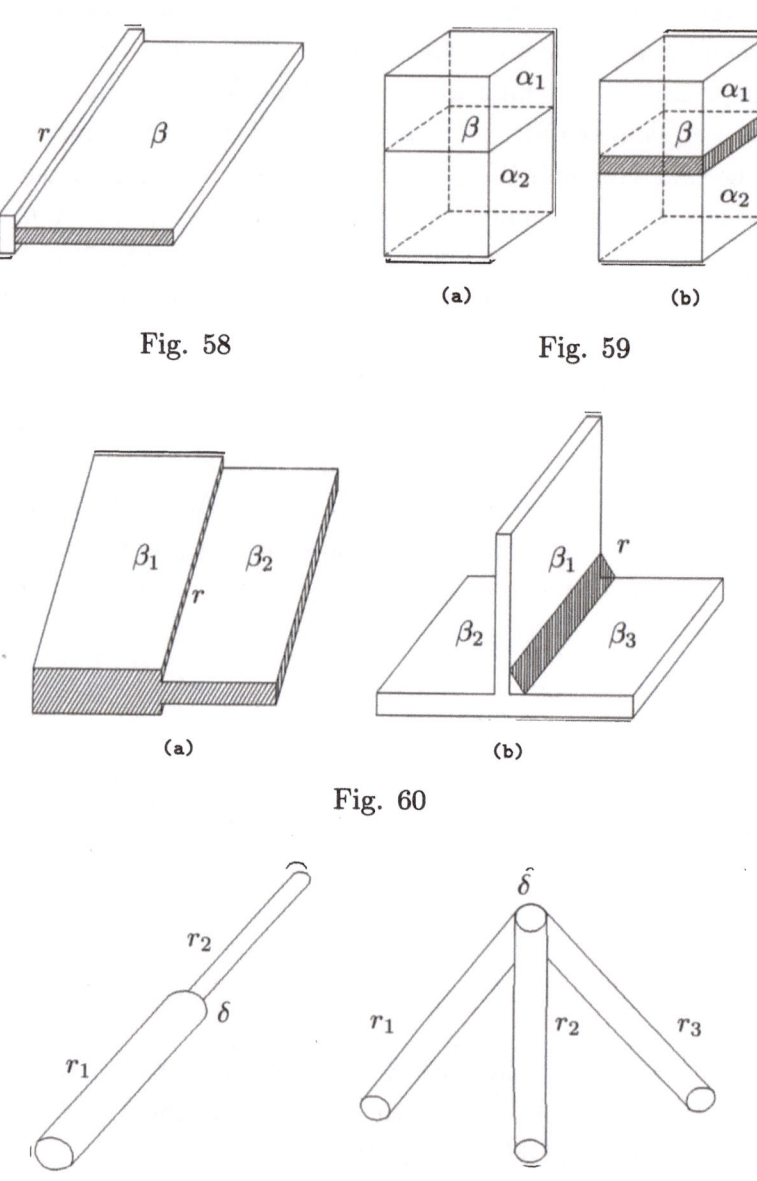

Fig. 58

(a)   (b)

Fig. 59

(a)

(b)

Fig. 60

Fig. 61

They compose a space structure Ω:

$$\Omega = \{\alpha_1, \cdots, \alpha_{N_3}; \beta_1, \cdots, \beta_{N_2}; \gamma_1, \cdots, \gamma_{N_1}\}.$$

The three dimensional, two dimensional and one dimensional parts of Ω are denoted, respectively, by

$$\Omega^3 = \{\alpha_1, \cdots, \alpha_{N_3}\}, \quad \Omega^2 = \{\beta_1, \cdots, \beta_{N_2}\},$$

$$\Omega^1 = \{\gamma_1, \cdots, \gamma_{N_1}\}.$$

As has been stated before, the boundary surface of a body member need not be a plate member, i.e., it may be a proper boundary area element. Denote such boundary area elements as $\beta_{N_2+1}, \cdots, \beta_{N_2'}$, and let

$$\Gamma^2 = \{\beta_{N_2+1}, \cdots, \beta_{N_2'}\}, \quad \overline{\Omega}^2 = \Omega^2 + \Gamma^2.$$

$\overline{\Omega}^2$ can be considered as the set of two dimensional boundaries of the composite body Ω. It contains not only the external boundaries of the body members but also the internal boundaries among the body members. Thus, the boundary ∂_α of an arbitrary body member α may not belong to Ω^2, but will surely belong to $\overline{\Omega}^2$, i.e.,

$$\alpha \in \Omega^3 \Longrightarrow \partial\alpha \in \overline{\Omega}^2.$$

Like wise, the boundary line of a plate member need not be a rod member but may be its own boundary line. Denote such boundary line elements as $\gamma_{N_1+1}, \cdots, \gamma_{N_1'}$, and let

$$\Gamma^1 = \{\gamma_{N_1+1}, \cdots, \gamma_{N_1'}\}, \quad \overline{\Omega} = \Omega^1 + \Gamma^1.$$

Then we have

$$\beta \in \Omega^2 \Longleftrightarrow \partial\beta \in \overline{\Omega}^1.$$

Γ^1 can be considered as the one dimensional boundaries of the composite body Ω, which contains the external boundaries of the plate members and the common edges among the plate members. Generally speaking, $\overline{\Omega}^1$ still does not contain all the line elements of the structure, e.g., there are still edge lines on proper boundaries of the body elements, but they are of no significance in the following mechanical model and are neglected.

Denote all the boundary points (a common point is counted only once) of the rod members $\gamma_1, \cdots, \gamma_{N_1}$, as $\delta_1, \cdots, \delta_{N_0}$, and let

$$\Gamma^0 = \{\delta_1, \cdots, \delta_{N_0}\},$$

then, we have

$$\gamma \in \Omega^1 \Longrightarrow \partial\gamma \in \Gamma^0.$$

Γ^0 can be considered as the null dimensional boundary of Ω. It contains the external boundary points of the rods as well as the intersecting points of the rods. Γ^0 still does not contain all the point elements of the structure, such as the corner points of the boundaries of the body members and the corner points of the proper boundaries of the plate members. The former are of no significance in the mechanical modal of the three dimensional elasticity, and can be neglected. However, the latter are significant in the mechanical model of thin plates in bending (see (4.30) in §4 of Chapter 3), and still should be taken into account for the sake of completeness. Let such corner points of proper boundaries of the plate members be $\delta_{N_0+1}, \cdots, \delta_{N_0'}$, and let

$$\overline{\Gamma}^0 = \{\delta_0, \cdots, \delta_{N_0}, \delta_{N_0+1}, \cdots, \delta_{N_0'}\} \supset \Gamma^0.$$

An element of $\overline{\Omega}^2$ is called an area element, an element of $\overline{\Omega}'$ is called a line element and an element of $\overline{\Gamma}^0$ is called a point element.

Inverse boundaries

For plane structure, we have already defined the inverse boundaries of line and point elements. These concepts are still valid for space structures. In addition, define the inverse boundary $\partial^{-1}\beta$ of an arbitrary area element $\beta \in \overline{\Omega}^2$ to be the set of all the body members having β as a boundary surface. For example, $\partial^{-1}\beta = \alpha$ in Fig. 57, and $\partial^{-1}\beta = \{\alpha_1, \alpha_2\}$ in Fig. 59. Clearly, we have

$$\alpha \in \partial^{-1}\beta \Longleftrightarrow \beta \in \partial\alpha.$$

Global coordinates and local coordinates

Fix a right hand orthogonal system (x_1, x_2, x_3) in the space, called the global coordinates. Further, define local right hand coordinate system for each body member α, area element β, line element γ, and point element δ of the structure Ω, respectively denoted as follows:

$$(x_1^\alpha, x_2^\alpha, x_3^\alpha), (x_1^\beta, x_2^\beta, x_3^\beta), (x_1^\gamma, x_2^\gamma, x^\gamma), (x_1^\delta, x_2^\delta, x_3^\delta).$$

In practice, the local coordinates of a body member α or a point element δ are taken to be just the global coordinates

$$(x_1^\alpha, x_2^\alpha, x_3^\alpha) = (x_1^\delta, x_2^\delta, x_3^\delta) = (x_1, x_2, x_3).$$

For the local coordinates $(x_1^\beta, x_2^\beta, x_3^\beta)$ of an area element β, we adopt the convention that x_1^β amd x_2^β are longitudinal directions of the plate, i.e., the coordinates in the neutral surface of the plate, and x_3^β is a direction transverse to the plate, i.e., normal to the neutral surface. Suppose (ξ_1, ξ_2, ξ_3) is the origin of the local coordinates, then the transformation relations between the global coordinates and the local coordinates have the form

$$x_i = \xi_i + \sum_{j=1}^{3} a_{ij}^\beta x_j^\beta, \quad \text{or} \quad \boldsymbol{x} = \boldsymbol{\xi} + A^\beta \boldsymbol{x}^\beta.$$

The transformation relations between the global components and the local components of the displacement are then

$$u_i = \sum_{j=1}^{3} a_{ij}^\beta u_j^\beta, \quad \text{or} \quad \boldsymbol{u} = A^\beta \boldsymbol{u}^\beta. \tag{3.1}$$

The transformation matrix A^β is an orthogonal matrix, which has the following concrete form:

$$A^\beta = (a_{ij}^\beta) = \begin{bmatrix} \cos(x_1, x_1^\beta) & \cos(x_1, x_2^\beta) & \cos(x_1, x_3^\beta) \\ \cos(x_2, x_1^\beta) & \cos(x_2, x_2^\beta) & \cos(x_2, x_3^\beta) \\ \cos(x_3, x_1^\beta) & \cos(x_3, x_2^\beta) & \cos(x_3, x_3^\beta) \end{bmatrix}. \tag{3.2}$$

Hence, (3.1) can be written in inverse form as

$$u_i^\beta = \sum_{j=1}^{3} a_{ji}^\beta u_j, \quad \text{or} \quad \boldsymbol{u}^\beta = (A^\beta)^T \boldsymbol{u}. \tag{3.3}$$

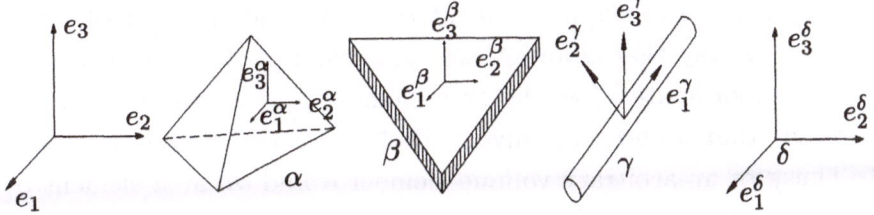

Fig. 62

For the local coordinates $(x_1^\gamma, x_2^\gamma, x_3^\gamma)$ of a line element γ, we adopt the convention that x_1^γ is the longitudinal direction, i.e., the straight line coordinate of the axis of rod, x_2^γ and x_3^γ are the transverse directions, i.e., normal to the axis of rod, and the origin of the local coordinates is located at an end point of the rod. The transformation formulas between the global coordinates and the local coordinates are identical with (3.1) and (3.3) except that the supercript β must be changed to γ.

Global coordinates and local coordinates are illustrated in Fig. 62.

Sometimes we want to perform the conversion between local displacement components of adjacent elements. For example, for $\gamma \in \partial\beta$, since

$$\boldsymbol{u} = A^\beta \boldsymbol{u}^\beta = A^\gamma \boldsymbol{u}^\gamma,$$

it follows that

$$\boldsymbol{u}^\beta = (A^\beta)^T A^\gamma \boldsymbol{u}^\gamma = A^{\beta,\gamma} \boldsymbol{u}^\gamma, \ A^{\beta,\gamma} = (A^\beta)^T A^\gamma,$$

or

$$u_i^\beta = \sum_{j=1}^3 a_{ij}^{\beta,\gamma} u_j^\gamma, \ a_{ij}^{\beta,\gamma} = \sum_{k=1}^3 a_{ki}^\beta a_{kj}^\gamma. \tag{3.4}$$

Oriented connections between members

In §2 the oriented connection relations $\varepsilon(\beta, \gamma)$ and $\varepsilon(\gamma, \delta)$ between the area element and the line element and between the line element and the point element were defined for plane structures. In the case of space structures, we must also say that the tangent $\boldsymbol{t}^\beta$ to the boundary $\partial\beta$ of area element β must be chosen so as to make $\{\boldsymbol{n}^\beta, \boldsymbol{t}^\beta, \boldsymbol{e}_3^\beta\}$ a right hand system. Here, $\boldsymbol{n}^\beta$ denotes the outward normal direction of the area element boundary $\partial\beta$ in the longitudinal plane, and $\boldsymbol{e}_3^\beta$ denotes the unit vector of the transverse coordinate of the area element.

Now define the oriented connection relations between volume elements and area elements. If $\beta \in \partial\alpha$, denote the outward normal direction to the boundary $\partial\alpha$ of the volume element α by $\boldsymbol{n}^\alpha$ and the part of $\boldsymbol{n}^\alpha$ on β by $[\boldsymbol{n}^\alpha]_\beta$. We say that α and β are conforming if the unit vector of the transverse coordinate $\boldsymbol{e}_3^\beta$ of the area element β is consistent with $[\boldsymbol{n}^\alpha]_\beta$, and we say that α and β are inverse if $\boldsymbol{e}_3^\beta$ is reverse to $[\boldsymbol{n}^\alpha]_\beta$ in sign (Fig. 63). Thus, for an arbitrary volume element α and ∂n area element β, we define the oriented connection relations by

$$\varepsilon(\alpha, \beta) = \begin{cases} 0, & \text{if } \beta \notin \partial\alpha, \\ 1, & \text{if } \beta \in \partial\alpha \text{ and } \alpha \text{ and } \beta \text{ are conforming}, \\ -1 & \text{if } \beta \in \partial\alpha \text{ and } \alpha \text{ and } \beta \text{ are inverse}, \end{cases}$$

consequently,

$$\beta \in \partial\alpha \Longrightarrow [\boldsymbol{n}^\alpha]_\beta = \varepsilon(\alpha, \beta)\boldsymbol{e}_3^\beta. \tag{3.5}$$

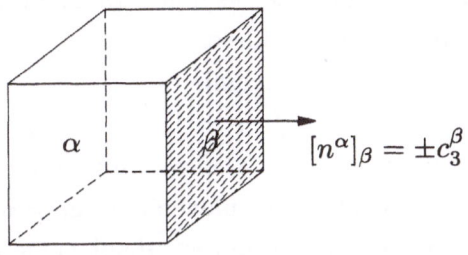

$$[n^\alpha]_\beta = \pm c_3^\beta$$

Fig. 63

3.2 Fundamental members

The following models, contain only three kinds of fundamental elastic members:

1°. One dimensional elastic members, having the deformation modes of slender rods in stretching, in bending, and in torsion (§3, §6, and §7 of Chapter 1).

2. Two dimensional plate members, having the deformation modes of thin plates in stretching and in bending (§1 and §4 of Chapter 3).

3. Three dimensional body members, having the general modes of three dimensional elastic deformations (§5 of Chapter 2). Those compose the general space elastic composite structure and are usually sufficient to deal with the stress problem in ordinary engineering structures.

As a special case, in addition to the plane structures in §2, we also have space plate-rod structures which only contain two-dimensional plate members and one-dimensional rod members, space frames which contain only one-dimensional rod members, and so on.

The geometrical and mechanical essentials of the three kinds of fundamental members are listed in the following. Global coordinates are adopted for body members, but all the variables of plate and rod members are expressed in terms of their local coordinates x_i^β and x_i^γ. At the same time, in order to simplify the notation, when there will be no confusion, we will omit the superscripts β and γ with respect to the local coordinates. The notation in the preceding chapters and sections will continue to be used, with minor modifications uniformity for the sake of uniformity. For example, for the displacement u and the rotation angle ω as well as the load f and the moment m, we shall use the parallel notations

$$\omega_i \equiv u_{3+i}, \; m_i \equiv f_{3+i}, \; i = 1, 2, 3.$$

Here, the rotation angle takes as a generalized displacement, and the moment takes as a generalized force.

1. One dimensional rod members.

Local coordinates: $x_i = x_i^\gamma$, $i = 1, 2, 3$. Longitudinal coordinate is x_1, transverse coordinates are x_2 and x_3, and $x_2 = x_3 = 0$ on the rod.

Fundamental variables: displacements $u_i = u_i^\gamma(x_1), i = 1, 2, 3$. Longitudinal displacement is u_1, transverse displacements are u_2 and u_3 rotation angles $\omega_i = \omega_i^\gamma = u_{3+i}^\gamma$, $i = 1, 2, 3$,

$$\omega_i = u_i^\gamma, \quad \omega_2 = -\frac{du_3}{dx_1}, \quad \omega_3 = \frac{du_2}{dx_1}.$$

Coordinate transformations:

$$\boldsymbol{u} = A^\gamma \boldsymbol{u}^\gamma, \quad \omega = A^\gamma \omega^\gamma, \quad A^\gamma = (a_{ij}^\gamma)_{i,j=1,2,3}.$$

Strains:

tension

$$\varepsilon_{11} = \varepsilon_{11}^\gamma(\boldsymbol{u}) = \frac{du_1}{dx_1},$$

bending

$$K_2 = K_2^\gamma(\boldsymbol{u}) = \frac{d\omega_2}{dx_1} = -\frac{d^2 u_3}{dx_1^2},$$

$$K_3 = K_3^\gamma(\boldsymbol{u}) = \frac{d\omega_3}{dx_1} = \frac{d^2 u_2}{dx_1^2},$$

torsion

$$K_1 = K_1^\gamma(\boldsymbol{u}) = \frac{d\omega_1}{dx_1} = \frac{du_4}{dx_1}.$$

Stresses:

tensile force

$$Q_1 = Q_1^\gamma(\boldsymbol{u}) = EA\varepsilon_{11}(\boldsymbol{u});$$

bending moments

$$M_2 = M_2^\gamma(\boldsymbol{u}) = EI_{22}K_3 + EI_{23}K_3,$$

$$M_3 = M_3^\gamma(\boldsymbol{u}) = EI_{32}K_2 + EI_{33}K_3;$$

twisting moment

$$M_1 = M_1^\gamma(\boldsymbol{u}) = \frac{E}{2(1+\nu)}JK_1 = GJK_1;$$

shearing forces

$$Q_2 = Q_2^{\gamma}(\boldsymbol{u}) = -\frac{dM_3(\boldsymbol{u})}{dx_1},$$

$$Q_3 = Q_3^{\gamma}(\boldsymbol{u}) = \frac{dM_2(\boldsymbol{u})}{dx_1},$$

where, A is the area of the cross section, I_{ij} is the moment of inertia of the cross section, and J is the geometric torsional rigidity of the cross section (Fig. 64).

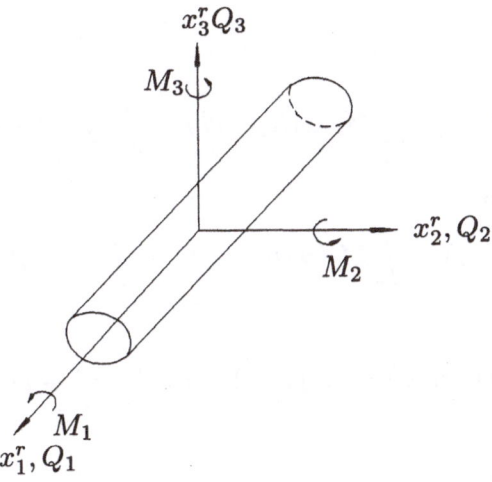

Fig. 64

Strain energies:

   tension

$$\frac{1}{2}D_1^{\gamma}(\boldsymbol{u}, \boldsymbol{u}) = \frac{1}{2}\int_{\gamma} Q_1(\boldsymbol{u})\varepsilon_{11}(\boldsymbol{u})d\gamma = \frac{1}{2}\int_{\gamma} EA\varepsilon_{11}^2(\boldsymbol{u})d\gamma,$$

which depends only on u_1:

   bending

$$\frac{1}{2}D_{23}^{\gamma}(\boldsymbol{u}, \boldsymbol{u}) = \frac{1}{2}\int_{\gamma} \sum_{i=2}^{3} M_i(\boldsymbol{u})K_i(\boldsymbol{u})d\gamma$$

$$= \frac{1}{2}\int_{\gamma} \sum_{i,j=2}^{3} EI_{ij}K_i(\boldsymbol{u})K_j(\boldsymbol{u})d\gamma,$$

which depends only on u_2 and u_3;

   torsion

$$\frac{1}{2}D_4^{\gamma}(\boldsymbol{u}, \boldsymbol{u}) = \frac{1}{2}\int_{\gamma} M_i(\boldsymbol{u})K_1(\boldsymbol{u})d\gamma = \frac{1}{2}\int_{\gamma} GJK_1^2(\boldsymbol{u})d\gamma,$$

which depends only on u_4;

total

$$\frac{1}{2}D^\gamma(\boldsymbol{u}, \boldsymbol{u}) = \frac{1}{2}[D_1^\gamma(\boldsymbol{u}, \boldsymbol{u}) + D_{23}^\gamma(\boldsymbol{u}, \boldsymbol{u}) + D^\gamma(\boldsymbol{u}, \boldsymbol{u})].$$

Functionals of virtual work:

tension

$$D_1^\gamma(\boldsymbol{u}, \boldsymbol{v}) = \int_\gamma Q_1(\boldsymbol{u})\varepsilon_{11}(\boldsymbol{v})d\gamma$$

bending

$$= \int_\gamma EA\varepsilon_{11}(\boldsymbol{u})\varepsilon_{11}(\boldsymbol{v})d\gamma;$$

$$D_{23}^\gamma(\boldsymbol{u}, \boldsymbol{v}) = \int_\gamma \sum_{i=2}^3 M_i(\boldsymbol{u})K_i(\boldsymbol{v})d\gamma = \int_\gamma \sum_{i,j=2}^3 EI_{ij}K_i(\boldsymbol{u})K_j(\boldsymbol{v})d\gamma;$$

torsion

$$D_4^\gamma(\boldsymbol{u}, \boldsymbol{v}) = \int_\gamma M_1(\boldsymbol{u})K_1(\boldsymbol{v})d\gamma = \int_\gamma GJK_1(\boldsymbol{u})K_1(\boldsymbol{v})d\gamma;$$

total

$$D^\gamma(\boldsymbol{u}, \boldsymbol{v}) = D_1^\gamma(\boldsymbol{u}, \boldsymbol{v}) + D_{23}^\gamma(\boldsymbol{u}, \boldsymbol{v}) + D_4^\gamma(\boldsymbol{u}, \boldsymbol{v}).$$

Strainless states:

tension

$$D_1^\gamma(\boldsymbol{u}, \boldsymbol{v}) = 0 \iff \varepsilon_{11} \equiv 0 \iff v_1 = a_1;$$

bending

$$D_{23}^\gamma(\boldsymbol{v}, \boldsymbol{v}) = 0 \iff K_2 = K_3 \equiv 0 \iff v_2 = a_2 + b_3 x_1,$$

$$v_3 = a_3 - b_2 x_1, \quad \omega_2 = b_2, \quad \omega_3 = b_3;$$

torsion

$$D_4^\gamma(\boldsymbol{v}, \boldsymbol{v}) = 0 \iff K_1 \equiv 0 \iff \omega_1 = v_4 = b_1;$$

total

$$D^\gamma(\boldsymbol{v}, \boldsymbol{v}) = 0 \iff v_1 = a_1, \quad v_2 = a_2 + b_3 x_1,$$

$$v_3 = a_3 - b_2 x_1, \quad \omega_1 = b_1, \quad \omega_2 = b_2, \quad \omega_3 = b_3,$$

i.e.,

$$\boldsymbol{v} = [\boldsymbol{a} + \boldsymbol{b} \wedge \boldsymbol{x}]_{x_2=x_3=0} = [\boldsymbol{a} + \boldsymbol{b} \wedge \boldsymbol{x}]_\gamma, \quad \boldsymbol{\omega} = \boldsymbol{b},$$

$$\boldsymbol{a} = (a_1, a_2, a_3)^T, \quad \boldsymbol{b} = (b_1, b_2, b_3)^T, \quad \boldsymbol{\omega} = (\omega_1, \omega_2, \omega_3)^T. \tag{3.6}$$

Loads:

    forces

$$f_i = f_i^\gamma(x_1), \quad i = 1, 2, 3,$$

    moments

$$m_1 = m_1^\gamma(x_1) = f_4^\gamma(x_1).$$

Potential energy of external work:

    tension

$$-F_1^\gamma(\boldsymbol{v}) = -\int_\gamma f_1 v_1 d\gamma,$$

which depends only on v_1;

    bending

$$-F_{23}^\gamma(\boldsymbol{v}) = -\int_\gamma \sum_{i=2}^3 f_i v_i d\gamma,$$

which depends only on v_2 and v_3;

    torsion

$$-F_4^\gamma(\boldsymbol{v}) = -\int_\gamma m_i \omega_1(\boldsymbol{v}) d\gamma = -\int_\gamma f_4 v_4 d\gamma,$$

which depends only on v_4;

    total

$$-F^\gamma(\boldsymbol{v}) = -[F_1^\gamma(\boldsymbol{v}) + F_{23}^\gamma(\boldsymbol{v}) + F_4^\gamma(\boldsymbol{v})] = -\int_\gamma \sum_{i=1}^4 f_i v_i d\gamma. \qquad (3.7)$$

Total potential energy:

$$J^\gamma(\boldsymbol{u}) = \frac{1}{2} D^\gamma(\boldsymbol{u}, \boldsymbol{u}) - F^\gamma(\boldsymbol{u}).$$

Green's formulas:

    tension

$$D_1^\gamma(\boldsymbol{u}, \boldsymbol{v}) = \int_\gamma \left(-\frac{dQ_1(\boldsymbol{u})}{dx_1}\right) v_1 d\gamma + \sum_{\delta \in \partial\gamma} \varepsilon(\gamma, \delta)[Q_1(\boldsymbol{u}) v_1]_\delta;$$

    bending

$$D_{23}^\gamma(\boldsymbol{u}, \boldsymbol{v}) = \int_\gamma \sum_{i=2}^3 \left(-\frac{dQ_i(\boldsymbol{u})}{dx_1}\right) v_i d\gamma$$

$$+ \sum_{\delta \in \partial\gamma} \varepsilon(\gamma, \delta)\left[\sum_{i=2}^2 Q_i(\boldsymbol{u}) v_i + \sum_{i=2}^3 M_i v_{3+i}\right]_\delta;$$

total

$$D^{\gamma}(\boldsymbol{u},\boldsymbol{v}) = \int_{\gamma}\Big[\sum_{i=1}^{3}\Big(-\frac{dQ_i(\boldsymbol{u})}{dx_1}\Big)v_i + \Big(-\frac{dM_1(\boldsymbol{u})}{dx_1}\Big)v_4\Big]d\gamma$$

$$+ \sum_{\delta\in\partial\gamma}\varepsilon(\partial,\delta)\Big[\sum_{i=1}^{3}Q_i(\boldsymbol{u})v_i + \sum_{i=1}^{3}M_i(\boldsymbol{u})v_{3+i}\Big]_{\delta}. \qquad (3.8)$$

Elastic support: When the rod member γ is elastically coupled with the external region, for example, the rod on elastic foundation. The elastic energy of the coupling must be included into the strain energy.

For strain energies:

$$\frac{1}{2}D_1^{\gamma}(\boldsymbol{u},\boldsymbol{u}) \text{ must be increased by } \frac{1}{2}\int_{\gamma}c_1 u_1^2 d\gamma,$$

$$\frac{1}{2}D_{23}^{\gamma}(\boldsymbol{u},\boldsymbol{u}) \text{ must be increased by } \frac{1}{2}\int_{\gamma}\sum_{i,j=2}^{3}c_{ij}u_i u_j d\gamma,$$

$$\frac{1}{2}D_4^{\gamma}(\boldsymbol{u},\boldsymbol{u}) \text{ must be increased by } \frac{1}{2}\int_{\gamma}c_4 u_4^2 d\gamma$$

For functionals of virtual work:

$$D^{\gamma}(\boldsymbol{u},\boldsymbol{v}) \text{ must be increased by } \int_{\gamma}c_1 u_1 v_1 d\gamma,$$

$$D_{23}^{\gamma}(\boldsymbol{u},\boldsymbol{v}) \text{ must be increased by } \int_{\gamma}\sum_{i,j=2}^{3}c_{ij}u_i v_j d\gamma,$$

$$D_4^{\gamma}(\boldsymbol{u},\boldsymbol{v}) \text{ must be increased by } \int_{\gamma}c_4 u_4 v_4 d\gamma,$$

For Green's formula:

The first term in the right side of $D_1^{\gamma}(\boldsymbol{u},\boldsymbol{v})$ must be changed to

$$\int_{\gamma}\Big(-\frac{dQ_1(\boldsymbol{u})}{dx_1}+c_1 u_1\Big)v_1 d\gamma;$$

the first term in the right side of $D_{23}^{\gamma}(\boldsymbol{u},\boldsymbol{v})$ must be changed to

$$\int_{\gamma}\sum_{i=2}^{3}\Big(-\frac{dQ_i(\boldsymbol{u})}{dx_1}+\sum_{j=2}^{3}c_{ij}u_j\Big)v_i d\gamma;$$

and the first term in the right side of $D_4^{\gamma}(\boldsymbol{u},\boldsymbol{v})$ must be changed to

$$\int_{\gamma}\Big(-\frac{dM_1(\boldsymbol{u})}{dx_1}+c_4 u_4\Big)v_4 d\gamma,$$

where, $c_1 = c_1^\gamma \geq 0$, $(c_{ij}) = (c_{ij}^\gamma)_{i,j=2,3}$ is a symmetric positive definite or positive semi-definite matrix, and $c_4 = c_4^\gamma \geq 0$, are the constants of elastic coupling.

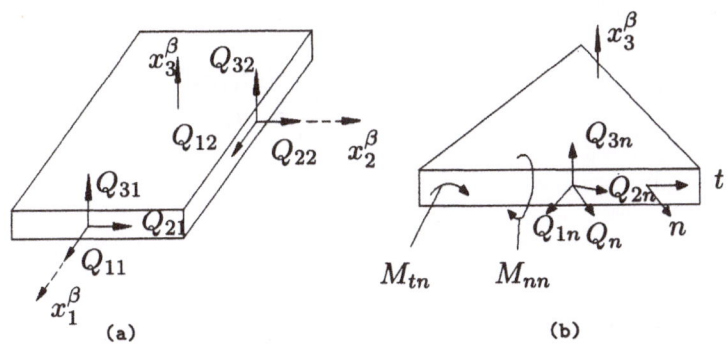

Fig. 65

2.   **Two dimensional plate members.**

Local coordinates: $x_i = x_i^\beta$, $i = 1, 2, 3$. The longitudinal coordinates are x_1 and x_2, the transverse coordinate is x_3, $x_3 = 0$ on the plate member. The fundamental variables: displacements $u_i = u_i^\beta(x_1, x_2)$, $i = 1, 2, 3$.

The longitudinal displacements are u_1 and u_2, the transverse displacement is u_3.

Rotation angles $\omega_i = \omega_i^\beta = u_{i+3}^\beta$, $i = 1, 2, 3$,

$$\omega_1^\beta = \frac{\partial u_3}{\partial x_2}, \quad \omega_2^\beta = -\frac{\partial u_3}{\partial x_1}, \quad \omega_3^\beta = \frac{1}{2}\left(\frac{\partial u_2}{\partial x_1} - \frac{\partial u_1}{\partial x_2}\right).$$

Coordinate transformations

$$\boldsymbol{u} = A^\beta \boldsymbol{u}^\beta, \quad \boldsymbol{\omega} = A^\beta \boldsymbol{\omega}^\beta, \quad A^\beta = (a_{ij}^\beta)_{i,j=1,2,3}.$$

Strains:

tension $\varepsilon_{ij} = \varepsilon_{ij}^\beta(\boldsymbol{u}) = \frac{1}{2}\left(\frac{\partial u_j}{\partial x_i} + \frac{\partial u_i}{\partial x_j}\right)$, $i, j = 1, 2$;

bending $K_{ij} = K_{ij}^\beta(\boldsymbol{u}) = -\frac{\partial^2 u_3}{\partial x_i \partial x_j}$, $i, j = 1, 2$.

Stresses:

tensile forces

$$Q_{ij} = Q_{ij}^\beta(\boldsymbol{u}) = h\sigma_{ij} = D'\left[(1-\nu)\varepsilon_{ij} + \nu\left(\sum_{k=1}^{2}\varepsilon_{kk}\right)\delta_{ij}\right], \quad i, j = 1, 2;$$

bending mements

$$M_{ij} = M_{ij}^{\beta}(\boldsymbol{u}) = D\Big[(1-\nu)K_{ij} + \nu\Big(\sum_{k=1}^{2} K_{kk}\Big)\delta_{ij}\Big], \ i,j = 1,2;$$

shearing force

$$Q_{3j} = Q_{3j}^{\beta}(\boldsymbol{u}) = \sum_{i=1}^{2} \frac{\partial M_{ij}(\boldsymbol{u})}{\partial x_i};$$

tensile rigidity

$$D' = \frac{Eh}{1-\nu^2},$$

flexural rigidity

$$D = \frac{Eh^3}{12(1-\nu^2)},$$

where h is the thickness (Fig. 41 and 65).

Strain energies:

tension

$$\frac{1}{2}D_{12}^{\beta}(\boldsymbol{u},\boldsymbol{u}) = \frac{1}{2}\iint_{\beta}\sum_{i,j=1}^{2} Q_{ij}(u)\varepsilon_{ij}(u)d\beta$$

$$= \frac{1}{2}\iint_{\beta} D'\Big[(1-\nu)\sum_{i,j=1}^{2}\varepsilon_{ij}^2 + \nu\Big(\sum_{k=1}^{2}\varepsilon_{kk}\Big)^2\Big]d\beta,$$

which depend only on $\boldsymbol{u}_1$ and $\boldsymbol{u}_2$;

bending

$$\frac{1}{2}D_{3}^{\beta}(\boldsymbol{u},\boldsymbol{u}) = \frac{1}{2}\iint_{\beta}\sum_{i,j=1}^{2} M_{ij}(\boldsymbol{u})K_{ij}(\boldsymbol{u})d\beta$$

$$= \frac{1}{2}\iint_{\beta} D\Big[(1-\nu)\sum_{i,j=1}^{2}K_{ij}^2 + \nu\Big(\sum_{k=1}^{2}K_{kk}\Big)^2\Big]d\beta,$$

which depend only on u_3;

total

$$\frac{1}{2}D^{\beta}(\boldsymbol{u},\boldsymbol{u}) = \frac{1}{2}[D_{12}^{\beta}(\boldsymbol{u},\boldsymbol{u}) + D_{3}^{\beta}(\boldsymbol{u},\boldsymbol{u})].$$

Functionals of virtual work:

tension

$$D_{12}^{\beta}(\boldsymbol{u}, \boldsymbol{v}) = \iint_{\beta} \sum_{i,j=1}^{2} Q_{ij}(\boldsymbol{u})\varepsilon_{ij}(\boldsymbol{v})d\beta$$

$$= \iint_{\beta} D'\Big[(1-\nu)\sum_{i,j=1}^{2}\varepsilon_{ij}(\boldsymbol{u})\varepsilon_{ij}(\boldsymbol{v})$$

$$+\nu\Big(\sum_{k=1}^{2}\varepsilon_{kk}(\boldsymbol{u})\Big)\Big(\sum_{k=1}^{2}\varepsilon_{kk}(\boldsymbol{v})\Big)\Big]d\beta;$$

bending

$$D_{3}^{\beta}(\boldsymbol{u}, \boldsymbol{v}) = \iint_{\beta} \sum_{i,j=1}^{2} M_{ij}(\boldsymbol{u})K_{ij}(\boldsymbol{v})d\beta$$

$$= \iint_{\beta} D\Big[(1-\nu)\sum_{i,j=1}^{2}K_{ij}(\boldsymbol{u})K_{ij}(\boldsymbol{v})$$

$$+\nu\Big(\sum_{k=1}^{2}K_{kk}(\boldsymbol{u})\Big)\Big(\sum_{k=1}^{2}K_{kk}(\boldsymbol{v})\Big)\Big]d\beta;$$

total

$$D^{\beta}(\boldsymbol{u}, \boldsymbol{v}) = D_{12}^{\beta}(\boldsymbol{u}, \boldsymbol{v}) + D_{3}^{\beta}(\boldsymbol{u}, \boldsymbol{v}).$$

Strainless states:

tension

$$D_{12}^{\beta}(\boldsymbol{v}, \boldsymbol{v}) = 0 \Longleftrightarrow \varepsilon_{ij} \equiv 0 \Longleftrightarrow v_1 = a_1 - b_3 x_2,$$

$$v_2 = a_2 + b_3 x_1, \ \omega_3 = b_3.$$

bending

$$D_{3}^{\beta}(\boldsymbol{v}, \boldsymbol{v}) = 0 \Longleftrightarrow K_{ij} \equiv 0 \Longleftrightarrow v_3 = a_3 - b_2 x_1 + b_1 x_2,$$

$$\omega_1 = b_1, \ \omega_1 = b_2;$$

total

$$D^{\beta}(\boldsymbol{v}, \boldsymbol{v}) = 0 \Longleftrightarrow v_1 = a_1 - b_3 x_2, \ v_2 = a_2 + b_3 x_1,$$

$$v_3 = a_3 - b_2 x_1 + b_1 x_2, \ \omega_1 = b_1, \ \omega_2 = b_2, \ \omega_3 = b_3,$$

i.e.,

$$\boldsymbol{v} = [\boldsymbol{a} + \boldsymbol{b} \wedge \boldsymbol{x}]_{x_3=0} = [\boldsymbol{a} + \boldsymbol{b} \wedge \boldsymbol{x}]_{\beta}, \boldsymbol{\omega} = \boldsymbol{b}. \qquad (3.9)$$

Loads:

   forces

$$f_i = f_i^\beta(x_1, x_2), \ i = 1, 2, 3.$$

Potential energies of external work:

   tension

$$-F_{12}^\beta(v) = -\iint_\beta \sum_{i=1}^2 f_i v_i d\beta;$$

   bending

$$-F_3^\beta(v) = -\iint_\beta f_3 v_3 d\beta;$$

   total

$$-F^\beta(v) = -(F_{12}^\beta(v) + F_3^\beta(v)) = -\iint_\beta \sum_{i=1}^3 f_i v_i d\beta. \qquad (3.10)$$

Total potential energy:

$$J^\beta(u) = \frac{1}{2} D^\beta(u, u) - F^\beta(u).$$

Green's formulas:

   tension

$$D_{12}^\beta(u, v) = -\iint_\beta \sum_{i=1}^2 \sum_{j=1}^2 \frac{\partial Q_{ij}(u)}{\partial x_j} v_i d\beta + \int_{\partial\beta} \sum_{i=1}^2 Q_{in}(u) v_i d\gamma$$

where $Q_{in}(u) = \sum_{j=1}^2 Q_{ij}(u)n_j$ and $n = (n_1, n_2, 0)^T$ is the outward normal direction to $\partial\beta$;

   bending

$$D_3^\beta(u, v) = -\iint_\beta \sum_{j=1}^2 \frac{\partial Q_{3j}(u)}{\partial x_j} v_3 d\beta + \int_{\partial\beta} \left(Q_{3n}(u)v_3 \right.$$

$$\left. - M_{nn}(u)\frac{\partial v_3}{\partial n} - M_{in}(u)\frac{\partial v_3}{\partial l} \right) d\gamma,$$

where

$$Q_{3n}(u) = \sum_{i=1}^2 Q_{3j}(u)n_j, \ M_{nn}(u) = \sum_{i,j=1}^2 M_{ij}(u)n_i n_j,$$

$$M_{in}(u) = \sum_{i,j=1}^2 M_{ij}(u)n_i t_j, \ n_1 = t_2, \ n_2 = -t_1;$$

total

$$D^\beta(\boldsymbol{u}, \boldsymbol{u}) = \iint_\beta \sum_{i=1}^3 \left(-\sum_{j=1}^2 \frac{\partial Q_{ij}(\boldsymbol{u})}{\partial x_j} \right) v_i d\beta$$

$$+ \int_{\partial\beta} \left(\sum_{i=1}^3 Q_{in}(\boldsymbol{u}) v_i - M_{nn}(\boldsymbol{u}) \frac{\partial v_3}{\partial n} - M_{ln}(\boldsymbol{u}) \frac{\partial v_3}{\partial l} \right) d\gamma.$$

$$(3.11)$$

Elastic support: the elastic energy of the coupling must be included into the strain energy when the plate member β is elastically coupled with the external region, as with plate on an elastic foundation.

Strain energies

$$\frac{1}{2} D_{12}^\beta(\boldsymbol{u}, \boldsymbol{u}) \text{ must be increased by } \frac{1}{2} \iint_\beta \sum_{i,j=1}^2 c_{ij} u_i u_j d\beta,$$

$$\frac{1}{2} D_3^\beta(\boldsymbol{u}, \boldsymbol{u}) \text{ must be increased by } \frac{1}{2} \iint_\beta c_3 u_3^2 d\beta.$$

Functionals of virtual work

$$D_{12}^\beta(\boldsymbol{u}, \boldsymbol{v}) \text{ must be increased by } \iint_\beta \sum_{i,j=1}^2 c_{ij} u_i v_j d\beta,$$

$$D_3^\beta(\boldsymbol{u}, \boldsymbol{v}) \text{ must be increased by } \iint_\beta c_3 u_3 v_3 d\beta.$$

Green's formula:

The first term on the right side of $D_{12}^\beta(\boldsymbol{u}, \boldsymbol{v})$ must be changed to

$$\iint_\beta \sum_{i=1}^2 \left[\sum_{j=1}^2 \left(-\frac{\partial Q_{ij}(\boldsymbol{u})}{\partial x_j} + c_{ij} u_j \right) \right] v_i d\beta,$$

the first term on the right side of $D_3^\beta(\boldsymbol{u}, \boldsymbol{v})$ must be changed to

$$\iint_\beta \left(-\sum_{j=1}^2 \frac{\partial Q_{3j}(\boldsymbol{u})}{\partial x_j} + c_3 u_3 \right) v_3 d\beta,$$

where, $(c_{ij}) = (c_{ij}^\beta)_{i,j=1,2}$ is a symmetric positive definite or positive semi-definite matrix, and $c_3 = c_3^\beta \geq 0$, are the constants of elastic coupling.

3.   The three dimensional body member

Local coordinates = global coordinates:

$$x_i = x_i^a, \ i = 1, 2, 3.$$

Fundamental variables: the displacements $u_i = u_i^a(x_1, x_2, x_3)$, $i = 1, 2, 3$; and the rotation angles $\omega_i = u_{3+i}$, $i = 1, 2, 3$;

$$\omega_1 = \omega_{23}, \quad \omega_2 = \omega_{31}, \quad \omega_3 = \omega_{12},$$

$$\omega_{ij} = \frac{1}{2}\left(\frac{\partial u_j}{\partial x_i} - \frac{\partial u_i}{\partial x_j}\right), \quad i, j = 1, 2, 3.$$

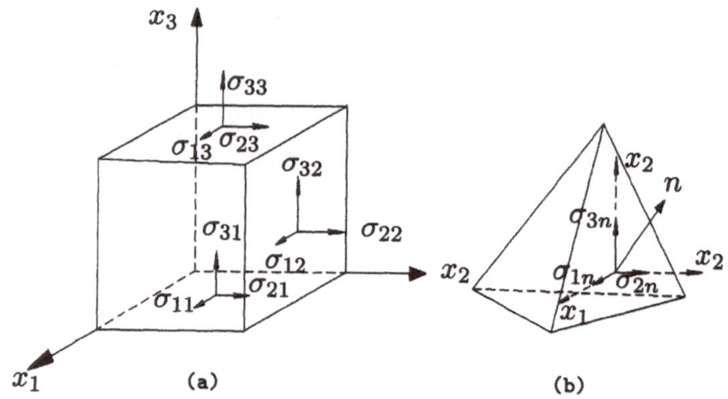

Fig. 66

Coordiante transformation: $\boldsymbol{u} = \boldsymbol{u}^a$.

Strains

$$\varepsilon_{ij} = \varepsilon_{ij}^\alpha(\boldsymbol{u}) = \frac{1}{2}\left(\frac{\partial u_j}{\partial x_i} + \frac{\partial u_i}{\partial x_j}\right), \quad i, j = 1, 2, 3.$$

Stresses

$$\sigma_{ij} = \sigma_{ij}^\alpha(\boldsymbol{u}) = \frac{E}{1+\nu}\varepsilon_{ij} + \frac{E\nu}{(1+\nu)(1-2\nu)}\left(\sum_{k=1}^{3}\varepsilon_{kk}\right)\delta_{ij},$$

$$i, j = 1, 2, 3.$$

(See Fig. 66)

Strain energy:

$$\frac{1}{2}D^\alpha(\boldsymbol{u}, \boldsymbol{u}) = \frac{1}{2}\iiint_\alpha \sum_{i,j=1}^{3} \sigma_{ij}(\boldsymbol{u})\varepsilon_{ij}(\boldsymbol{u})d\alpha$$

$$= \frac{1}{2}\iiint_\alpha \Big[\frac{E}{1+\nu}\sum_{i,j=1}^{3}\varepsilon_{ij}^2(\boldsymbol{u})$$

$$+ \frac{E\nu}{(1+\nu)(1-2\nu)}\left(\sum_{k=1}^{3}\varepsilon_{kk}(\boldsymbol{u})\right)^2\Big]d\alpha.$$

Functional of virtual work:

$$
D^\alpha(\boldsymbol{u}, \boldsymbol{v}) = \iiint_\alpha \sum_{i,j=1}^3 \sigma_{ij}(\boldsymbol{u})\varepsilon_{ij}(\boldsymbol{v})d\alpha
$$

$$
= \iiint_\alpha \Big[\frac{E}{1+\nu} \sum_{i,j=1}^3 \varepsilon_{ij}(\boldsymbol{u})\varepsilon_{ij}(\boldsymbol{v})
$$

$$
+ \frac{E\nu}{(1+\nu)(1-2\nu)} \Big(\sum_{k=1}^3 \varepsilon_{kk}(\boldsymbol{u}) \Big) \times \Big(\sum_{k=1}^3 \varepsilon_{kk}(\boldsymbol{v}) \Big) \Big] d\alpha.
$$

Strainless states:

$$
D^\alpha(\boldsymbol{v}, \boldsymbol{v}) = 0 \iff \varepsilon_{ij} \equiv 0 \iff v_1 = a_1 + b_2 x_3 - b_3 x_2,
$$

$$
v_2 = a_2 + b_3 x_1 - b_1 x_3, \quad v_3 = a_3 + b_1 x_2 - b_2 x_1,
$$

$$
\omega_1 = b_1, \ \omega_2 = b_2, \ \omega_3 = b_3.
$$

i.e.,

$$
\boldsymbol{v} = \boldsymbol{a} + \boldsymbol{b} \wedge \boldsymbol{x}, \ \boldsymbol{a} = (a_1, a_2, a_3)^T,
$$

$$
\boldsymbol{b} = (b_1, b_2, b_3)^T, \quad \boldsymbol{\omega} = \boldsymbol{b}. \tag{3.12}
$$

Loads:

   forces

$$
f_i = f_i^\alpha(x_1, x_2, x_3), \ i = 1, 2, 3.
$$

Potential energy of external work:

$$
-F^\alpha(\boldsymbol{v}) = - \iiint \sum_{i=1}^3 f_i v_i d\alpha. \tag{3.13}
$$

Total potential energy:

$$
J^\alpha(\boldsymbol{u}) = \frac{1}{2} D^\alpha(\boldsymbol{u}, \boldsymbol{u}) - F^\alpha(\boldsymbol{u}).
$$

Green's formulas:

$$
D^\alpha(\boldsymbol{u}, \boldsymbol{v}) = \iiint_\alpha \sum_{i=1}^3 \Big(- \sum_{j=1}^3 \frac{\partial \sigma_{ij}(\boldsymbol{u})}{\partial x_j} \Big) v_i d\alpha + \iint_{\partial\alpha} \sum_{i=1}^3 \sigma_{in}(\boldsymbol{u}) v_i d\beta, \tag{3.14}
$$

where $\sigma_{in}(\boldsymbol{u}) = \sum_{j=1}^3 \sigma_{ij}(\boldsymbol{u}) n_j$ and $\boldsymbol{n} = (n_1, n_2, n_3)^T$ is the outward normal direction to $\partial\alpha$.

3.3   The rigid connection

Rigidly connected space structures, like rigidly connected plane structures, exhibit two kinds of continuity: continuity of displacements and continuity of rotation angles. The continuity of displacements is identical in principle with that of plane structures. However, the continuity of rotation angles is somewhat more complicated than for plane structures. These two kinds of continuity will be illustrated for the various types of members, as follows.

Continuity of displacements (to transfer forces)

1.   Continuity of displacements at a joint of members with the same dimension.

(i) The displacement is continuous where two rod members γ and γ' are connected at the point δ:

$$[u]_{\gamma,\delta} = [u]_{\gamma',\delta}.$$

(ii) The displacement is continuous where two plate members β and β' are connected on the line element γ (not a rod member):

$$[u]_{\beta,\gamma} = [u]_{\beta',\gamma}.$$

(iii) The displacement is continuous when two body members α and α' are connected on the area element β (not a plate member):

$$[u]_{\alpha,\beta} = [u]_{\alpha',\beta}.$$

Accordingly, there exist single-valued displacements $[u]_\delta, [u]_\gamma$, and $[u]_\beta$ on the boundary elements $\delta \in \Gamma^0, \gamma \in \Gamma^1$ and $\beta \in \Gamma^2$, which satisfy:

$[u]_\delta = [u]_{\gamma,\delta}$ for any rod member γ intersecting
                 at the point element δ;

$[u]_\gamma = [u]_{\beta,\gamma}$ for any plate member β intersecting
                 on the line element γ;

$[u]_\beta = [u]_{\alpha,\beta}$ for any body element α intersecting
                 on the area element β.

2.   Continuity of displacements at joints of members with different dimensions.

(i) The displacement is continuous where a rod member γ and a plate member β are connected (i.e., $\gamma \in \partial\beta$):

$$[u]_\gamma = [u]_{\beta,\gamma}.$$

(ii) The displacement is continuous where a plate member β and a body member α are connected (i.e., $\beta \in \partial\alpha$):

$$[\boldsymbol{u}]_\beta = [\boldsymbol{u}]_{\alpha,\beta}.$$

According, there exist the single-valued displacements $[\boldsymbol{u}]_\gamma$ and $[\boldsymbol{u}]_\beta$ on every rod member $\gamma \in \Omega^1$ and plate member $\beta \in \Omega^2$.

Continuity of rotation angles (to transfer moments)

1. ˙ The rotation angle is continuous where two rod members γ and γ' are connected at the point element δ:

$$[\boldsymbol{\omega}]_{\gamma,\delta} = [\boldsymbol{\omega}]_{\gamma',\delta}.$$

Accordingly, at every point element $\delta \in \Gamma^0$, there exists a single-valued rotation angle $[\boldsymbol{\omega}]_\delta$ satisfying

$[\boldsymbol{\omega}]_\delta = [\boldsymbol{\omega}]_{\gamma,\delta}$, for any rod member γ intersecting at point δ.

Note that all three components of the rotation angle ω_i, $i = 1, 2, 3$ are required to be continuous here, while in the plane structure only the component ω_3 is required to be continuous.

2.   The tangential angle of rotation (i.e., the angle of rotation about the unit coordinate vector $\boldsymbol{e}_1^\gamma$ of the line element γ) is continuous where two plate members β and β' intersect on the line element γ (not a rod member):

$$[\boldsymbol{\omega}]_{\beta,\gamma} \cdot \boldsymbol{e}_1^\gamma = [\boldsymbol{\omega}]_{\beta',\gamma} \cdot \boldsymbol{e}_1^\gamma,$$

where the notation "·" denotes the "dot" or inner product of these two vectors, i.e., the length of projection of the vector $[\omega]_{\beta,\gamma}$ on the direction of $\boldsymbol{e}_1^\gamma$. By (2.6) in §2, we have

$$[\boldsymbol{\omega}]_{\beta,\gamma} \cdot \boldsymbol{e}_1^\gamma = \varepsilon(\beta,\gamma)[\boldsymbol{\omega}]_{\beta,\gamma} \cdot [\boldsymbol{t}^\beta]_\gamma.$$

Moreover, we know from (4.20) in §4 of Chapter 3 that the tangential angle of rotation with respect to the boundary $\partial\beta$ on γ satisfies

$$[\boldsymbol{\omega}]_{\beta,\gamma} \cdot [\boldsymbol{t}^\beta]_\gamma = [\omega_\gamma^\beta]_\gamma = -\left[\frac{\partial u_3^\beta}{\partial n^\beta}\right]_\gamma,$$

so that

$$\varepsilon(\beta,\gamma)\left[\frac{\partial u_3^\beta}{\partial n^\beta}\right]_\gamma = \varepsilon(\beta',\gamma)\left[\frac{\partial u_3^{\beta'}}{\partial n^{\beta'}}\right]_\gamma.$$

Accordingly, on the boundary line element $\gamma \in \Gamma^1$, there exists a single-valued tangential angle of rotation ω_1^γ satisfying

$$\omega_1^\gamma = [\boldsymbol{\omega}]_{\beta,\gamma} \cdot \boldsymbol{e}_1^\gamma = -\varepsilon(\beta,\gamma)\left[\frac{\partial u_3^\beta}{\partial n^\beta}\right]_\gamma,$$

$$(3.15)$$

for any plate member β intersecting at γ.

3. The tangential angle of rotation is continuous where a rod member γ and a plate member β are connected (i.e., $\gamma \in \partial\beta$):

$$\omega_1^\gamma = [\boldsymbol{\omega}]_{\beta,\gamma} \cdot \boldsymbol{e}_1^\gamma = -\varepsilon(\beta,\gamma)\left[\frac{\partial u_3^\beta}{\partial n^\beta}\right]_\gamma.$$

Accordingly, there exists a single-valued tangential angle of rotation ω_1^γ on the rod $\gamma \in \Omega^1$.

Note that, in the plane structure, the continuity of rotation angles is need not hold ingeneral at a joint of plates and plates or of plates and rods (see the two examples in Section 2.3 of §2). In the space structure, according to 4 and 5 above, the tangential angle of rotation at a joint of plates and plates or of plates and rods must be continuous. The rotation angles in other two directions need not be continuous. Continuity of the rotation angle at a joint of bodies and bodies as well as of bodies and plates also need not hold in general for the same reason as in the corresponding examples for plane structures. The rigid connection conditions 1–5 are themselves sufficient to guarantee that the strainless state of the whole composite structure is just a global rigid displacement.

3.4   Boundary conditions

Boundary conditions can be imposed on the boundary of a composite structure Ω no matter whether it is a two dimensional Γ^2, a one-dimensional Γ^1, or a zero dimensional Γ^0. In general, they are of three kinds: The first kind is the geometric boundary condition, which prescribes the displacement and/or the rotation angle. The second kind is the mechanical boundary condition. The third kind is the elastic support. The second and the third kinds may exist simultaneously. In the variational principle formulation, only the first kind of the boundary condition must be listed out as an essential geometric constraint, while the second and the third kinds are natural boundary conditions and it is only necessary to absorb the corresponding loads and elastic energy of couplings into the potential energy of external work and into the strain energy, respectively.

We illustrate boundaries of various dimensions as follows.

1.   The two dimensional boundary Γ^2, $\beta \in \Gamma^2$.

According to the theory of three dimensional elasticity, the displacements or only a subset of them can be prescribed as geometric boundary conditions:

$$\beta : u_i^\beta = \bar{u}_i^\beta, \ i = 1, 2, 3.$$

β may also be subjected to area force loads f_i^β, $i = 1, 2, 3$. Then the potential energy of external work is

$$-F^\beta(v) = -\iint_\beta \sum_{i=1}^3 f_i^\beta v_i^\beta d\beta = -\iint_\beta \sum_{i=1}^3 [f_i^\beta]_\beta [v_i^\beta]_\beta d\beta. \qquad (3.16)$$

If β has an elastic support and elastic reaction forces per unit area is

$$-\sum_{j=1}^3 c_{ij}^\beta u_j^\beta, \ i = 1, 2, 3, \text{ then}$$

the strain energy

$$\frac{1}{2} D^\beta(u, u) \text{ is increased by } \frac{1}{2} \iint_\beta \sum_{i,j=1}^3 c_{ij}^\beta u_i^\beta u_j^\beta d\beta,$$

the functional of virtual work

$$D^\beta(u, v) \text{ is increased by } \iint_\beta \sum_{i,j=1}^3 c_{ij}^\beta u_i^\beta v^\beta d\beta.$$

Due to the continuity of displacements for rigid connections, the displacement is also single-valued even if β is an internal interface of a body member. Hence the foregoing statement about boundary conditions is still correct. However, the internal interface of a body member in three-dimensional space is unreachable, and the conditions listed above are in practice difficult to realize at such an interface, so these boundary conditions are generally imposed only on the external boundary of a body member.

2.   The one dimensional boundary Γ^1, $\gamma \in \Gamma^1$.

According to the theories of thin plates in stretching and in bending, the displacements and the tangential angle of rotation or only a subset of them can be prescribed as geometric boundary conditions:

$$\gamma : u_i^\gamma = \bar{u}_i^\gamma, \ i = 1, 2, 3, \ \omega_1^\gamma = \bar{\omega}_1^\gamma.$$

If γ is subjected to line force loads f_i^γ, $i = 1, 2, 3$, then the potential energy of external work is

$$-F_{123}^\gamma(\varepsilon) = -\int_\gamma \sum_{i=1}^3 f_i^\gamma v_i^\gamma d\gamma = -\int_\gamma \sum_{i=1}^3 [f_i^\gamma]_\gamma [v_i^\gamma]_\gamma d\gamma. \qquad (3.17)$$

If γ is also subjected to an axial moment load $m_1^\gamma = f_4^\gamma$, then the potential energy of external work is

$$-F_4^\gamma(v) = -\int_\gamma f_4^\gamma v_4^\gamma d\gamma = \int_\gamma [f_4^\gamma]_\gamma [v_4^\gamma]_\gamma d\gamma. \qquad (3.18)$$

If γ has an elastic support and elastic reaction forces per unit length are $-\sum_{j=1}^3 c_{ij}^\gamma u_j^\gamma$, $i = 1, 2, 3$, then

the strain energy

$$\frac{1}{2} D_{123}^\gamma(u, u) \text{ is increased by } \frac{1}{2} \int_\gamma \sum_{i,j=1}^3 c_{ij}^\gamma u_i^\gamma u_j^\gamma d\gamma,$$

the functional of virtual work

$$D_{123}^\gamma(u, v) \text{ is increased by } \int_\gamma \sum_{i,j=1}^3 c_{ij} u_i^\gamma v_j^\gamma d\gamma.$$

If an axial elastic reaction moment subjected to a unit length is $-c_4^\gamma u_4^\gamma$, then

the strain energy

$$\frac{1}{2} D_4^\gamma(u, u) \text{ is increased by } \frac{1}{2} \int_\gamma c_4^\gamma u_4^2 d\gamma,$$

the functional of virtual work

$$D_4^\gamma(u, v) \text{ is increased by } \int_\gamma c_4^\gamma u_4 v_4 d\gamma.$$

If γ is a common interface line of several plates, since the rigid connection condition guarantees that the displacement and the rotation angle are single-valued, the foregoing statement about boundary condition is still correct for this case. In the there dimensional space, the interface line of plates is generally reachable so long as it is not enclosed in three dimensional body members, so that the conditions listed above are realizable.

3.   The null-dimensional boundary γ^0 and $\delta \in \Gamma^0$.

According to the theories of slender rods in stretching, in bending, and in torsion, the displacements and the rotation angles or only a part of them can be prescribed as geometric boundary conditions:

$$\delta: \ u_i = \bar{u}_i^\delta, \ i = 1, 2, 3, \ \omega_i = \bar{\omega}_i^\delta, \ i = 1, 2, 3.$$

If δ is subjected to a point force load $[f_i]_\delta$ and moment loads $[m_i]_\delta = [f_{3+i}]_\delta, \ i = 1, 2, 3,$ then the potential energies of external work are

$$-F_{123}^\delta(v) = -\sum_{i=1}^{3} [f_i]_\delta [v_i]_\delta, \tag{3.19}$$

$$-F_{456}^\delta(v) = -\sum_{i=1}^{3} [f_{3+i}]_\delta [v_{3+i}]_\delta. \tag{3.20}$$

If δ also has an elastic support with the elastic reaction forces

$$-\sum_{j=1}^{3} c_{ij}^\delta [u_j]_\delta, \ i = 1, 2, 3,$$

and the elastic reaction moments

$$-\sum_{i=1}^{3} c_{3+i,3+j}^\delta [u_{3+j}]_\delta, \ i = 1, 2, 3,$$

then for the strain energies:

$$\frac{1}{2} D_{123}^\delta(u, u) \text{ is increased by } \frac{1}{2} \sum_{i,j=1}^{3} c_{ij}^\delta [u_i]_\delta [u_j]_\delta,$$

$$\frac{1}{2} D_{456}^\delta(u, u) \text{ is increased by } \frac{1}{2} \sum_{i,j=1}^{3} c_{3+i,3+j}^\delta [u_{3+i}]_\delta [u_{3+j}]_\delta,$$

and for the functionals of virtual:

$$D_{123}^\delta(u, v) \text{ is increased by } \sum_{i,j=1}^{3} c_{ij}^\delta [u_i]_\delta [v_i]_\delta,$$

$$D_{456}^\delta(u, v) \text{ is increased by } \sum_{i,j=1}^{3} c_{3+i,3+j}^\delta [u_{3+i}]_\delta [v_{3+j}]_\delta.$$

Since the rigid connection guarantees the monodromy of the displacement and of the rotation angle at a common intersecting point of rods, the above conditions also hold for a common intersecting point δ of several rods and are realizable.

In all of the above mentioned cases, whenever the geometric constraint u_i (including the displacement and the rotation angle) is prescribed, the corresponding load component f_i is out of action.

3.5   The pinned connection

The pinned connection is a kind of elastic connection which relaxes the continuity of the rotation angle. For space structures, we see from the continuity conditions 3–5 for the rotation angle in Section 3.3 that the pinned connection may exist at a common line element of plate and plate as well as of plate and rod and at a common point element of rod and rod. At such a pin-connected line or point, the rotation angle will not be single-valued and consequently only the force but not the moment will be transferred.

Suppose γ is a pin-connected line element, and $\partial^{-1}\gamma = \{\beta_1, \cdots, \beta_4\}$ is a set of all the plate members with γ as the common boundary line. We discuss the following three cases.

1.   The intersecting line γ is not a rod member, i.e., $\gamma \in \gamma \in \Gamma^1$.

Suppose the plate members in $\partial^{-1}\gamma$ are divided into p sets of rigidly connected members $\partial_j^{-1}\gamma, j = 1, 2, \cdots, p$:

$$\partial^{-1}\gamma = \{\partial_1^{-1}\gamma, \cdots, \partial_p^{-1}\gamma\} = \{\beta_1, \cdots, \beta_q\}.$$

This corresponds to a completely rigid connection when $p = 1$, and to a completely pinned connection when $p = q$. The plates are mutually rigid-connected in each group $\partial_j^{-1}\gamma$, so that they have a common value of the tangential angle of rotation (expressed in terms of the local coordinates of γ) on the boundary γ

$$[\omega_1^\gamma]_{\partial_j^{-1}\gamma,\gamma} \equiv [v_4^\gamma]_{\partial_i^{-1}\gamma,\gamma} = -\varepsilon(\beta,\gamma)\left(\frac{\partial v_3^\beta}{\partial n^\beta}\right)_\gamma, \quad \text{for any } \beta \in \partial_j^{-1}\gamma.$$

Since the rotation angles are free among the various sets, there are altogether p independent values of the tangential angle of rotation

$$[v_4^\gamma]_{\partial_j^{-1}\gamma,\gamma}, \quad j = 1, \cdots, p$$

instead of a single angle of rotation $[v_4^\gamma]_\gamma$ for the rigid connection. In that case, there is also no single tangential moment load $[m_1^\gamma]_\gamma \equiv [f_4^\gamma]_\gamma$ Rather, there is a distinct load for each rigidly connected set:

$$[m_1^\gamma]_{\partial_j^{-1}\gamma,\gamma} \equiv [f_4^\gamma]_{\partial_j^{-1}\gamma,\gamma}, \quad j = 1, 2, \cdots, p.$$

Consequently, instead of (3.18), the potential energy of external work takes the form

$$-F_4^\gamma(v) = -\int_\gamma \sum_{j=1}^p [f_4^\gamma]_{\partial_j^{-1}\gamma,\gamma}[v_4^\gamma]_{\partial_j^{-1}\gamma,\gamma}d\gamma, \qquad (3.21)$$

where

$$[v_4^\gamma]_{\partial_j^{-1}\gamma,\gamma} = -v(\beta,\gamma)\left[\frac{\partial v_3^\beta}{\partial n^\beta}\right]_\gamma, \quad \text{for any } \beta \in \partial_j^{-1}\gamma.$$

2.   The line of intersection γ is a rod member $\gamma \in \Omega^1$ and the rod member and the intersecting plates are all pin-connected.

The difference between this case and 1 is that the rod γ has its own independent value of the tangential angle of rotation v_1^γ. Hence, it can be independently subjected to a load f_4^γ whose potential energy of external work is $-\int_\gamma [f_4^\gamma]_\gamma[v_4^\gamma]_\gamma dl$, then the total potential energy of external work is

$$-F_4^\gamma(v) = -\int_\gamma \left([f_4^\gamma]_\gamma[v_4^\gamma]_\gamma + \sum_{j=1}^p [f_4^\gamma]_{\partial_j^{-1}\gamma,\gamma}[v_4^\gamma]_{\partial_j^{-1}\gamma,\gamma}\right)d\gamma. \qquad (3.22)$$

3.   The line of intersection γ is a rod member and is rigid-connected with at least one of the intersecting plates.

For the sake of simplicity, assume that $\partial^{-1}\gamma$ is divided into a total of $p+1$ sets rigidly connected members, while the rod member γ is rigidly connected with the last set of the rigid connection $\partial_{p+1}^{-1}\gamma$. Then both of them have a common value of the rotation angle $[v_4^\gamma]_\gamma = [v_4^\gamma]_{\partial_{p+1}^{-1}\gamma,\gamma}$, and are jointly subjected to the load $[f_4^\gamma]_\gamma$. Thus, the case is identical with 2, and consequently the form of the potential energy of external work is identical with (3.32).

For the above three cases, the moment loads is all vanish, so the potential energy of external work is

$$-F_4^\gamma(v) = 0.$$

The values of the tangential angle of rotation $[v_4^\gamma]_{\partial_j^{-1}\gamma,\gamma}$, $j = 1,\cdots,p$ can in principle also be prescribed as geometric constraint conditions.

For the pinned connection among the rod members, the situation becomes even more complicated. Suppose δ is a pin-connected point, and $\partial^{-1}\delta = \{\gamma_1,\cdots,\gamma_q\}$ is the set of the rod members which contain the point δ. Each rod member γ_i has its own boundary values for the three components of rotation angle vector at the point δ (expressed uniformly in global coordinates at point δ), hence there are altogether $3q$ components at the point δ. In the case of a completely rigid connection, all of the corresponding components are equal, and there are only three independent values. For the general pinned connection, however, there may be various combinations of rigid and pinned connections.

In the following, only a simple case will be discussed. Suppose the rod members in $\partial^{-1}\delta$ are divided into p sets $\partial_j^{-1}\cdots$, $j = 1,\cdots,p$:

$$\partial^{-1}\delta = \{\partial_1^{-1}\delta,\cdots,\partial_p^{-1}\delta\} = \{\gamma_1,\cdots,\gamma_4\}.$$

The rod members within each set $\partial_j^{-1}\delta$ are all mutually rigid-connected, so that they have a common rotation angle vector $[\boldsymbol{\omega}]_{\partial_j^{-1}\delta,\delta}$ at the point δ. The various sets are mutually pin connected, i.e., the rotation angle vectors are independent of one another. Hence, at the point δ there are altogether γ rotation angle vectors $[\boldsymbol{\omega}]_{\partial_j^{-1}\delta,\delta}$, $j = 1,\cdots,p$. Correspondingly, at the point δ there may be p moment loads $[\boldsymbol{m}]_{\partial_j^{-1}\delta,\delta}$ imposed on the individual rigidly connected groups $\partial_j^{-1}\delta$, $j = 1,\cdots,p$, respectively. Therefore, at the point δ, the potential energy of external work is

$$-F_{456}^\delta(\boldsymbol{v}) = -\sum_{i=1}^{3}\sum_{j=1}^{p}[f_{3+i}]_{\partial_j^{-1}\delta,\delta}[v_{3+i}]_{\partial_j^{-1}\delta,\delta} \qquad (3.23)$$

instead of the (3.20) for the rigid connection.

Usually these moment loads all vanish, hence $F_{456}^\delta(\boldsymbol{v}) = 0$. In principle, the values of the rotation angle $[\boldsymbol{\omega}]_{\partial_j^{-1}\delta,\delta}$, $j = 1,\cdots,p$ can also be prescribed as geometric constraint conditions at the pin-connected point δ.

3.6   Variational principles

Suppose Ω is a rigidly connected elastic composite structure. Let Ω^3, Ω^2 and Ω^1 stand for the sets of three, two, and one dimensional members, and let Γ^2, Γ^1 and Γ^0 stand for its two, one, and null dimensional boundaries, respectively.

In Section 3.2, the functionals of strain energy were established for individual members, as well as functional of the elastic energy of coupling for an elastic support on the boundary. Suppose that loads have been imposed on the individual members and on the boundary elements; these induce a corresponding potential energy of external work. The total strain energy $\frac{1}{2}D(\boldsymbol{u},\boldsymbol{u})$ and the total potential energy of external work $-F(\boldsymbol{u})$ of the structure Ω are the sums of the individual local quantities, respectively,

$$
\frac{1}{2}D(u, u) = \frac{1}{2}\Big[\sum_{\alpha \in \Omega^3} D^\alpha(u, u) + \sum_{\beta \in \Omega^2 + \Gamma^2} D^\beta(u, u)
$$

$$
+ \sum_{\gamma \in \Omega^1 + \Gamma^1} D^\gamma(u, u) + \sum_{\delta \in \Gamma^0} D^\delta(u, u) \Big], \tag{3.24}
$$

$$
D(u, v) = \sum_{\alpha \in \Omega^3} D^\alpha(u, v) + \sum_{\beta \in \Omega^2 + \Gamma^2} D^\beta(u, v)
$$

$$
+ \sum_{\gamma \in \Omega^1 + \Gamma^1} D^\gamma(u, v) + \sum_{\delta \in \Gamma^0} D^\delta(u, v), \tag{3.25}
$$

$$
F(u) = \sum_{\alpha \in \Omega^3} F^\alpha(u) + \sum_{\beta \in \Omega^2 + \Gamma^2} F^\beta(u) + \sum_{\gamma \in \Omega^1 + \Gamma^1} F^\gamma(u)
$$

$$
+ \sum_{\delta \in \Gamma^0} F^\delta(u). \tag{3.26}
$$

The total potential energy of the whole structure Ω is

$$
J(u) = \frac{1}{2}D(u, u) - F(u).
$$

The displacements u and v denote compatible generalized displacement fields defined on the whole structure Ω, respectively. Their local values on the volume, area, line, and point elements are just the displacements and the rotation angles in the local coordinates of the individual kinds of members. By a generalized displacement we mean the displacement and the rotation angle components of the different kinds of members (for example, a rod element has three local displacement components and one component of the tangential angle of rotation). By compatibility we mean that the continuity conditions 1–5 for rigid connection in Section 3.3 are satisfied.

We can still impose geometric boundary conditions on the generalized displacement of various members on the boundaries $\Gamma = \{\Gamma^2, \Gamma^1, \Gamma^0\}$ of structures. However, the total strain energy is clearly nonnegative no matter what the prescribed geometric constraint is.

According to the variational principle, the displacement field of the equilibrium configuration u under the prescribed geometric constraint must make the total potential energy $J(u)$ a minimum

$$
J(u) = \frac{1}{2}D(u, u) - F(u) = \text{Min.} \tag{3.27}
$$

This is equivalent to the principle of virtual work

$$D(\boldsymbol{u}, \boldsymbol{v}) - F(\boldsymbol{v}) = 0, \text{ for any virtual displacement } \boldsymbol{v}. \qquad (3.28)$$

3.7   Equilibrium equations

The equations of elastic equilibrium inside the individual members of the structure and on its boundary can, as usual, be derived from the variational principle using Green's formula (3.28). Since the elastic coupling among the members occupies an important place in the composite structure, it is necessary to modify the Green's formula for various members into the form more suitable to the coupling.

For a three-dimensional body member α, from (3.14), Green's formula becomes

$$D^\alpha(\boldsymbol{u}, \boldsymbol{v}) = \iiint_\alpha \sum_{i=1}^3 \left(-\sum_{j=1}^3 \frac{\partial \sigma_{ij}(\boldsymbol{u})}{\partial x_j} \right) v_i d\alpha + \iint_{\partial \alpha} \sum_{j=1}^3 \sigma_{jn}(\boldsymbol{u}) v_j d\beta.$$

Expressing the boundary integral in terms of the sum of the integrals of individual area elements on the boundary and considering the transformation of local coordinates between the body element and the area element,

$$v_j = \sum_{i=1}^3 a_{ji}^\beta v_i^\beta,$$

we have

$$D^\alpha(\boldsymbol{u}, \boldsymbol{v}) = \iiint_\alpha \sum_{i=1}^3 \left(-\sum_{j=1}^2 \frac{\partial \sigma_{ij}(\boldsymbol{u})}{\partial x_j} \right) [v_i]_\alpha d\alpha$$

$$+ \sum_{\beta \in \partial \alpha} \iint_\beta \sum_{i=1}^3 \left(\sum_{j=1}^3 \sigma_{jn}(\boldsymbol{u}) a_{ji}^\beta \right) [v_i]_{\alpha, \beta}^\beta d\beta. \qquad (3.29)$$

For a two-dimensional plate member β, Green's formula (3.11) is

$$D^\beta(\boldsymbol{u}, \boldsymbol{v}) = \iint_\beta \sum_{i=1}^3 \left(-\sum_{j=1}^2 \frac{\partial Q_{ij}^\beta(\boldsymbol{u})}{\partial x_j^\beta} \right) v_i^\beta d\beta$$

$$+ \int_{\partial \beta} \left(\sum_{j=1}^3 Q_{jn}^\beta(\boldsymbol{u}) v_j^\beta - M_{nn}^\beta \frac{\partial v_3^\beta}{\partial n^\beta} - M_{jn}^\beta \frac{\partial v_3^\beta}{\partial l} \right) d\gamma.$$

Integrating the third term of the line integral in the above expression once by parts and transforming the local displacements on the plate member β into the global displacements at the point element δ, we have

$$-\int_{\partial\beta} M_{tn}^\beta \frac{\partial v_3^\beta}{\partial l}d\gamma = -\sum_{\delta\in\partial\gamma}\sum_{\gamma\in\partial\beta}\varepsilon(\beta,\gamma)\varepsilon(\gamma,\delta)[M_{tn}^\beta v_3^\beta]_{\gamma,\delta} + \int_{\partial\beta}\frac{\partial M_{tn}^\beta}{\partial l}v_3^\beta d\gamma$$

$$= -\sum_{\delta\in\partial\gamma}\sum_{\gamma\in\partial\beta}\varepsilon(\beta,\gamma)\varepsilon(\gamma,\delta)\times\sum_{i=1}^{3}[M_{tn}^\beta]_{r,\delta}a_{i3}^\beta[v_i]_{\beta,\gamma,\delta}$$

$$+\sum_{\gamma\in\partial\beta}\int_\gamma\frac{\partial M_{tn}^\beta}{\partial l}v_3^\beta d\gamma.$$

Using the transformation formulas

$$v_j^\beta = \sum_{i=1}^{3}\sum_{k=1}^{3}a_{kj}^\beta a_{ki}^\gamma v_i^\gamma,$$

which transform the local displacements on the plate element p contained in the individual terms of the line integral into the local displacements on the line element γ, and using (3.15), i.e.

$$\left[\frac{\partial v_3^\beta}{\partial n^\beta}\right] = -\varepsilon(\beta,\gamma)\omega_1^\gamma = -\varepsilon(\beta,\gamma)v_4^\gamma,$$

we obtain

$$D^\beta(\boldsymbol{u},\boldsymbol{v}) = \iint_\beta\sum_{i=1}^{3}\left(-\sum_{j=1}^{2}\frac{\partial Q_{ij}^\beta(\boldsymbol{u})}{\partial x_j^\beta}\right)[v_i^\beta]_\beta d\beta$$

$$+\sum_{\gamma\in\partial\beta}\int_\gamma\Big[\sum_{i=1}^{3}\Big(\sum_{j=1}^{3}\sum_{k=1}^{3}Q_{jn}^\beta(\boldsymbol{u})a_{kj}^\beta a_{ki}^\gamma$$

$$+\sum_{k=1}^{3}\frac{\partial M_{tn}^\beta}{\partial l}a_{k3}^\beta a_{ki}^\gamma\Big)[v_i^\gamma]_{\beta,\gamma}$$

$$+\varepsilon(\beta,\gamma)M_{nn}^\beta[v_4^\gamma]_{\beta,\gamma}\Big]d\gamma - \sum_{\delta\in\partial\gamma}\sum_{\gamma\in\partial\beta}\varepsilon(\beta,\gamma)$$

$$\times\varepsilon(\gamma,\delta)\sum_{i=1}^{3}[M_{tn}^\beta]_{\gamma,\delta}a_{i3}^\beta[v_i]_{\beta,\gamma,\delta}. \tag{3.30}$$

For the one dimensional rod member γ, (3.8) Green's formula (3.8) is

$$D^\gamma(\boldsymbol{u},\boldsymbol{v}) = \int_\gamma\Big[\sum_{i=1}^{3}\Big(-\frac{dQ_i^\gamma(\boldsymbol{u})}{dx_1^\gamma}\Big)v_i^\gamma + \Big(-\frac{dM_1^\gamma(\boldsymbol{u})}{dx_1^\gamma}\Big)v_4^\gamma\Big]d\gamma$$

$$+\sum_{\delta\in\partial\gamma}\varepsilon(\gamma,\delta)\Big[\sum_{j=1}^{3}Q_j^\gamma(\boldsymbol{u})v_j^\gamma + \sum_{j=1}^{3}M_j^\gamma(\boldsymbol{u})v_{3+j}^\gamma\Big]_i.$$

Transforming the local displacements on the line element γ in the second term into the global displacements at the point element δ

$$v_j^\gamma = \sum_{i=1}^3 a_{ij}^\gamma v_i, \quad v_{3+j}^\gamma = \omega_j^\gamma = \sum_{i=1}^3 a_{ij}^\gamma \omega_i = \sum_{i=1}^3 a_{ij}^\gamma v_{3+i},$$

we have

$$D^\gamma(\boldsymbol{u}, \boldsymbol{v}) = \int_\gamma \left[\sum_{i=1}^3 \left(-\frac{dQ_i^\gamma(\boldsymbol{u})}{dx_1^\gamma} \right) [v_i^\gamma]_\gamma + \left(-\frac{dM_1^\gamma(\boldsymbol{u})}{dx_1^\gamma} \right) \right.$$

$$\times [v_4^\gamma]_\gamma \bigg] d\gamma + \sum_{\delta \in \partial \gamma} \varepsilon(\gamma, \delta) \bigg[\sum_{i=1}^3 \sum_{j=1}^3 Q_i^\gamma(\boldsymbol{u})$$

$$\left. \times a_{ij}^\gamma [v_i]_{\gamma,\delta} + \sum_{i=1}^3 \sum_{j=1}^3 M_j^\gamma(\boldsymbol{u}) a_{ij}^\gamma [v_{3+i}]_{\gamma,\delta} \right]. \tag{3.31}$$

Adding up the Green's formulas (3.29)–(3.31) for the above various members and the corresponding potential energies of external work, and noting the rigid connection conditions 1–5, finally we obtain

$$D(\boldsymbol{u}, \boldsymbol{v}) - F(\boldsymbol{v})$$

$$= \sum_{\alpha \in \Omega^3} \iiint_\alpha \sum_{i=1}^3 \left[-\sum_{j=1}^3 \frac{\partial \sigma_{ij}(\boldsymbol{u})}{\partial x_j} - [f_i]_\alpha \right] [v_i]_\alpha d\alpha$$

$$+ \sum_{\beta \in \Omega^2} \iint_\beta \sum_{i=1}^3 \left[-\sum_{j=1}^2 \frac{\partial Q_{ij}^\beta(\boldsymbol{u})}{\partial x_j^\beta} \right.$$

$$+ \sum_{\alpha \in \partial^{-1}\beta} \sum_{j=1}^3 \sigma_{jn}(\boldsymbol{u}) a_{ji}^\beta - [f_i^\beta]_\beta \bigg] [v_i^\beta]_\beta d\beta \quad .$$

$$+ \sum_{\beta \in \Gamma^2} \iint_\beta \sum_{i=1}^3 \left[\sum_{\alpha \in \partial^{-1}\beta} \sum_{j=1}^3 \sigma_{jn}(\boldsymbol{u}) a_{ji}^\beta - [f_i^\beta]_\beta \right] [v_i^\beta]_\beta d\beta \quad .$$

$$+ \sum_{\gamma \in \Omega^1} \int_\gamma \left\{ \sum_{i=1}^3 \left[-\frac{dQ_j^\gamma(\boldsymbol{u})}{dx_1^\gamma} + \sum_{\beta \in \partial^{-1}\gamma} \left(\sum_{j=1}^3 \sum_{k=1}^3 Q_{jn}^\beta(\boldsymbol{u}) a_{kj}^\beta a_{ki}^\gamma \right. \right. \right.$$

$$\left. + \sum_{k=1}^3 \frac{\partial M_{tn}^\beta(\boldsymbol{u})}{\partial l} a_{k3}^\beta a_{ki}^\gamma \right) - [f_i^\gamma]_\gamma \bigg] [v_i^\gamma]_\gamma$$

$$+\left[-\frac{dM_1^\gamma(u)}{dx_1^\gamma}+\sum_{\beta\in\partial^{-1}\gamma}\varepsilon(\beta,\gamma)M_{nn}^\beta(u)-[f_4^\gamma])\gamma\right][v_4^\gamma]_\gamma\Big\}d\gamma$$

$$+\sum_{\gamma\in\Gamma^1}\int_\gamma\Big\{\sum_{i=1}^3\Big[\sum_{\beta\in\partial^{-1}\gamma}\Big(\sum_{j=1}^3\sum_{k=1}^3 Q_{jn}^\beta(u)a_{kj}^\beta a_{ki}^\gamma$$

$$+\sum_{k=1}^3\frac{\partial M_{tn}^\beta(u)}{\partial l}a_{k3}^\beta a_{ki}^\gamma\Big)-[f_i^\gamma]_\gamma\Big][v_i^\gamma]_\gamma$$

$$+\Big[\sum_{\beta\in\partial^{-1}\gamma}\varepsilon(\beta,\gamma)M_{nn}^\beta(u)-[f_4^\gamma]_\gamma\Big][v_4^\gamma]_\gamma\Big\}d\gamma$$

$$+\sum_{\delta\in\Gamma^0}\sum_{i=1}^3\Big[\sum_{\gamma\in\partial^{-1}\delta}\varepsilon(\gamma,\delta)\sum_{j=1}^3 Q_j^\gamma(u)a_{ij}^\gamma$$

$$-\sum_{\gamma\in\partial^{-1}\delta}\sum_{\beta\in\partial^{-1}\gamma}\varepsilon(\beta,\gamma)\varepsilon(\gamma,\delta)M_{tn}^\beta(u)a_{i3}^\beta-[f_i]_\delta\Big][v_i]_\delta$$

$$+\sum_{i=1}^3\Big[\sum_{\gamma\in\partial^{-1}\delta}\varepsilon(\gamma,\delta)\sum_{j=1}^3 M_j^\gamma(u)a_{ij}^\gamma-[f_{3+i}]_\delta\Big][v_{3+i}]_\delta$$

$$-\sum_{\delta\in\bar\Gamma^0-\Gamma^0}\sum_{i=1}^3\Big[\sum_{\gamma\in\partial^{-1}\delta}\sum_{\beta\in\partial^{-1}\gamma}\varepsilon(\beta,\gamma)\varepsilon(\gamma,\delta)M_{tn}^\beta(u)a_{i3}^\beta[v_i]_\delta\Big]\qquad(3.32)$$

Now we consider the equilibrium of a geometrically unrestrained, free elastically supported and completely rigid-connected composite structure. According to the variational principle, u of the equilibrium configuration makes $D(u,v)-F(v)=0$ for any v. In particular, this having holds for any $[v_i]_\alpha$, $[v_i^\beta]_\beta$, $[v_i^\gamma]_\gamma$, and $[v_i]_\delta$, hence it is equivalent having all of the brackets at the right hand of (3.32) vanish, $[\cdots]=0$. The equilibrium equations on the various members and on the boundaries of various dimensions can now be written as follows:

$$-\sum_{j=1}^3\frac{\partial\sigma_{ij}(u)}{\partial x_j}=[f_i]_\alpha,\ i=1,2,3,\ \alpha\in\Omega^3,\qquad(3.33)$$

$$-\sum_{j=1}^2\frac{\partial Q_{ij}^\beta(u)}{\partial x_j^\beta}+\sum_{\alpha\in\partial^{-1}\beta}\sum_{j=1}^3[\sigma_{jn}(u)]_{\alpha,\beta}a_{ji}^\beta$$

$$=[f_i^\beta]_\beta,\ i=1,2,3,\ \beta\in\Omega^2,\qquad(3.34)$$

$$\sum_{\alpha\in\partial^{-1}\beta}\sum_{j=1}^{3}[\sigma_{jn}(\boldsymbol{u})]_{\alpha,\beta}a_{ji}^{\beta}=[f_i^{\beta}]_{\beta},\ i=1,2,3,\ \beta\in\Gamma^2, \tag{3.35}$$

$$-\frac{dQ_i^{\gamma}(\boldsymbol{u})}{dx_1^{\gamma}}+\sum_{\beta\in\partial^{-1}\gamma}\Big(\sum_{j=1}^{3}\sum_{k=1}^{3}[Q_{jn}^{\beta}(\boldsymbol{u})]_{\beta,\gamma}a_{kj}^{\beta}a_{ki}^{\gamma}$$

$$+\sum_{k=1}^{3}\Big[\frac{\partial M_{tn}^{\beta}(\boldsymbol{u})}{\partial l}\Big]_{\beta,\gamma}a_{k3}^{\beta}a_{ki}^{\gamma}\Big)$$

$$=[f_i^{\gamma}]_{\gamma},\quad i=1,2,3, \tag{3.36}$$

$$-\frac{dM_1^{\gamma}(\boldsymbol{u})}{dx_1^{\gamma}}+\sum_{\beta\in\partial^{-1}\gamma}\varepsilon(\beta,\gamma)[M_{nn}^{\beta}(\boldsymbol{u})]_{\beta,\gamma}=[f_4^{\gamma}]_{\gamma},\ \gamma\in\Omega^1, \tag{3.37}$$

$$\sum_{\beta\in\partial^{-1}\gamma}\Big(\sum_{j=1}^{3}\sum_{k=1}^{3}[Q_{jn}^{\beta}(\boldsymbol{u})]_{\beta,\gamma}a_{kj}^{\beta}a_{ki}^{\gamma}+\sum_{k=1}^{3}\Big[\frac{\partial M_{tn}^{\beta}(\boldsymbol{u})}{\partial l}\Big]_{\beta,\gamma}a_{k3}^{\beta}a_{ki}^{\gamma}\Big)$$

$$=[f_i^{\gamma}]_{\gamma},\quad i=1,2,3, \tag{3.38}$$

$$\sum_{\beta\in\partial^{-1}\gamma}\varepsilon(\beta,\gamma)[M_{nn}^{\beta}(\boldsymbol{u})]_{\beta,\gamma}=[f_4^{\gamma}]_{\gamma},\ \gamma\in\Gamma^1 \tag{3.39}$$

$$\sum_{\gamma\in\partial^{-1}\delta}\varepsilon(\gamma,\delta)\sum_{j=1}^{3}[Q_j^{\gamma}](\boldsymbol{u})]_{\gamma,\delta}a_{ij}^{\gamma}$$

$$-\sum_{\gamma\in\partial^{-1}\delta}\sum_{\beta\in\partial^{-1}\gamma}\varepsilon(\beta,\gamma)\varepsilon(\gamma,\delta)[M_{tn}^{\beta}(\boldsymbol{u})]_{\gamma,\delta}a_{i3}^{\beta}$$

$$=[f_i]_{\delta},\quad i=1,2,3,\ \delta\in\Gamma^0, \tag{3.40}$$

$$\sum_{\gamma\in\partial^{-1}\delta}\varepsilon(\gamma,\delta)\sum_{j=1}^{3}[M_j^{\gamma}(\boldsymbol{u})]_{\gamma,\delta}a_{ij}^{\gamma}=[f_{3+i}]_{\delta},\quad i=1,2,3,\ \delta\in\Gamma^0, \tag{3.41}$$

$$\sum_{\gamma\in\partial^{-1}\delta}\sum_{\beta\in\partial^{-1}\gamma}\varepsilon(\beta,\gamma)v(\gamma,\delta)[M_{tn}^{\beta}(\boldsymbol{u})]_{\gamma,\delta}a_{i3}^{\beta}=0,$$

$$i=1,2,3,\quad \delta\in\Gamma^0-\Gamma^0. \tag{3.42}$$

Note that the sums $\displaystyle\sum_{\alpha\in\partial^{-1}\beta}$, $\displaystyle\sum_{\beta\in\partial^{-1}\gamma}$, $\displaystyle\sum_{\gamma\in\partial^{-1}\delta}$, etc. stand for the elastic coupling terms of higher dimensional members with respect to the lower dimensional members. This is an important property of the equilibrium equations of the composite structure.

If there exist geometric constraints, then whenever some generalized displacement u_i has been prescribed at some place, the equation containing the load f_i of the same subscript is displaced locally. If there are elastic supports, then the strain energy must absorb the contributions of the elastic support. The Green's formula (3.32) must be modified to some extent, and the resulting equilibrium equations must be increased by the relevant elastic reaction force terms. We will not repeat the details here.

Equilibrium equations at a pin-connected line (point)

Since the rotation angle at a pin-connected line (point) is multiple-valued, the form of the equilibrium equation must be changed at that place.

For a pin-connected line γ, we shall investigate the three cases discussed in Section 3.5.

1. γ is not a rod member but a boundary line element $\gamma \in \Gamma^1$.

Divide $\partial^{-1}\gamma$ into p sets of rigidly connected members $\partial_j^{-1}\gamma$, $j = 1, \cdots, p$. Each set has its own rotation angle $[v_4^\gamma]_{\partial_j^{-1}\gamma, \gamma}$—there is no common rotation angle v_4^γ on γ. From (3.21), the corresponding potential energy of external work is

$$-F_4^\gamma(\boldsymbol{v}) = -\int_\gamma \sum_{j=1}^p [f_4^\gamma]_{\partial_j^{-1}\gamma,\gamma} [v_4^\gamma]_{\partial_j^{-1}\gamma,\gamma} d\gamma.$$

Hence

$$\left[\sum_{\beta \in \partial^{-1}\gamma} \varepsilon(\beta, \gamma) M_{nn}^\beta(\boldsymbol{u}) - [f_4^\gamma]_\gamma \right] [v_4^\gamma]_\gamma.$$

The boundary line element term $\displaystyle\sum_{\gamma \in \Gamma^1}$ in Green's formula (3.32) must be modified to

$$\sum_{j=1}^p \left[\sum_{\beta \in \partial_j^{-1}\gamma} \varepsilon(\beta, \gamma) M_{nn}^\beta(\boldsymbol{u}) - [f_4^\gamma]_{\partial_{j,\gamma,\gamma}^{-1}} \right] [v_4^\gamma]_{\partial_j^{-1}\gamma,\gamma}.$$

Consequently, the equilibrium equations on γ become

$$\sum_{\beta \in \partial_j^{-1}\gamma} \varepsilon(\beta, \gamma)[M_{nn}^\beta(\boldsymbol{u})]_{\beta,\gamma} = [f_4^\gamma]_{\partial_j^{-1}\gamma,\gamma}, \quad j = 1, \cdots, p, \ \gamma \in \Gamma^1 \qquad (3.43)$$

instead of the single equation (3.39).

2. γ is a rod member, i.e., $\gamma \in \Omega^1$, and this rod member and the plates intersecting are all pin-connected.

There are $p+1$ independent rotation angles on γ, one of them, v_4^γ, belongs to the rod member itself, $[v_4^\gamma]_{\partial_j^{-1}\gamma,\gamma}$, $j = 1,\cdots,p$ belong to the p sets of rigid connection $\partial_j^{-1}\gamma$, respectively, and the corresponding potential energy of external work is (3.22), i.e.,

$$-F_4^\gamma(v) = -\int_\gamma \left([f_4^\gamma]_\gamma [v_4^\gamma]_\gamma + \sum_{j=1}^p [f_4^\gamma]_{\partial_j^{-1}\gamma,\gamma}[v_4^\gamma]_{\partial_j^{-1}\gamma,\gamma}\right)d\gamma.$$

Hence

$$\left[-\frac{dM_1^\gamma(u)}{dx_1^\gamma} + \sum_{\beta\in\partial^{-1}\gamma}\varepsilon(\beta,\gamma)M_{nn}^\beta(u) - [f_4^\gamma]_\gamma\right][v_4^\gamma]_\gamma.$$

The rod member term $\sum\limits_{\gamma\in\Omega^1}$ in Green's formulas (3.32) must be modified to

$$\left[\left(-\frac{dM_1^\gamma(u)}{dx_1^\gamma} - [f_4^\gamma]_\gamma\right)[v_4^\gamma]_\gamma\right.$$
$$\left.+\sum_{j=1}^p\left(\sum_{\beta\in\partial_j^{-1}\gamma}\varepsilon(\beta,\gamma)M_{nn}^\beta(u) - [f_4^\gamma]_{\partial_j^{-1}\gamma,\gamma}\right)[v_4^\gamma]_{\partial_j^{-1}\gamma,\gamma}\right]. \tag{3.44}$$

We then obtain the equilibrium equations on γ

$$\gamma\in\Omega^1:\begin{cases} -\dfrac{dM_1^\gamma(u)}{dx_1^\gamma} = [f_4^\gamma]_\gamma, \\[2mm] \displaystyle\sum_{\beta\in\partial_j^{-1}\gamma}\varepsilon(\beta,\gamma)[M_{nn}^\beta(u)]_{\beta,\gamma} = [f_4^\gamma]_{\partial_j^{-1}\gamma,\gamma},\ j=1,\cdots,p \end{cases} \tag{3.45}$$

instead of (3.37).

3. γ is a rod member and is rigidly connected with at least one of the intersecting plates.

Then according to Section 3.5, $\partial^{-1}\gamma$ is divided into $p+1$ rigidly connected sets $\partial_j^{-1}\gamma$, the rod member γ being rigidly connected with the members of last group $\partial_{p+1}^{-1}\gamma$. Hence on γ there are $p+1$ independent values of the rotation angle

$$[v_4^\gamma]_\gamma = [v_4^\gamma]_{\partial_{p+1}^{-1}\gamma,\gamma},\ [v_4^\gamma]_{\partial_j^{-1}\gamma,\gamma},\ j=1,\cdots,p,$$

while the form of the potential energy of external work is still (3.22).

However, (3.44) must be further modified to

$$\left[\left(-\frac{dM_1^\gamma(\boldsymbol{u})}{dx_1^\gamma} + \sum_{\beta\in\partial_{p+1}^{-1}\gamma} \varepsilon(\beta,\gamma)M_{nn}^\beta(\boldsymbol{u}) - [f_4^\gamma]_\gamma\right)[v_4^\gamma]_\gamma\right.$$

$$\left.+\sum_{j=1}^p\left(\sum_{\beta\in\partial_j^{-1}\gamma} \varepsilon(\beta,\gamma)M_{nn}^\beta(\boldsymbol{u}) - [f_4^\gamma]_{\partial_j^{-1}\gamma,\gamma}\right)[v_4^\gamma]_{\partial_j^{-1}\gamma,\gamma}\right].$$

From this, the equilibrium equations on γ can be obtained:

$$\begin{cases} -\dfrac{dM_1^\gamma(\boldsymbol{u})}{dx_1^\gamma} + \displaystyle\sum_{\beta\in\partial_{p+1}^{-1}\gamma} \varepsilon(\beta,\gamma)[M_{nn}^\beta(\boldsymbol{u})]_{\beta,\gamma} = [f_4^\gamma]_\gamma, \\[3mm] \displaystyle\sum_{\beta\in\partial_j^{-1}\gamma} \varepsilon(\beta,\gamma)[M_{nn}^\beta(\boldsymbol{u})]_{\beta,\gamma} = [f_4^\gamma]_{\partial_j^{-1}\gamma,\gamma}, \ j=1,\cdots,p, \end{cases} \quad \gamma\in\Omega^1. \quad (3.46)$$

These equations replace (3.37).

For a pin-connected point δ, we shall investigate only the simple case mentioned in Section 3.5.

In this special case, $\partial^{-1}\delta$ is divided into p sets rigidly connected $\partial_j^{-1}\delta$, $j=1,\cdots,p$. On δ, each set has its own independent vector of rotation angle $[\omega]_{\partial_j^{-1}\delta,\delta}$, and the corresponding potential energy of external work is (3.23):

$$-F_{456}^\delta(\boldsymbol{v}) = -\sum_{i=1}^3\sum_{j=1}^p [f_{3+i}]_{\partial_j^{-1}\delta,\delta}[v_{3+i}]_{\partial_j^{-1}\delta,\delta}.$$

Hence

$$\sum_{i=1}^3\left[\sum_{\gamma\in\partial^{-1}\delta} \varepsilon(\gamma,\delta)\sum_{k=1}^3 M_k^\gamma(\boldsymbol{u})a_{ik}^\gamma - [f_{3+i}]_\delta\right][v_{3+i}]_\delta.$$

The point term $\displaystyle\sum_{\delta\in\Gamma^0}$ in Green's formula (3.32) must be changed to

$$\sum_{i=1}^3\sum_{j=1}^p\left[\sum_{\gamma\in\partial_j^{-1}\gamma} \varepsilon(\gamma,\delta)\sum_{k=1}^3 M_k^\gamma(\boldsymbol{u})a_{ik}^\gamma - [f_{3+i}]_{\partial_j^{-1}\delta,\delta}\right][v_{3+i}]_{\partial_j^{-1}\delta,\delta}.$$

Consequently, the equilibrium equations on δ are

$$\delta\in\Gamma^0: \quad \sum_{\gamma\in\partial_j^{-1}\delta} \varepsilon(\gamma,\delta)\sum_{k=1}^3 [M_k^\gamma(\boldsymbol{u})]_{\gamma,\delta}a_{ik}^\gamma = [f_{3+i}]_{\partial_j^{-1}\delta,\delta}, \quad (3.47)$$

$$i=1,2,3, \ j=1,\cdots,p$$

instead of (3.41).

3.8   Strainless states

For a completely rigid-connected composite structure Ω which is not subject to any geometric constraint or elastic support, the strainless states are given by

$$D(\boldsymbol{v}, \boldsymbol{v}) = 0 \Longleftrightarrow D^\alpha(\boldsymbol{v}, \boldsymbol{v}) = D^\beta(\boldsymbol{v}, \boldsymbol{v}) = D^\gamma(\boldsymbol{v}, \boldsymbol{v}) = 0,$$

for any α, β, γ.

We can see from (3.6), (3.9) and (3.12) in Section 3.2 that

1. $D^\alpha(\boldsymbol{v}, \boldsymbol{v}) = 0 \Longleftrightarrow \boldsymbol{v} = (\boldsymbol{a}^\alpha + \boldsymbol{b}^\alpha \wedge \boldsymbol{x})_\alpha, \ \boldsymbol{\omega} = \boldsymbol{b}^\alpha.$

This is a rigid displacement.

2. $D^\beta(\boldsymbol{v}, \boldsymbol{v}) = 0 \Longleftrightarrow \boldsymbol{v}^\beta = (\bar{\boldsymbol{a}} + \bar{\boldsymbol{b}}^\beta \wedge \boldsymbol{x}^\beta)_{x_3^\beta = 0}, \ \boldsymbol{\omega}^\beta = \bar{\boldsymbol{b}}^\beta.$

This is also a rigid displacement. Transformed into global coordinates, it becomes

$$\boldsymbol{v} = (\boldsymbol{a}^\beta, \boldsymbol{b}\beta \wedge \boldsymbol{x})_\beta, \ \boldsymbol{\omega} = \boldsymbol{b}^\beta.$$

3. $D^\gamma(\boldsymbol{v}, \boldsymbol{v}) = 0 \Longleftrightarrow \boldsymbol{v}^\gamma = (\bar{\boldsymbol{a}}^\gamma + \bar{\boldsymbol{b}}^\gamma \wedge \boldsymbol{x}^\gamma)_{x_2^\gamma = x_3^\gamma = 0}, \ \boldsymbol{\omega}^\gamma = \bar{\boldsymbol{b}}^\gamma.$

This is also a rigid displacement. Transformed into the global coordinates, it becomes

$$\boldsymbol{v} = (\boldsymbol{a}^\gamma + \boldsymbol{b}^\gamma \wedge \boldsymbol{x})_\gamma, \ \boldsymbol{\omega} = \boldsymbol{b}^\gamma.$$

Therefore, on every member of the structure there exist constant vectors $\boldsymbol{a}^\alpha, \boldsymbol{b}^\alpha; \boldsymbol{a}^\beta, \boldsymbol{b}^\beta$ and $\boldsymbol{a}^\gamma, \boldsymbol{b}^\gamma$, which represent the individual rigid translations and rotation angles, respectively. Using the continuity conditions 1–5 for the displacements and rotation angles of the rigid connection, one can prove that the respective displacement vectors and rotation angles must be equal whenever two members are connected by a joint. One can see from the foregoing and from the connectedness of the structure that, there exist constant vectors $\boldsymbol{a}$ and $\boldsymbol{b}$ such that

$$\boldsymbol{a} = \boldsymbol{a}^\alpha = \boldsymbol{a}^\beta = \boldsymbol{a}^\gamma, \ \boldsymbol{b} = \boldsymbol{b}^\alpha = \boldsymbol{b}^\beta = \boldsymbol{b}^\gamma, \text{ for any } \alpha, \beta, \gamma.$$

Therefore, as expected, the strainless state of the whole composite structure is a global rigid displacement

$$\boldsymbol{v} = (\boldsymbol{a} + \boldsymbol{b} \wedge \boldsymbol{x})_\Omega,$$

which contain six parameters $a_i, b_i, i = 1, 2, 3$. Any strainless state $\boldsymbol{v}$ can be expressed as

$$\boldsymbol{v} = \sum_{i=1}^3 a_i \boldsymbol{v}^{(i)} + \sum_{i=1}^3 b_i \boldsymbol{v}^{(3+i)}, \ \boldsymbol{\omega} = \sum_{i=1}^3 b_i \boldsymbol{v}^{(i)}, \tag{3.48}$$

where

$$
\boldsymbol{v}^{(1)} = \begin{bmatrix} 1 \\ 0 \\ 0 \end{bmatrix}, \quad
\boldsymbol{v}^{(2)} = \begin{bmatrix} 0 \\ 1 \\ 0 \end{bmatrix}, \quad
\boldsymbol{v}^{(3)} = \begin{bmatrix} 0 \\ 0 \\ 1 \end{bmatrix}, \quad
\boldsymbol{v}^{(4)} = \begin{bmatrix} 0 \\ -x_3 \\ x_2 \end{bmatrix},
$$

$$
\boldsymbol{v}^{(5)} = \begin{bmatrix} x_3 \\ 0 \\ -x_1 \end{bmatrix}, \quad
\boldsymbol{v}^{(6)} = \begin{bmatrix} -x_2 \\ x_1 \\ 0 \end{bmatrix},
\tag{3.49}
$$

while

$$
\boldsymbol{\omega}^{(1)} = \boldsymbol{\omega}^{(2)} = \boldsymbol{\omega}^{(3)} = 0, \quad \boldsymbol{\omega}^{(4)} = \boldsymbol{v}^{(1)}, \quad \boldsymbol{\omega}^{(5)} = \boldsymbol{v}^{(2)}, \quad \boldsymbol{\omega}^{(6)} = \boldsymbol{v}^{(3)}.
\tag{3.50}
$$

In this case, the equilibrium problem without geometric constraints and elastic supports is degenerate, and necessary and sufficient conditions for solvability are

$$
F(\boldsymbol{v}^{(k)}) = 0, \quad k = 1, \cdots, 6,
\tag{3.51}
$$

where

$$
F(\boldsymbol{v}) = \sum_{\alpha \in \Omega^1} \iiint_\alpha \sum_{i=1}^3 [f_i]_\alpha [v_i]_\alpha d\alpha + \sum_{\beta \in \Omega^2 + \Gamma^2} \iint_\beta \sum_{i=1}^3 [f_i^\beta]_\beta [v_i^\beta]_\beta d\beta
$$
$$
+ \sum_{\gamma \in \Omega^1 + \Gamma^1} \int_\gamma \sum_{i=1}^4 [f_i^\gamma]_\gamma [v_i^\gamma]_\gamma d\gamma + \sum_{\delta \in \Gamma^0} \sum_{i=1}^6 [f_i]_\delta [v_i]_\delta.
$$

Transforming the local coordinates on β and γ into the global coordinates, this equation becomes

$$
F(\boldsymbol{v}) = \sum_{\alpha \in \Omega^3} \iiint_\alpha \sum_{i=1}^3 [f_i]_\alpha [v_i]_\alpha d\alpha + \sum_{\beta \in \Omega^2 + \Gamma^2} \iint_\beta \sum_{i=1}^3 [f_i]_\beta [v_i]_\beta d\beta
$$
$$
+ \sum_{\gamma \in \Omega^1 + \Gamma^1} \int_\gamma \sum_{i=1}^4 [f_i]_\gamma [v_i]_\gamma d\gamma + \sum_{\delta \in \Gamma^0} \sum_{i=1}^6 [f_i]_\delta [v_i]_\delta.
$$

Using (3.49) and (3.50) and noting that $f_{3+i} \equiv m_i, v_{3+i} \equiv \omega_i, i = 1, 2, 3$, the necessary and sufficient conditions (3.51) become

$$
\sum_{\alpha \in \Omega^3} \iiint_\alpha [f]_\alpha d\alpha + \sum_{\beta \in \Omega^2 + \Gamma^2} \iint_\beta [f]_\beta d\beta
$$
$$
+ \sum_{\gamma \in \Omega^1 + \Gamma^1} \int_\gamma [f]_\gamma d\gamma + \sum_{\delta \in \Gamma^0} [f]_\delta = 0,
\tag{3.52}
$$

when $k = 1, 2, 3$. If we let $[m]_\gamma = [m_1^\gamma, 0, 0]^T$, then the conditions (3.51) become

$$\sum_{\alpha \in \Omega^3} \iiint_\alpha [x \wedge f]_\alpha d\alpha + \sum_{\beta \in \Omega^2 + \Gamma^2} \iint_\beta [x \wedge f]_\beta d\beta$$

$$+ \sum_{\gamma \in \Omega^1 + \Gamma} \int_\gamma [x \wedge f]_\gamma d\gamma + \sum_{\gamma \in \Omega^1 + \Gamma^1} \int_\gamma [m]_\gamma d\gamma$$

$$+ \sum_{\delta \in \Gamma^0} [x \wedge f]_\delta + \sum_{\delta \in \Gamma^0} [m]_\delta = 0 \qquad (3.53)$$

when $k = 4, 5, 6$. These are two vector equations. They have a total of six component equations, which state that the resultant force and moment of the loads must vanish, respectively. As expected, when the displacement and the rotation angle are unrestrained and there is no elastic support, the self-equilibrium of the loads is the prerequisite to make the structure achieve equilibrium.

When the prerequisites (3.51) have been satisfied, there exists a solution of the above equilibrium problem without constraints and elastic supports, but the displacement solution is not unique and may differ by a strainless state. However, for the strainless state v, the strains

$$\varepsilon_{ij}^\alpha(v) \equiv \varepsilon_{ij}^\beta(v) \equiv K_{ij}^\beta(v) \equiv \varepsilon_{11}^\gamma(v) \equiv K_i^\gamma(v) \equiv 0,$$

correspondingly, the stresses

$$\sigma_{ij}^\alpha(v) \equiv Q_{ij}^\beta(v) \equiv M_{ij}^\beta(v) \equiv Q_1^\gamma(v) = M_i^\gamma(v) \equiv 0.$$

Hence the stress has still a unique solution.

As is usual in elastic equilibrium, whenever a geometric constraint or an elastic support is imposed, it always reduces the degrees of freedom of the strainless state, and raise the positive definiteness of the strain energy $D(v, v)$. When there are enough geometric constraints or elastic supports to make $D(v, v)$ positive definite, the equilibrium problem has a unique displacement solution.

Suppose the displacements u_1, u_2 and u_3 are fixed on some two dimensional boundary area element $\beta \in \Gamma^2$. Then, under this prescribed constraint, D is positive definite, this is $D(v, v) = 0$ implies $v \equiv 0$. This is because

$$D(v, v) = 0 \Longleftrightarrow v = a + b \wedge x,$$

which corresponds to a rigid displacement, while the virtual displacement v should satisfy the annihilating constraint condition $v|_\beta = 0$ under the prescribed constraint which corresponds to fixing the rigid body on a plane block β, and consequently, its displacement $\boldsymbol{v} \equiv 0$. Such a conclusion is consistent with that in the case of the single three dimensional elastic body.

Next, suppose no geometric constraint is imposed on the two dimensional boundary Γ^2, while the displacements u_1, u_2 and u_3 and the rotation angle ω_1^γ about γ are fixed only on some one dimensional boundary line element, i.e., on the boundary γ of the plate element. D is also positive definite under this prescribed constraint. This is because there must be $\boldsymbol{v} \equiv 0$ when the rigid body is located on a segment of straight line and the rotation angle about the straight line is fixed. This case is different from that of the single three dimensional body. Under the condition of the single three dimensional body, fixing the displacement and the tangential angle of rotation only on a segment of straight line of the boundary surface, it is still not enough to guarantee the positive definiteness of the strain energy, i.e., not enough to guarantee the existence of a unique solution. However in the three dimensional composite structure, it is enough to guarantee the positive definiteness of the strain energy provided that the segment fixing the displacements is located on the boundary of a plate member.

Furthermore, suppose no constraint is imposed on either the two dimensional or the one dimensional boundaries, but take three arbitrary non-collinear points among the null dimensional boundaries Γ^0 (i.e., the end points of the rod members) and fix the displacements there, or only take two points and fix the displacements and the rotation angle there. Then positive definiteness is guaranteed. This is because the rigid displacement $\boldsymbol{v} \equiv 0$ can be made under such fixing conditions, but the positive definiteness still can not be guaranteed in the single three dimensional body.

The geometric constraints to guarantee the positive definiteness of the strain energy for a composite structure and for a single elastic body, are thus quite different. For the composite structure, through the elastic coupling between the members with different dimensions, the geometric constraints necessary to guarantee positive definiteness can be applied to the boundary with the lowest dimension down to two or three individual points. It is, however, impossible to do this for a single elastic body. This is an important peculiarity of composite structures.

The mechanical background of the composite structure, of course, is the three dimensional elastic body. For the linear three dimensional elastic body, we know that a strainless state is just a rigid displacement while the prerequisite for achieving equilibrium under the condition without displacement constraints and elastic supports is a self-equilibrium of external loads. The composite structure model discussed in the present chapter is on the one hand the extension of the three dimensional elasticity to the composite body with different dimensions and on the other hand is the simplification and modification of three dimensional elasticity owing to the introduction of the specified synthesis of some deformation modes (plate, beam). However, under the case of this model, it is still possible to guarantee the validity of the conclusions about the strainless state and the unrestrained elastic equilibrium (some modes are only satisfied approximately). This shows that the composite structure model is in some sense a reasonable one.

3.9   Treatment of the offset distance

Although the composite structure model employed in the present chapter is reasonable, it is in the end simplified model, in which a thin plate is simplified to be an elastic plane (neutral surface) without the thickness and depending only on two plane coordinates, and a slender rod is simplified to be an elastic line (neutral axis) without cross-sectional area cross section and depending only on one straight line coordinate. For a single elastic body, just such deformation modes are used in Chapter 3 to discuss plates and rods. This is consistent with the actual situation. For a composite structure, however, such simplified modes are often to some degree discrepant with the actual situation in elastic coupling between members. This is because during the discussion of interconnections, the joints between a body member and a plate member (three dimensional and two dimensional), between a plate member and a plate member (two dimensional and two dimensional), between a plate member and a rod member (two dimensional and one dimensional), as well as between a rod member and a rod member (one dimensional and one dimensional), are not the actual joints between the two members mentioned above e.g. thin rod and thin plate, thin plate and body. The deviation between the joint in the model and the actual joint is generally called the offset distance.

When Green's formula is used to derive the elastic relation between adjacent members and consequently to obtain the equilibrium equation, there will be some inaccuracy in the equations due to the offset distance between the members. Generally speaking, since the plate is comparatively thin and the rod is comparatively slender, the influence of the offset distance is small and can be neglected consequently. However, the influence of the offset distance must still be taken into account under some conditions. One method of dealing with the offset distance will be given in the following through an example plane structural problem.

Suppose a plate member β and a rod γ are rigidly connected on the boundary (Fig. 67). The neutral axis of the rod γ is denoted by γ and is taken as the x_1-axis. The local coordinates of the rod γ are also the global coordinates. Let us first write out the Green's formulas and the equilibrium equations for both plate β and rod γ according to the plane structure model in §2. We assume that the line of intersection of plate β and rod γ is γ in this model.

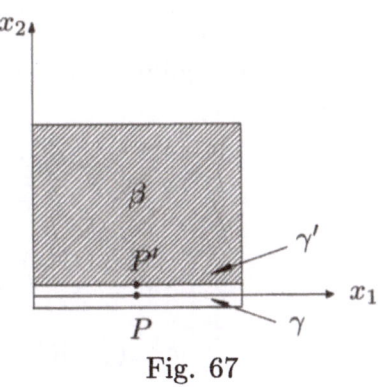

Fig. 67

The Green's formula for the plate β (see (2.22)) is

$$D^{\beta}(\boldsymbol{u}, \boldsymbol{v}) = \iint_{\beta} \sum_{i=1}^{2} \left(-\sum_{i=1}^{2} \frac{\partial Q_{ij}(\boldsymbol{u})}{\partial x_j} \right) v_i d\beta + \int_{\gamma} \sum_{i=1}^{2} Q_{in}(\boldsymbol{u}) v_i d\gamma$$

+ line integral terms on the other three boundaries of plate β;

the Green's formula for rod γ (see (2.33)) is

$$D^{\gamma}(\boldsymbol{u}, \boldsymbol{v}) = \int_{\Gamma} \sum_{i=1}^{2} \left(-\frac{dQ_1(\boldsymbol{u})}{dx_1} \right) v_i dx_1$$

$$+ \sum_{\delta \in \partial \gamma} \varepsilon(\gamma, \delta) \left[\sum_{i=1}^{2} Q_i(\boldsymbol{u}) v_i + M_3(\boldsymbol{u}) \frac{dv_2}{dx_1} \right]_{\delta};$$

hence

$$D(\boldsymbol{u}, \boldsymbol{v}) - F(\boldsymbol{v}) = D^\beta(\boldsymbol{u}, \boldsymbol{v}) - F^\beta(\boldsymbol{v}) + D^\gamma(\boldsymbol{u}, \boldsymbol{v}) - F^\gamma(\boldsymbol{v})$$

$$= \iint_\beta \sum_{i=1}^2 \left[-\sum_{j=1}^2 \frac{\partial Q_{ij}(\boldsymbol{u})}{\partial x_j} - [f_i]_\beta \right] v_i d\beta$$

$$+ \int_\gamma \sum_{i=1}^2 \left[-\frac{dQ_i(\boldsymbol{u})}{dx_1} + Q_{in}(\boldsymbol{u}) - [f_i]_\gamma \right] v_i d\gamma$$

$$+ \sum_{\delta \in \partial\gamma} \left[\sum_{i=1}^2 (\varepsilon(\gamma, \delta) Q_i(\boldsymbol{u}) - [f_i]_\delta)[v_i]_\delta \right.$$

$$\left. + (\varepsilon(\gamma, \delta) M_3(\boldsymbol{u}) - [m_3]_\delta) \left[\frac{dv_2}{dx_1} \right]_\delta \right].$$

The following equilibrium equations can be obtained from the principle of virtual work

$$\beta : -\sum_{j=1}^2 \frac{\partial Q_{ij}(\boldsymbol{u})}{\partial x_j} = [f_i]_\beta, \ i = 1, 2,$$

$$\gamma : -\frac{dQ_i(\boldsymbol{u})}{dx_1} + Q_{in}(\boldsymbol{u}) = [f_i]_\gamma, \ i = 1, 2,$$

$$\delta : \varepsilon(\gamma, \delta) Q_i(\boldsymbol{u}) = [f_i]_\delta, \ i = 1, 2,$$

$$\varepsilon(\gamma, \delta) M_3(\boldsymbol{u}) = [m_3]_\delta.$$

But the model mentioned above does not exactly describe the actual situation because the actual intersection line between the plate β and the rod γ is someline γ', not γ. Suppose γ' is a straight line parallel to γ, and the distance between these two lines is h. Take a point P' on γ' and another point P on γ, and let the displacement vectors at points P' and P be $\boldsymbol{u}'$ and $\boldsymbol{u}$, and the rotation angle vectors be Ω' and Ω, respectively. Suppose there is a rigid displacement from point P to point P', i.e.,

$$\boldsymbol{u}'(P') = \boldsymbol{u}(P) + \Omega \wedge \overrightarrow{PP'}, \tag{3.54}$$

where $\overrightarrow{PP'}$ stands for the position vector from P to P'.

In the case of plane structures, the displacement vector $\boldsymbol{u}' = (u_1', u_2', 0)^T$ at point P', the displacement vector $\boldsymbol{u} = (u_1, u_2, 0)^T$ at point P, and the rotation angle vector $\Omega = \left(0, 0, \dfrac{du_2}{dx_1}\right)^T$. Suppose the position vector $\overrightarrow{PP'} = (0, h, 0)$, i.e., points P' and P have a common coordinate in the x_1-direction while their coordinates in the x_2-direction differ by a constant *h*. Then, by (3.54), we get

$$u_1' = u_1 - h\frac{du_2}{dx_1}, \quad u_2' = u_2. \tag{3.55}$$

Since the actual boundary line of the plate β is γ' and not γ, the Green's formula for plate β must be changed to

$$D^\beta(\boldsymbol{u}, \boldsymbol{v}) = \iint_\beta \sum_{i=1}^2 \left(-\sum_{j=1}^2 \frac{\partial Q_{ij}(\boldsymbol{u})}{\partial x_j} \right) v_i d\beta + \int_{\gamma'} \sum_{i=1}^2 Q_{in}(\boldsymbol{u}') v_i' d\gamma$$

+ line integral terms on the other three boundaries.

Substituting (3.55) into the second term and then integrating once by parts, we obtain

$$D^\beta(\boldsymbol{u}, \boldsymbol{v}) = \iint_\beta \sum_{i=1}^2 \left(-\sum_{j=1}^2 \frac{\partial Q_{ij}(\boldsymbol{u})}{\partial x_j} \right) v_i d\beta$$

$$+ \int_\gamma \left[Q_{in}(\boldsymbol{u}') v_1 + \left(Q_{2n}(\boldsymbol{u}') + h\frac{dQ_{in}(\boldsymbol{u}')}{dx_1} \right) v_2 \right] d\gamma$$

$$- \sum_{\delta \in \partial\gamma} \varepsilon(\gamma, \delta) h Q_{in}(\boldsymbol{u}') v_2$$

+ line integral terms on the other three boundaries.

Hence the equilibrium equations should be

$$\beta : -\sum_{j=1}^2 \frac{\partial Q_{ij}(\boldsymbol{u})}{\partial x_j} = [f_i]_\beta, \ i = 1, 2,$$

$$\gamma : -\frac{dQ_1(\boldsymbol{u})}{dx_1} + Q_{in}(\boldsymbol{u}') = [f_1]_\gamma,$$

$$-\frac{dQ_2(\boldsymbol{u})}{dx_2} + Q_{2n}(\boldsymbol{u}') + h\frac{dQ_{1n}(\boldsymbol{u}')}{dx_1} = [f_2]_\gamma,$$

$$\delta : \varepsilon(\gamma, \delta) Q_1(\boldsymbol{u}) = [f_1]_\delta, \ \varepsilon(\gamma, \delta)(Q_2(\boldsymbol{u}) - h Q_{1n}(\boldsymbol{u}')) = [f_2]_\delta,$$

$$\varepsilon(\gamma, \delta) M_3(\boldsymbol{u}) = [m_3]_\delta.$$

We can see from comparing these two systems of equilibrium equations that the equilibrium equations for the plate β remain unchanged in the form. However, because of the offset distance, the x_2-directional equilibrium equation for the rod γ has changed and an extra $h\dfrac{dQ_{1n}(\boldsymbol{u}')}{dx_1}$ which is a shearing force term induced by the pair of forces $hQ_{1n} = -hQ_{12}$, and the x_2-directional equilibrium equation for the end point δ also has an extra term induced by a pair of forces $-hQ_{1n}(\boldsymbol{u}')$.

This method of dealing with the offset distance is likewise suitable for general space structure. The main point is: use the rigid connection relation (3.54) between the points P' and P to express the displacement and rotation angle at the corresponding point of the adjacent member of higher dimension in terms of the displacement and the rotation angle on the member of lower dimension and substitute it into the relevant energy functional and the corresponding Green's formula.

Chapter 5

Finite Element Methods

§1   Introduction

More than ten years ago, the finite element method, a modern, systematic
numerical method for solving differential equations, was created and developed independently and along different lines in China[6] and the West.
Its original purpose was to solve for equilibria and stable configurations,
i.e. to solve elliptic equations. It has stood a large number of practical
tests, and, in particular, has been used widely in the field of elastic structures with remarkable success. Recently, with the aid of computers, the
finite element method has been applied to almost all fileds of engineering
and to many fields of science and technology, and has become a routine
means for modern engineering analysis. It is an important achievement of
modern computational mathematics.

The mathematical bases of the finite element method are the variational principle and triangulated or piecewise approximation. As is well
known, a round well can be built up with rectangular bricks, a plate
can be sawed along a curved path with a straight saw, and a continuous
curve can be approximated by broken lines. The finer the subdivision, the
closer is the approximation. This is known as subdivision approximation
or piecewise interpolation. Starting from the variational principle, the finite element method uses the subdivision approximation approach to find
a solution. It subdivides a complex integrated structure into a finite number of fundamental elements, i.e., point, line, area, and volume elements.
The undetermined function was performed the piecewise interpolation,
which is generally a simple linear or low order polynomial interpolation.
Thus the global energy functional can reasonably be approximated by a
sum of element energies. This process discretizes an extremum problem of
quadratic functionals with infinite degrees of freedom into an extremum

problem of multivariate quadratic functions with a finite number of degrees of freedom. The latter is equivalent to a system of linear algebraic equations. It is linear system that one finally solves. To sum up, the guiding idea of the finite element method can be summarized in the following formula: "break up the whole into parts, substitute the straight for the bent, drive out complexity with simplicity, and turn the difficult into the easy".

Owing to the beautiful combination of the variational principle and the piecewise interpolation, the finite element method successfully combines the advantages of the traditional energy method and of the traditional finite difference method. Being monotonic and mechanical in form, it is easy to program on computers. Being very essence, it is especially suitable for problems with great geometrical and physical complexity. Furthermore, being based on the combination of the variational principle and the subdivision approximation, it can simultaneously treat the infinite and the finite and continuity and discontinuity, so as to form a complete theoretical basis and to provide theoretical assurance that meets current demands for reliability of the method and closes the longstanding gap between theory and practice[5]. In comparison with traditional methods, therefore, the finite element method has many different advantages, providing a highly efficient solution method, a firm theoretical basis, a straight forward, well-defined method and a wide range of applications.

In the present chapter, we will introduce the basic idea and the discretization process of the finite element method via several typical problems in elasticity. We will, however, leave aside the solution of the system of linear algebraic equations, the problem of convergence of the finite element solution to the exact solution, and the problem of error estimates.

§2   Stretching and Torsion of Rods

2.1   Variational problems

According to §3 and §6 of Chapter 1, though the mechanical backgrounds of the rod in tension and the rod in torsion are quite different, the mathematical forms of their equilibrium equations are identical and can be expressed as the second order ordinary differential equation

$$-\frac{d}{dx}\left(p\frac{du}{dx}\right) = f, \quad a < x < b, \tag{2.1}$$

where the coefficient $p = p(x) > 0$. In the case of the rod in tension, u denotes the longitudinal displacement, p denotes the tensile rigidity of the cross section, and f denotes the force load distribution in the longitudinal direction; while in the case of the rod in torsion, u denotes the twist angle, p denotes the torsional rigidity of the cross section, and f denotes the torque load distribution in the longitudinal direction. There are, besides, many other practical problems, such as the one-dimensional heat transfer problem, which can also be summarized in the form (2.1).

In addition to the equilibrium equation (2.1), we must consider boundary conditions which are generally classified into three kinds:

1. $u(a) = \bar{u}(a), \; u(b) = \bar{u}(b)$.

2. $-p(a)u'(a) + C_a u(a) = g_a$,

3. $-p(a)u'(a) + C_a u(a) = g_a$,

$p(b)u'(b) + C_b u(b) = g_b, \; C_a, C_b \geq 0$.

Various combinations of boundary conditions can be imposed at the two end points. The second kind of boundary condition can be considered as a special case of the third kind with $C_a = C_b = 0$.

We showed in §3 of Chapter 1 that (2.1) and the boundary conditions 1, 2, and 3 are, respectively, equivalent to the following three variational problems:

$$1° \quad \begin{cases} J(u) = \dfrac{1}{2}\displaystyle\int_a^b pu'^2 dx - \int_a^b fu\,dx = \text{Min}, \\[2mm] u(a) = \bar{u}(a), \; u(b) = \bar{u}(b). \end{cases} \tag{2.2}$$

$$2° \quad J(u) = \frac{1}{2}\int_a^b pu'^2 dx - \int_a^b fu\,dx - g_a u(a) - g_b u(b)$$

$$= \text{Min}, \tag{2.3}$$

$$3° \quad J(u) = \frac{1}{2}\int_a^b pu'^2 dx + \frac{1}{2}(C_a u^2(a) + C_b u^2(b))$$

$$- \int_a^b fu\,dx - g_a u(a) - g_b u(b) = \text{Min}. \tag{2.4}$$

In what follows, we take the variational problems (2.4) and (2.2) as examples to illustrate the fundamental idea and the discretization process of the finite element method.

2.2   Subdivision and interpolation
(1) Geometric subdivision

First subdivide the interval $[a, b]$ on which a solution is to be found into a finite number of subintervals (which need not be of equal length). we call these line elements and denote them by $\gamma_1, \cdots, \gamma_{N_1}$. The two end points of the line elements γ_i i.e., the subdividing points or the nodal points, are called point elements and are denoted $\delta_1, \cdots, \delta_{N_0}$. We see that $N_0 = N_1 + 1$. The line and point elements are both just called elements on the straight line. Thus, the interval $[a, b]$ is subdivided into a finite number of elements.

Note that if the coefficient p or the right hand side f of the equation is discontinuous in $[a, b]$, then the discontinuity point should be taken as a nodal point of subdivision.

(2) Linear interpolation

Consider an arbitrary line element γ. Suppose its two end points are δ_1 and δ_2. Let s be the variable of the chord length from δ_1 toward δ_2, and let l denote the length of γ. Linearly interpolate the function $u(s)$ on γ by

$$\begin{cases} Q(s) = u_1 \lambda_1(s) + u_2 \lambda_2(s), \\ Q(0) = u_1, \quad Q(l) = u_2, \end{cases} \tag{2.5}$$

where, u_1 and u_2 are the values of $u(s)$ at the end points δ_1 and δ_2, corresponding to $s = 0$ and $s = l$, respectively, and $\lambda_1(s)$ and $\lambda_2(s)$ are linear functions of s satisfying the relations

$$\lambda_1(0) = 1, \quad \lambda_1(l) = 0;$$
$$\lambda_2(0) = 10, \quad \lambda_2(l) = 1.$$

Clearly, we have

$$\lambda_1(s) = 1 - \frac{s}{l}, \quad \lambda_2(s) = \frac{s}{l}, \tag{2.6}$$

and

$$\lambda_1 + \lambda_2 = 1. \tag{2.7}$$

λ_1 and λ_2 are called the basis functions of the linear interpotion on γ.

The following table of integrals will be used later on.

Table 1 $\int_\gamma \varphi\psi ds$

ψ \\ φ	1	λ_j	$\partial\lambda_j/\partial s$
1	l	$l/2$	$(-1)^f$
λ_i		$l(1+\delta_{ij})/6$	$(-1)^{j/2}$
$\partial\lambda_i/\partial s$			$(-1)^{i+j/l}$

$$\delta_{ij} = \begin{cases} 1, & i=j, \\ 0, & i \neq j. \end{cases}$$

(3) Quadratic interpolation

We will interpolate the function $u(s)$ on γ by a quadratic polynomial. Since a quadratic polynomial has three coefficients, a third interpolating point is required to determine it. The mid point of γ is generally taken as the third point. It is denoted δ_3 after the two end points δ_1 and δ_2. Let

$$Q(s) = u_1\mu_1(s) + u_2\mu_2(s) + u_3\mu_3(s),$$
$$Q(s_i) = u_i, \quad s_1 = 0, \quad s_2 = l, \quad s_3 = l/2, \tag{2.8}$$

where, $\mu_i, i = 1, 2, 3$ are quadratic functions of s satisfying the relations

$$\mu_i(s_j) = \delta_{ij}, \quad i, j = 1, 2, 3.$$

From the foregoing, one can derive:

$$\begin{cases} \mu_1(s) = \dfrac{2}{l^2}(s - l)\left(s - \dfrac{l}{2}\right) = 2\lambda_1^2 - \lambda_1, \\[2mm] \mu_2(s) = \dfrac{2}{l^2}s\left(s - \dfrac{l}{2}\right) = 2\lambda_2^2 - \lambda_2, \\[2mm] \mu_3(s) = \dfrac{4}{l^2}s(l - s) = 4\lambda_1\lambda_2, \end{cases} \tag{2.9}$$

where λ_1 and λ_2 are the basis functions of the linear interpolation (2.6). μ_1, μ_2 and μ_3 are called basis functions of the quadratic interpolation on γ.

Using Euler's formula,

$$\int_0^1 t^{p_1}(1-t)^{p_2} dt = p_1!p_2!/(p_1 + p_2 + 1)!, \tag{2.10}$$

we know that

$$\int_\gamma \lambda_1^{p_1}\lambda_2^{p_2} ds = p_1!p_2!/(p_1 + p_2 + 1)!. \tag{2.11}$$

We have the following table of integrals.

$$\textbf{Table 2}\quad \int_{\gamma}\varphi\psi ds$$

ψ \ φ	1	μ_1	μ_2	μ_3	$\partial\mu_1/\partial s$	$\partial\mu_2/\partial s$	$\partial\mu_3/\partial s$
1	l	$l/6$	$l/6$	$2l/3$	-1	1	0
μ_1		$2l/15$	$-l/30$	$l/15$	$-1/2$	$-1/6$	$2/3$
μ_2				$8l/15$	$-2/3$	$2/3$	0
$\partial\mu_1/\partial s$					$7/31$	$1/3l$	$-8/31$
$\partial\mu_2/\partial s$						$7/3l$	$-8/3l$
$\partial\mu_3/\partial s$							$16/3l$

2.3　Analysis of elements (linear elements)

After subdividing the interval $[a, b]$ into line elements and point elements, the energy expression $J(u)$ (2.4) can be approximated by the sum of energies on the line and point elements:

$$J(u) = \sum_{i=1}^{N_1} J_{\gamma_i}(u) + J_{\delta_1}(u) + J_{\delta_{N_0}}(u), \tag{2.12}$$

$$J_{\gamma}(u) = \frac{1}{2}\int_{\gamma} p u'^2 dx - \int_{\gamma} f u dx, \tag{2.13}$$

$$J_{\delta_1}(u) = \frac{1}{2}C_a u^2(a) - g_a u(a),$$

$$J_{\delta_{N_0}}(u) = \frac{1}{2}C_b u^2(b) - g_b u(b). \tag{2.14}$$

This step breaks up the whole into parts. The next goal is to complete the discretization of $J(u)$ by approximating the energy on each element by a quadratic function of nodal variables. This step is called the analysis of elements.

(1) Analysis of line elements

Consider an arbitrary line element $\gamma = (\delta_1, \delta_2)$, and the two integrals over it

$$\frac{1}{2}\int_{\gamma} p u'^2 dx, \quad \int_{\gamma} f u dx.$$

Generally speaking, the undetermined function $u(x)$ is a rather complicated distribution function on the whole interval $[a, b]$. But if we subdivide the interval into very small intervals, that is, if the length of the line element γ is very small, then $u(x)$ on each line element γ can be treated as linear, i.e., the linear interpolation function in (2.5)

$$Q(x) = \lambda_1 u_1 + \lambda_2 u_2$$

can be used to approximate $u(x)$. This is the step to substitute the straight for curve. Thus, the integral becomes

$$\frac{1}{2}\int_\gamma pu'^2 dx \approx \frac{1}{2}\int_\gamma pQ'^2 dx = \frac{1}{2}\int_\gamma p(\lambda_1' u_1 + \lambda_2' u_2)^2 dx$$

$$= \frac{1}{2}(k_{11}u_1^2 + 2k_{12}u_1 u_2 + k_{22}u_2^2)$$

$$= \frac{1}{2}U_\gamma^T K_\gamma U_\gamma, \tag{2.15}$$

where, $U_\gamma = (u_1, u_2)^T$ is a two dimensional column vector, and

$$K_\gamma = \begin{bmatrix} k_{11} & k_{12} \\ k_{21} & k_{22} \end{bmatrix}$$

is a second order symmetric matrix called the element stiffness matrix, its entries are

$$k_{11} = \int_\gamma p\lambda_1'^2 dx = \frac{1}{l^2}P_\gamma,$$

$$k_{12} = k_{21} = \int_\gamma p\lambda_1'\lambda_2' dx = -\frac{1}{l^2}P_\gamma,$$

$$k_{22} = \int_\gamma p\lambda_1'^2 dx = \frac{1}{l^2}P_\gamma,$$

where,

$$P_\gamma = \int_\gamma pdx,$$

so that

$$K_\gamma = \frac{P_\gamma}{l^2}\begin{bmatrix} 1 & -1 \\ -1 & 1 \end{bmatrix}. \tag{2.16}$$

In particular, if p is a constant on γ, then

$$P_\gamma = \int_\gamma pdx = pl,$$

$$K_\gamma = \frac{p}{l}\begin{pmatrix} 1 & -1 \\ -1 & 1 \end{pmatrix} \tag{2.17}$$

For a general function p, quadrature formulas can be used to evaluate p_γ (see Table 5).

Another integral over γ is

$$\int_\gamma fu\,dx \approx \int_\gamma fQ\,dx = \int_\gamma f(\lambda_1 u_1 + \lambda_2 u_2)\,dx$$

$$= f_1^{(\gamma)}u_1 + f_2^{(\gamma)}u_2 = U_\gamma^T F_\gamma, \tag{2.18}$$

where,

$$F_\gamma = (f_1, f_2)^T, \quad f_i^{(\gamma)} = \int_\gamma f\lambda_i\,dx, \quad i = 1, 2. \tag{2.19}$$

If f is a constant on γ, then

$$f_1^{(\gamma)} = f_2^{(\gamma)} = \frac{1}{2}fl, \quad F_\gamma = \frac{1}{2}fl(1, 1)^T. \tag{2.20}$$

For a general f, quadrature formulas can also be used to evaluate $f_1^{(\gamma)}$ and $f_2^{(\gamma)}$.

The energy integral over the line element γ is thus discretized as

$$J_\gamma(u) \approx J_\gamma(Q) = \frac{1}{2}U_\gamma^T K_\gamma U_\gamma - U_\gamma^T F_\gamma. \tag{2.21}$$

This is a quadratic function of the nodal values u_1 and u_2. This discretization completes the analysis of line elements.

In the above, we have approximated $u(x)$ by a linear interpolation on γ, called the first order linear element, or just the linear element. We can also approximate $u(x)$ using a quadratic interpolation, we will discuss quadratic interpolation later in §2.6.

(2) Analysis of point elements

The discretization of energy on a point element δ is trivial, because a point is already discrete. There are two terms:

$$J_{\delta_1} = \frac{1}{2}u_1 C_a u_1 - g_a u_1, \quad J_{\delta_{N_0}} = \frac{1}{2}u_{N_0} C_b u_{N_0} - g_b u_{N_0}. \tag{2.22}$$

The element stiffness matrix is of the first order, i.e., it is just a scalar quantity C_a or C_b.

2.4 Assembly

Piecing together all of the linear interpolation functions on the individual line elements, we obtain a piecewise linear function $Q(x)$ on $[a, b]$, which is determined by the nodal values $u_1, \cdots, u_{N_0}$ of $u(x)$. $Q(x)$ is continuous over the whole interval, but its first derivatives are discontinuous at the nodal points.

After discretizing the energy expression something missing using a quadratic function on each element, the global energy will have been discretized as

$$
\begin{aligned}
J(u) \approx J(Q) &= J(u_1, \cdots, u_{N_0}) \\
&= \sum_{i=1}^{N_1} J_{ri}(Q) J_{\delta_1} + J_{\delta_{N_0}} \\
&= \sum_{i=1}^{N_1} \left(\frac{1}{2} U_{\gamma_i}^T K_{\gamma_i} U_{\gamma_i} - U_{\gamma_i}^T F_{\gamma_i} \right) \\
&\quad + \left(\frac{1}{2} u_1 C_a u_1 - g_a u_1 \right) \\
&\quad + \left(\frac{1}{2} u_{N_0} C_b u_{N_0} - g_b u_{N_0} \right) \\
&= \frac{1}{2} U^T K U - U^T F.
\end{aligned}
\tag{2.23}
$$

This is a quadratic function of the variables $u_1, \cdots, u_{N_0}$, and the coefficient matrix K is called the global stiffness matrix. K is symmetric and can be obtained by superposing the stiffness matrices of the individual line and point elements (being, respectively, of the second and the first orders) in the appropriate way. The vector F is called the load vector. F can be obtained in a similar manner by superimposing F_γ and F_δ of the individual line and point elements. This is the assembly step.

Note that the superposition producing K and F is performed on the existed basis, and that different elements can contribute to a given entry location of K and F. We illustrate this for the case $N_1 = 3$ as follows.

First construct a table, whose first four rows and four columns stand for the matrix K, while the last column stands for the vector F, and initially set all entries of K and F to zero. Superposing K_{γ_i} and F_{γ_i} ($i = 1, 2, 3$) for each line element γ_i, we obtain in turn

$$K_{\gamma_1} \qquad\qquad F_{\gamma_1}$$

	u_1	u_2	u_3	u_4	F
u_1	p_1/l_1^2	$-p_1/l_1^2$	0	0	$\frac{1}{2}f_1^{(1)}l_1$
u_2		p_1/l_1^2	0	0	$\frac{1}{2}f_2^{(1)}l_1$
u_3	sym.		0	0	0
u_4				0	0

$$K_{\gamma_1} + K_{\gamma_2} \qquad\qquad F_{\gamma_1} + F_{\gamma_2}$$

	u_1	u_2	u_3	u_4	F
u_1	p_1/l_1^2	$-p_1/l_1^2$	0	0	$\frac{1}{2}f_1^{(1)}l_1$
u_2		$p_1/l_1^2 + p_2/l_2^2$	$-p_2/l_2^2$	0	$\frac{1}{2}(f_2^{(1)}l_1 + f_1^{(2)}l_2)$
u_3	sym.		p_2/l_2^2	0	$\frac{1}{2}f_2^{(2)}l_2$
u_4				0	0

$$K_{\gamma_1} + K_{\gamma_2} + K_{\gamma_3} \qquad\qquad F_{\gamma_1} + F_{\gamma_2} + F_{\gamma_3}$$

	u_1	u_2	u_3	u_4	F
u_1	p_1/l_1^2	$-p_1/l_1^2$	0	0	$\frac{1}{2}f_1^{(1)}l_1$
u_2		$\dfrac{p_1}{l_1^2}+\dfrac{p_2}{l_2^2}$	$-p_2/l_2^2$	0	$\frac{1}{2}(f_2^{(1)}l_1 + f_1^{(2)}l_2)$
u_3	sym.		$\dfrac{p_2}{l_2^2}+\dfrac{p_3}{l_3^2}$	$-p_3/l_3^2$	$\frac{1}{2}(f_2^{(2)}l_2 + f_1^{(3)}l_3)$
u_4				p_3/l_3^2	$\frac{1}{2}f_2^{(3)}l_3$

Further superposing K_{δ_i} and F_{δ_i} $(i = 1, 4)$ for each point element δ_i, we finally obtain

$$
K = \begin{bmatrix}
\dfrac{p_1}{l_1^2} + C_a & -\dfrac{p_1}{l_1^2} & 0 & 0 \\[2mm]
 & \dfrac{p_1}{l_1^2} + \dfrac{p_2}{l_2^2} & \dfrac{-p_2}{l_2^2} & 0 \\[2mm]
 & & \dfrac{p-2}{l_2^2} + \dfrac{p_3}{l_3^2} & -\dfrac{p_3}{l_3^2} \\[2mm]
 & \text{sym.} & & \dfrac{p_3}{l_3^2} + C_b
\end{bmatrix},
$$

$$
F = \begin{bmatrix}
\dfrac{1}{2} f_1^{(1)} l_1 + g_a \\[2mm]
\dfrac{1}{2}(f_2^{(1)} l_1 + f_1^{(2)} l_2) \\[2mm]
\dfrac{1}{2}(f_2^{(2)} l_2 + f_1^{(3)} l_3) \\[2mm]
\dfrac{1}{2} f_2^{(3)} l_3 + g_b
\end{bmatrix}. \tag{2.24}
$$

The matrix K is symmetric, and it is easy to verify that K is positive definite as well, provided $C_a(\geq 0)$ and $C_b(\geq 0)$ do not vanish simultaneously. Furthermore, K is tridiagonal, its nonvanishing entries being concentrated on the two sides of the diagonal.

In (2.23), the quadratic functional of energy $J(u)$ is discretized into a multivariate function $J(u_1, \cdots, u_{N_0})$. The variational problem (2.4) is correspondingly discretized into an unconditional minimum problem for multivariate functions

$$
J(u_1, \cdots, u_{N_0}) = \frac{1}{2} U^T K U - U^T F = \text{Min.} \tag{2.25}
$$

According to the variational principle of linear algebra, the latter is equivalent to the system of linear equations

$$
KU = F. \tag{2.26}
$$

This is the final form to be solved. So far the discretizing process of the finite element method has been finished in the case without essential conditions.

The coefficient matrix of the system of equations (2.26) is symmetric, positive definite and banded. There are many approaches to solving such systems of equations. Among those commonly used at present are the frontal method (the direct method) and the overrelaxation method (a combination of the direct method and the iterative method).

We can see from the process of superposition used to form the global stiffness matrix K that product terms $u_i u_j$ do not arise in the discretized quadratic function of energy $J(u_1, \cdots, u_{N_0})$ when the point elements δ_i and δ_j are not adjacent, i.e., do not belong to the same line element. As a result, the entries k_{ij} of the corresponding matrix K must vanish. It follows that K is a sparse matrix, that is, the vast majority of its entries vanish. To save computer memory and computational effort, programs need only store the nonvanishing entries of K. This is called the method of condensation.

We can solve the system of equations (2.26) for $U = (u_1, \cdots, u_{N_0})^T$, which is just the discretized solution of the unknown function $u(x)$ at the nodal points $\delta_1, \cdots, \delta_{N_0}$. Sometimes in practical problems we also need to know the value of the derivative $\dfrac{du}{dx}$. This value can be obtained as follows. Since the interpolation function $Q(x)$ is piecewise linear, its first derivative is piecewise constant and can be taken as the value of $\dfrac{du}{dx}$ on the line element γ. If we need the value of $\dfrac{du}{dx}$ at a nodal point, then we can use the mean value of the derivatives on the two adjacent line elements.

2.5   Treatment of essential conditions

If the variational problem has further constraints, then after obtaining the coefficient matrix K and the right hand side F, we still need more work to obtain the system of linear equations which we must finally solve. As an example, we discuss the variational problem (2.2).

Following the steps given in the last section, we discretize the variational problem of (2.2) into

$$\begin{cases} J(u_1, \cdots, u_{N_0}) = \dfrac{1}{2} U^T K U - U^T F = \text{Min}, \\ u_1 = \bar{u}(a), \ u_{N_0} = \bar{u}(b). \end{cases} \qquad (2.27)$$

This is a constrained extremum problem, where the variables u_1 and u_{N_0} in $J(u_1, \cdots, u_{N_0})$ take prescribed values. Hence J can be regarded as a function of the other variables $u_2, \cdots, u_{N_0-1}$, making the conditional extremum problem into an unconditional extremum problem in variables. Thus, we should delete the first and N_0-th equations in the system

$$KU = F$$

and substitute the values of $u_1 = \bar{u}(a)$ and $u_{N_0} = \bar{u}(b)$ into the remaining equations after transposing u_1 and from the left hand to the right hand side. We then obtain $N_0 - 2$ equations in $N_0 - 2$ variables $u_2, \cdots, u_{N_0-1}$. The order of the coefficient matrix of the new system of equations is $N_0 - 2$, but the matrix is still symmetric, positive definite, and banded.

An equivalent procedure is to make the following modifications in the original coefficient matrix K and in the right hand member F:

Change the first entry of F to $\bar{u}(a)$, the N_0-th entry to $\bar{u}(b)$, and the remaining entries to

$$F_i - k_{i1}\bar{u}(a) - k_{iN_0}\bar{u}(b), \quad i = 2, \cdots, N_0 - 1.$$

The diagonal entries on the first and the N_0-th row are set to be 1 and all the other entries in the first row, the first column, the N_0-th row and the N_0-th column are set to zero.

Thus we obtain a system of equations, of the same order as the original one, whose coefficient matrix is still symmetric, positive definite, and banded. This procedure may be somewhat simpler to program, because it avoids the trouble of renumbering due to the deletion of equations and variables, but it requires somewhat more memory and computational effor.

Note that even if the coefficient p or f' in the right-hand of (2.1) is discontinuous, so long as the discontinuity point is taken as the nodal point and both p and f take the different values on the two sides of the discontinuity point during the evaluation of the line integrals, then the remaining steps of discretization are exactly the same as those in the case without discontinuity.

2.6 Applications of the quadratic element

In the preceding subsections, we used is linear interpolation on the linear elements. Its advantages are simplicity and convenience. Its disadvantages are low accuracy and that it requires many nodal points, resulting in a rather large coefficient matrix. In this section, we perform the element analysis by interpolating with quadratic functions, called quadratic elements. This is also quite simple, but is a tangible improvement over the method of linear elements.

Choose an arbitrary line element $\gamma = (\delta_1, \delta_2)$. By substituting the quadratic interpolation function $Q(x)$ in (2.8) for $u(x)$, the energy integral over γ is discretized into

$$J_\gamma(u) \approx J_\gamma(Q) = \frac{1}{2} \int_\gamma pQ'^2 dx - \int_\gamma fQ dx$$

$$= \frac{1}{2} U_\gamma^T K_\gamma U_\gamma - U_\gamma^T F_\gamma$$

where $U_\gamma = (u_1, u_2, u_3)^T$, while u_1, u_2, and u_3 are, respectively, the values of $u(x)$ at the two end points δ_1 and δ_2 and at the midpoint δ_3 of γ;

$$F_\gamma = (f_1, f_2, f_3)^T, \quad f_i = \int_\gamma f\mu_i dx, \quad i = 1, 2, 3; \tag{2.28}$$

$$K_\gamma = \begin{bmatrix} k_{11} & k_{12} & k_{13} \\ & k_{22} & k_{23} \\ \text{sym.} & & \\ & & k_{33} \end{bmatrix},$$

$$k_{ij} = \int_\gamma pu_i' u_j' dx, \quad i, j = 1, 2, 3. \tag{2.29}$$

K_γ is just the element stiffness matrix of the quadratic element. When p and f are constant on γ, according to Table 2 we obtain

$$K_\gamma = \frac{p}{3l} \begin{bmatrix} 7 & 1-8 \\ & 7-8 \\ \text{sym.} & \\ & 16 \end{bmatrix}, \tag{2.30}$$

$$F_\gamma = \frac{fl}{6}(1, 1, 4)^T. \tag{2.31}$$

Note that, the order of the three nodal points is $\delta_1, \delta_2, \delta_3$ in the evaluation of K_γ and F_γ. If we change the numbering so that δ_2 is the midpoint and δ_3 is the end point, then the K_γ and F_γ will become

$$K_\gamma = \frac{p}{3l}\begin{bmatrix} 7 & -8 & 1 \\ & 16 & -8 \\ \text{sym.} & & \\ & & 7 \end{bmatrix},$$

$$F_\gamma = \frac{fl}{6}(1,4,1)^T.$$

For quadratic elements, the analysis of point elements and the assembly step are just the same as for linear elements, so we will not repeat them here.

In order to compare the quadratic element with the linear element method, we give a solution using quadratic elements of the variational problem (2.4). When $N_1 = 3$, the coefficient matrix K and the right-hand vector F (assuming that p and f are constant on the individual line elements) are,

$$K = \frac{1}{3}\begin{bmatrix} \frac{7p_1}{l_1}+C_a & \frac{-8p_1}{l_1} & \frac{p_1}{l_1} & & & & \\ & \frac{16p_1}{l_1} & \frac{-8p_1}{l_1} & & & & \\ & & \frac{7p_1}{l_1}+\frac{7p_2}{l_2} & \frac{-8p_2}{l_2} & \frac{p_2}{l_2} & & 0 \\ & & & \frac{16p_2}{l_2} & \frac{-8p_2}{l_2} & & \\ & & \text{sym.} & & \frac{7p_2}{l_2}+\frac{7p_3}{l_3} & \frac{-8p_3}{l_3} & \frac{p_3}{l_3} \\ & & & & & \frac{16p_3}{l_3} & \frac{-8p_3}{l_3} \\ & & & & & & \frac{7p_3}{l_3}+C_b \end{bmatrix}_{7\times 7} \qquad (2.32)$$

$$F_\gamma = \frac{1}{6}(f_1l_1 + g_a, 4f_1l_1, f_1l_1 + f_2l_2, 4f_2l_2, f_2l_2 + f_3l_3, 4f_3l_3, f_3l_3 + g_b)^T. \qquad (2.33)$$

The solution vector is

$$U = (u_1, u_2, u_3, u_4, u_5, u_6, u_7)^T,$$

where, u_2, u_4, and u_6 are respectively the displacements at the midpoints of the line elements γ_1, γ_2, and γ_3.

It is easy to verify that the coefficient matrix of the quadratic element is still symmetric, positive definite, and sparse. The proof of positive definiteness is as follows. Since the quadratic functional

$$D(u,u) = \frac{1}{2}\int_a^b pu'^2 dx + \frac{1}{2}C_a u^2(a) + \frac{1}{2}C_b u^2(b)$$

is positive definite, i.e.

$$D(u,u) = 0 \Longleftrightarrow u \equiv 0,$$

and we know from the process of assembling K that

$$\frac{1}{2}U^T KU = D(Q,Q),$$

we have

$$U^T KU = 0 \Longleftrightarrow D(Q,Q) = 0 \rightarrow Q \equiv 0 \rightarrow U = 0,$$

Hence K is positive definite.

It is a general property of the finite element method that a positive definite variational problem gives rise to a positive definite coefficient matrix.

§3   Bending of Beams

3.1   Variational problems

According to §7 of Chapter 1, the equilibrium equation for plane bending of beams is

$$\frac{d^2}{dx^2}\left(D\frac{d^2 u}{dx^2}\right) = f, \quad a < x < b, \tag{3.1}$$

where $D = EI_z$ denotes the flexural rigidity of the cross section while f denotes the transverse load distribution.

As boundary conditions, we take the left end to be fixed and the right end to be free, i.e.,

$$\begin{cases} u(a) = u'(a) = 0, & x = a, \tag{3.2} \\ u''(b) = (Du''(x))'_b = 0, & x = b. \tag{3.3} \end{cases}$$

we assume that the flexural rigidity D has a discontinity at $x = \xi$, Consequently, we have the interface condition

$$[Du'']_{\varepsilon-0}^{\varepsilon+0} = 0, \tag{3.4}$$

and the continuity conditions for the displacement and the rotation angle:

$$[u]_{\varepsilon-0}^{\varepsilon+0} = 0, \quad [u']_{\varepsilon-0}^{\varepsilon+0} = 0. \tag{3.5}$$

More generally, let us consider the problem of beams on elastic foundation. Then the equilibrium equation (3.1) must be replaced by

$$\frac{d^2}{dx^2}\left(D\frac{d^2u}{dx^2}\right) + cu = f, \quad a < x < b, \tag{3.6}$$

where $c > 0$ is a constant of elastic coupling with the foundation.

By the variational principle, the equilibrium equation (3.6), the boundary condtiions (3.2) and (3.3), and the interface conditions (3.4) and (3.5) are, taken together, equivalent to the following variational problem

$$\begin{cases} J(u) = \dfrac{1}{2}\displaystyle\int_a^b Du''^2 dx + \dfrac{1}{2}\displaystyle\int_a^b cu^2 dx - \displaystyle\int_a^b fu\,dx = \text{Min}, \\ u(a) = u'(a) = 0. \end{cases} \tag{3.7}$$

The boundary conditions (3.3) and the interface condition (3.4) are all the natural boundary conditions of the variational problem (3.7), which are automatically satisfied by the extremum function and hence donot need to be listed explicitly. The continuity conditions (3.5) at the discontinuity point of the medium also don't need to be listed explicitly as constraint conditions of the variational problem, so long as we prescribe that u and u' are single-valued at $x = \xi$.

3.2 The cubic hermite element

(1) The interpolation polynomial

Since the second derivative of the displacement u is contained in the energy integral of beams in bending, the first derivative of u must be continuous. The linear or quadratic elements used so far for rods in tension can only assure the continuity of the interpolation function itself. The first derivative of the interpolation function is only is piecewise continuous, so that these two kinds of elements are not generally used for beams in bending.

In order to obtain an interpolation function with continuous first derivatives, we must prescribe the value of the first derivatives (the rotation angle) in addition to the value of the displacement u at the nodal point. This is called Hermite interpolation.

Consider an arbitrary line element γ with end points δ_1 and δ_2. Let s be the variable of the chord length from δ_1 to δ_2, and let l be the length of γ. We require a cubic polynomial $Q(s)$ inter polating the function $u(s)$ on γ such that

$$\begin{cases} Q(s) = a_1 s^3 + a_2 s^2 + a_3 s + a_4, \\ Q(0) = u_1, \ Q(l) = u_2, \ Q'(0) = u_1', \ Q'(l) = u_2'. \end{cases}$$

The four coefficients a_1, a_2, a_3 and a_4 of $Q(s)$ are determined by the values u_1, u_2, u_1', and u_2' at the end points of γ. Rewrite the interpolation function as

$$Q(s) = u_1 \varphi(s) + u_2 \varphi_2(s) + u_1' \psi_1(s) + u_2' \psi_2(s), \qquad (3.8)$$

where, $\varphi_1, \varphi_2, \psi_1$ and ψ_2 are cubic polynomials of s satisfying the relations

$$\begin{cases} \varphi_1(0) = 1, \ \varphi_1(l) = \varphi_1'(0) = \varphi_1'(l) = 0, \\ \varphi_2(l) = 1, \ \varphi(0) = \varphi_2'(0) = \varphi_2'(l) = 0, \\ \psi_1'(0) = 1, \ \psi_1(0) = \psi_1(l) = \psi_1'(l) = 0, \\ \psi_2'(l) = 1, \ \psi_2(0) = \psi_2(l) = \psi_2'(0) = 0. \end{cases} \qquad (3.9)$$

Accordingly, it is easy to show that

$$\begin{cases} \varphi_1(s) = 1 - 3\left(\dfrac{s}{l}\right)^2 + 2\left(\dfrac{s}{l}\right)^3 = \lambda_1^2(3 - 2\lambda_1), \\ \varphi_2(s) = 3\left(\dfrac{s}{l}\right)^2 - 2\left(\dfrac{s}{l}\right)^3 = \lambda_2^2(3 - 2\lambda_2), \\ \psi_1(s) = l\left[\dfrac{s}{l} - 2\left(\dfrac{s}{l}\right)^2 + \left(\dfrac{s}{l}\right)^3\right] = \lambda_1^2(\lambda_1 - 1), \\ \psi_2(s) = -l\left[\left(\dfrac{s}{l}\right)^2 - \left(\dfrac{s}{l}\right)^3\right] = \lambda_2^2(\lambda_2 - 1), \end{cases}$$

where λ_1 and λ_2 are the basis functions of the linear interpolation on γ. The functions $\varphi_1, \varphi_2, \psi_1$ and ψ_2 are called the basis functions of the cubic Hermite interpolation on γ.

(2) Analysis of elements

As in §2, divide the interval $[a, b]$ into line elements $\gamma_1, \cdots, \gamma_{N_1}$ and point elements $\delta_1, \cdots, \delta_{N_0}$, and suppose the discontinuity point ξ of D is one of the point elements. Substituting the cubic Hermite interpolation polynomial $Q(x)$ for the displacement $u(x)$ on each line element γ, we obtain a piecewise cubic function $Q(x)$ on the whole interval $[a, b]$, whose first derivative is continuous and whose second derivative is piecewise continuous. Consequently, $Q(x)$ is an admissible interpolation function for the variational problem (3.7).

The energy integral over the line element γ is

$$J_\gamma(u) = \frac{1}{2} \int_\gamma D u''^2 dx + \frac{1}{2} \int_\gamma c u^2 dx - \int_\gamma f u dx \approx J_\gamma(Q).$$

For convenience, we introduce the notation:

$$U_\gamma = (u_1, u_1', u_2, u_2')^T,$$
$$N(x) = (\varphi_1(x), \psi_1(x), \varphi_2(x), \psi_2(x))^T.$$

(3.8) can then be abbreviated as $Q(x) = U_\gamma^T N(x)$. The various terms of the energy integral $J_\gamma(Q)$ can be expressed as

$$\frac{1}{2} \int_\gamma D Q''^2 dx = \frac{1}{2} \int_\gamma D U_\gamma^T N'' N''^T U_\gamma dx = \frac{1}{2} U_\gamma^T K_\gamma U_\gamma,$$
$$\frac{1}{2} \int_\gamma c Q^2 dx = \frac{1}{2} \int_\gamma c U_\gamma^T N N^T U_\gamma dx = \frac{1}{2} U_\gamma^T G_\gamma U_\gamma,$$
$$\int_\gamma f Q dx = \int_\gamma f U_\gamma^T N dx = U_\gamma^T F_\gamma,$$

so that

$$J_\gamma(u) \approx J_\gamma(Q) = \frac{1}{2} U_\gamma^T (K_\gamma + G_\gamma) U_\gamma - U_\gamma^T F_\gamma, \tag{3.11}$$

where

$$K_\gamma = \int_\gamma D N'' N''^T dx, \tag{3.12}$$

$$G_\gamma = \int_\gamma c N N^T dx, \tag{3.13}$$

$$F_\gamma = \int_\gamma f N dx. \tag{3.14}$$

If D and c are constant on γ, then we can use formula (2.11) to evaluate

$$K_\gamma = D \begin{bmatrix} \int_\gamma \varphi_1''^2 dx & \int_\gamma \varphi_1'' \psi_1'' dx & \int_\gamma \varphi_1'' \varphi_2''^2 dx & \int_\gamma \varphi_1'' \psi_2'' dx \\ & \int_\gamma \psi_1''^2 dx & \int_\gamma \psi_1'' \varphi_2'' dx & \int_\gamma \psi_1'' \psi_2'' dx \\ & & \int_\gamma \varphi_2''^2 dx & \int_\gamma \varphi_2'' \psi_2'' dx \\ & \text{sym.} & & \int_\gamma \psi_2''^2 dx \end{bmatrix}$$

$$= D \begin{bmatrix} \dfrac{12}{l^3} & \dfrac{6}{l^2} & \dfrac{-12}{l^3} & \dfrac{6}{l^2} \\[2mm] & \dfrac{4}{l} & \dfrac{-6}{l^2} & \dfrac{2}{l} \\[2mm] & & \dfrac{12}{l^3} & \dfrac{-6}{l^2} \\[2mm] \text{sym.} & & & \dfrac{4}{l} \end{bmatrix}, \tag{3.15}$$

$$G_\gamma = c \begin{bmatrix} \int_\gamma x_1^2 dx & \int_\gamma \varphi \psi_1 dx & \int_\gamma \varphi_1 \varphi_2 dx & \int_\gamma \varphi_1 \psi_2 dx \\ & \int_\gamma \psi_1^2 dx & \int_\gamma \psi_1 \varphi_2 dx & \int_\gamma \psi_1 \psi_2 dx \\ & \text{sym.} & \int_\gamma \varphi_2^2 dx & \int_\gamma \varphi_2 \psi_2 dx \\ & & & \int_\gamma \psi_2^2 dx \end{bmatrix}$$

$$= c \begin{bmatrix} \dfrac{13l}{35} & \dfrac{11l^2}{210} & \dfrac{9l}{70} & \dfrac{-13l^2}{420} \\[2mm] & \dfrac{l^3}{105} & \dfrac{13l}{420} & \dfrac{-l^3}{140} \\[2mm] & & \dfrac{13l}{35} & \dfrac{-11l^2}{210} \\[2mm] \text{sym.} & & & \dfrac{l^3}{105} \end{bmatrix}. \tag{3.16}$$

The following are concrete forms of F_γ for several commonly used load distributions f:

1. f is constant on γ. Then

$$F_\gamma = f \left(\frac{l}{2}, \frac{l^2}{12}, \frac{l}{2}, -\frac{l^2}{12} \right)^T. \tag{3.17}$$

2. f is linearly distributed over γ. Suppose at the left end point $f = f_1$, while at the right end point $f = f_2$, then

$$F_\gamma = \left(\frac{l}{20}(7f_1 + 3f_2), \frac{l^2}{10}\left(\frac{f_1}{2} + \frac{f_2}{3}\right), \frac{l}{20}(3f_1 + 7f_2),\right.$$
$$\left.\frac{-l^2}{10}\left(\frac{f_1}{3} + \frac{f_2}{2}\right)\right)^T. \tag{3.18}$$

3. f is a concentrated force of intensity q acting at point η on γ. Suppose the distance between η and the left end point is $sl, 0 \le s \le 1$, then

$$F_\gamma = q(\varphi_1(sl), \psi_1(sl), \varphi_2(sl), \psi_2(sl))^T$$
$$= q(1 - 3s^2 + 2s^2, l(s - 2s^2 + s^3), 3s^2 - 2s^3, -l(s^2 - s^3))^T. \tag{3.19}$$

After the nodal values U of the displacement u have been found by solving the system of linear equations

$$KU = F,$$

we must still evaluate the rotation angle θ and the bending moment M. On γ we can take

$$\theta = \frac{du}{dx} \approx dQ/dx = \varphi_1' u_1 + \psi_1' u_1' + \varphi_2' u_2 + \psi_2' u_2' \quad \text{(including the nodal points)}, \tag{3.20}$$

$$\frac{1}{D}M = \frac{d^2u}{dx^2} \approx \frac{d^2Q}{dx^2} = \varphi_1'' u_1 + \psi_1'' u_1' + \varphi_2'' u_2 + \psi_2'' u_2'$$

(not including the nodal points). $\tag{3.21}$

Since $\dfrac{d^2Q}{dx^2}$ is generally discontinuous at nodal points, if we need the value

of $\dfrac{d^2u}{dx^2}$ at the nodal point, then we can approximate it using the average

of the values of $\dfrac{d^2Q}{dx^2}$ at the midpoints of the two adjacent elements.

§4   Poisson Equation

4.1   Variational problems

So far, we have introduced the finite element method for one-dimensional problems. In this section we shall start to discuss the two-dimensional problem. Consider first the classical second order elliptic partial differential equation—Poissons equation

$$-\left[\frac{\partial}{\partial x}\left(p\frac{\partial u}{\partial x}\right) + \frac{\partial}{\partial y}\left(p\frac{\partial u}{\partial y}\right)\right] = f, \ (x,y) \in \Omega, \tag{4.1}$$

where Ω is a bounded domain on the (x,y)-plane, $p = p(x,y) > 0$ is the prescribed coefficient, and $f = f(x,y)$ is the prescribed right-hand distribution. For the torsion of cylinders, both the torsion function ψ and the stress function φ satisfy (4.1), has already been discussed in §3 of Chapter 3. Some other problems, such as the bending of membranes and plane heat transfer, can also be summarized this equation.

Since (4.1) is a second order equation, to determine a unique solution we need to prescribe on $\partial\Omega$ boundary conditions of the following three kinds:

1. $u = \bar{u}$;

2. $p\dfrac{\partial u}{\partial n} = g$;

3. $p\dfrac{\partial u}{\partial n} + cu = g, \ c \geq 0.$

Here, n denotes the outward normal direction to the boundary $\partial\Omega$. The second kind of boundary condition can be considered as a special case of the third kind with $c = 0$. The different kinds of boundary conditions can be prescribed on different segments of the boundary.

To be definite, we assume that Γ_1 and Γ_3 are two complementary parts of the boundary $\partial\Omega$, i.e.,

$$\partial\Omega = \Gamma_1 + \Gamma_3, \ \Gamma_1 \cap \Gamma_3 = 0,$$

and prescribe

$$\begin{cases} \Gamma_1 : u = \bar{u}, \\ \Gamma_3 : p\dfrac{\partial u}{\partial n} + cu = g. \end{cases} \tag{4.2}$$

Equation (4.1) and the boundary conditions (4.2) are together equivalent to the variational problem

$$
\begin{cases}
J(u) = \displaystyle\iint_\Omega \left\{ \frac{1}{2}p\left[\left(\frac{\partial u}{\partial x}\right)^2 + \left(\frac{\partial u}{\partial y}\right)^2\right] - fu \right\} dx\,dy \\[2mm]
\qquad + \displaystyle\int_{\Gamma_3} \left[\frac{1}{2}cu^2 - gu\right] ds \\[2mm]
= \min, \\[2mm]
\Gamma_1 : u = \bar{u}.
\end{cases}
\tag{4.3}
$$

When the coefficient p of the medium has a discontinuity in the interior of Ω, let the discontinuity occur on a line L that subdivides the domain Ω into several smaller pieces. For simplicity, suppose it is subdivided into two pieces $\Omega = \Omega^+ + \Omega^-$. We prescribe a positive direction on L, for instance, the normal direction ν pointing to Ω^+ from Ω^-. Then, using Green's formula on Ω^- and Ω^+, respectively, we can derive the interface condition

$$
L : \left(p\frac{\partial u}{\partial \nu}\right)^- - \left(p\frac{\partial u}{\partial \nu}\right)^+ = 0,
$$

on the discontinuity line L. This condition needed to determine a unique solution to the differential equation (4.1), but it is automatically satisfied in the variational problem (4.3) and need not be listed as a constraint.

4.2   Subdivision and interpolation

The two dimensional problem is much more complicated than the one dimensional problem. Nevertheless, the fundamental idea and the process of discretization of the finite element method are still roughly similar to that of the one-dimensional problem.

(1) Geometric subdivision

First subdivide the domain Ω of undetermined solutions. The fundamental element could a triangle, rectangle, quadrangle, etc. Among these, the trianglular subdivision is the simplest and the most often used one, so we will introduce it first.

The domain Ω may be either simply connected or multiply-connected, i.e., with holes. If there exist q holes, then the boundary $\partial\Omega$ is composed of $q + 1$ closed curves.

Using a broken line to approximate the boundary $\partial\Omega$ of Ω, we obtain a polygonal domain which we still denote Ω, temporarily ignoring the error caused by the substitution of straight for curved. Divide the polygonal domain Ω into triangles more precisely subdivide Ω into

area elements: $\beta_1, \cdots, \beta_{N_1}$,

line elements: $\gamma_1 \cdots, \gamma_{N_1}$, and

point elements: $\delta_1, \cdots, \delta_{N_0}$.

We call all of these elements. The area element is just the triangle, the line elements are just the edges of the triangles, and the point elements are just the vertices of the triangles.

We point out, by the way, that the numbers of the point, line, and area elements N_0, N_1 and N_2 satisfy Euler's formula

$$N_0 - N_1 + N_2 = 1 - q, \qquad (4.4)$$

where, q is just the number of the holes in Ω. Formula (4.4) can be used to check the information in programs.

Having numbered the point, line, and area elements, and having given the following information:

1.   the coordinates of the point elements $(x_k, y_k), k = 1, 2, \cdots, N_0$;
2.   the numbers of the two end points of the line elements;
3.   the number of the three vertices of the area elements;

the geometric subdivision of Ω is entirely determined.

If the coefficient p has discontinuities, then the discontinuity line L should be approximated as closely as possible by line elements of the subdivision. To sum up, the subdivision is basically arbitrary, and it may be denser in some places and sparser in others according to practical demands.

(2) Linear interpolation on triangles

Consider an arbitrary triangle β. Suppose that the three vertices of β are δ_1, δ_2, and δ_3 taken in anti-clockwise order (Fig. 68), and that the coordinates of δ_i are $(x_i, y_i), i = 1, 2, 3$. Linearly interpolate the function $u(x, y)$ by

$$\begin{cases} Q(x, y) = ax + by + c, \\ Q(x_i, y_i) = u(x_i, y_i) = u_i, \quad i = 1, 2, 3, \end{cases}$$

where the coefficients a, b, and c are determined by three interpolation conditions. We can also rewrite the interpolation function as

$$Q(x, y) = u_1 \lambda_1 + u_2 \lambda_2 + u_3 \lambda_3, \qquad (4.5)$$

where, λ_i $(i = 1, 2, 3)$ are all linear polynomials of x and y satisfying the relations

$$\lambda_i(x_j, y_j) = \delta_{ij}, \ i, j = 1, 2, 3.$$

Accordingly, we obtain

$$
\begin{cases}
\lambda_1(x,y) = \dfrac{1}{2\Delta}(\eta_1 x - \xi_1 y + \omega_1), \\[2mm]
\lambda_2(x,y) = \dfrac{1}{2\Delta}(\eta_2 x - \xi_2 y + \omega_2), \\[2mm]
\lambda_3(x,y) = \dfrac{1}{2\Delta}(\eta_3 x - \xi_3 y + \omega_3),
\end{cases}
\tag{4.6}
$$

where,

$$
\begin{aligned}
&\eta_1 = y_2 - y_3,\ \eta_2 = y_3 - y_1,\ \eta_3 = y_1 - y_2, \\
&\xi_1 = x_2 - x_3,\ \xi_2 = x_3 - x_1,\ \xi_3 = x_1 - x_2, \\
&\omega_1 = x_2 y_3 - x_3 y_2,\ \omega_2 = x_3 y_1 - x_1 y_3,\ \omega_3 = x_1 y_2 - x_2 y_1
\end{aligned}
\tag{4.7}
$$

$$
2\Delta =
\begin{vmatrix}
x_1 & y_1 & 1 \\
x_2 & y_2 & 1 \\
x_3 & y_3 & 1
\end{vmatrix}
> 0,
\tag{4.8}
$$

Δ is the area of the triangle β. Note that when the three vertices of the triangle are ordered in the clockwise direction, the determinant

$$
\begin{vmatrix}
x_1 & y_1 & 1 \\
x_2 & y_2 & 1 \\
x_3 & y_3 & 1
\end{vmatrix}
= -2\Delta < 0.
$$

The functions λ_1, λ_2, and λ_3 are the basis functions of linear interpolation on the triangle. From (4.6),

$$
\frac{\partial \lambda_i}{\partial x} = \eta_i/2\Delta, \quad \frac{\partial \lambda_i}{\partial y} = -\xi_i/2\Delta.
\tag{4.9}
$$

When the interpolated function $u(x,y)$ itself is just a linear function, we obviously have

$$
u(x,y) \equiv Q(x,y).
$$

Hence, letting $u(x,y)$ be $1, x$, and y in turn, we immediately obtain the following identities

$$
\begin{cases}
1 = \lambda_1 + \lambda_2 + \lambda_3, \\
x \equiv x_1\lambda_1 + x_2\lambda_2 + x_3\lambda_3, \\
y \equiv y_1\lambda_1 + y_2\lambda_2 + y_3\lambda_3.
\end{cases}
\tag{4.10}
$$

(3) Centroid coordinates and numerical quadrature

The linear interpolation basis functions λ_1, λ_2, and λ_3 occupy a very important place in interpolation and in calculus operations on the triangle $\beta = (\delta_1, \delta_2, \delta_3)$. We shall discuss these three functions in more detail in the following.

Consider an arbitrary point δ. Let its rectangular coordinates be (x, y). The corresponding functions $(\lambda_1, \lambda_2, \lambda_3)$ can be evaluated from formula (4.6). Only two of these three functions, however, are independent, because the three functions satisfy identity

$$\lambda_1 + \lambda_2 + \lambda_3 \equiv 1.$$

Conversely, give $(\lambda_1, \lambda_2, \lambda_3)$ satisfying the above identity, the corresponding (x, y) can be evaluated from the formula (4.10). Hence, like $(x, y), (\lambda_1, \lambda_2, \lambda_3)$ can also be used as coordinates for points in the plane. These are called centroid coordinates. The reason is that if we place three masses denoted λ_1, λ_2, and λ_3, at the vertices δ_1, δ_2 and δ_3, then the centroid $\delta = (x, y)$ of this system of masses will be

$$x = (x_1\lambda_1 + x_2\lambda_2 + x_3\lambda_3)/(\lambda_1 + \lambda_2 + \lambda_3) = x_1\lambda_1 + x_2\lambda_2 + x_3\lambda_3,$$

$$y = (y_1\lambda_1 + y_2\lambda_2 + y_3\lambda_3)/(\lambda_1 + \lambda_2 + \lambda_3) = y_1\lambda_1 + y_2\lambda_2 + y_3\lambda_3.$$

These are just the formulas (4.9).

On area elements, the centroid coordinates are sometimes called area coordinates. For consider a point $\delta = (x, y)$ inside the triangle $(\delta_1, \delta_2, \delta_3)$. $\delta\delta_2\delta_3, \delta\delta_3\delta_1$, and $\delta\delta_1\delta_2$ are all anticlockwise because δ_1, δ_2 and δ_3 are anticlockwise. We have the following area formulas (Fig. 68):

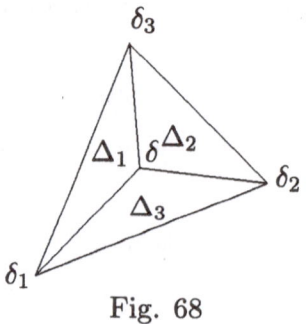

Fig. 68

$$\Delta = \text{the area of the triangle } \delta_1\delta_2\delta_3 = \frac{1}{2}\begin{vmatrix} x_1 & y_1 & 1 \\ x_2 & y_2 & 1, \\ x_3 & y_3 & 1 \end{vmatrix},$$

$$\Delta_1 = \text{the area of the triangle } \delta\delta_2\delta_3 = \frac{1}{2}\begin{vmatrix} x & y & 1 \\ x_2 & y_2 & 1, \\ x_3 & y_3 & 1 \end{vmatrix}$$

$$= \frac{1}{2}(\eta_1 x - \xi_1 y + \omega_1) = \lambda_1\Delta,$$

$$\Delta_2 = \text{the area of the triangle } \delta\delta_3\delta_1 = \frac{1}{2}\begin{vmatrix} x & y & 1 \\ x_3 & y_3 & 1, \\ x_1 & y_1 & 1 \end{vmatrix}$$

$$= \frac{1}{2}(\eta_2 x - \xi_2 y + \omega_2) = \lambda_2\Delta,$$

$$\Delta_3 = \text{the area of the triangle } \delta\delta_1\delta_2 = \frac{1}{2}\begin{vmatrix} x & y & 1 \\ x_1 & y_1 & 1, \\ x_2 & y_2 & 1 \end{vmatrix}$$

$$= \frac{1}{2}(\eta_3 x - \xi_3 y + \omega_3) = \lambda_3\Delta.$$

Hence, the controid coordinates $(\lambda_1, \lambda_2, \lambda_3)$ of the point $\delta = (x, y)$ are just the ratios of the area of the triangles $\dfrac{\Delta_1}{\Delta}, \dfrac{\Delta_2}{\Delta}, \dfrac{\Delta_3}{\Delta}$.

As we vary the point $\delta = (x, y)$ in the triangle, the values of the centroid coordinates $(\lambda_1, \lambda_2, \lambda_3)$ vary through

$$0 \le \lambda_1, \lambda_2, \lambda_3 \le 1, \lambda_1 + \lambda_2 + \lambda_3 = 1.$$

The centroid coordinates of the three vertices δ_1, δ_2 and δ_3 are $(1, 0, 0)$, $(0, 1, 0)$, and $(0, 0, 1)$ in turn. Furthermore, $\lambda_i = 0$ $(i = 1, 2, 3)$ on the edge opposite to the vertex δ_i.

When analyzing area elements, we often want to evaluate integrals of the form

$$\iint_\beta \lambda_1^{p_1} \lambda_2^{p_2} \lambda_3^{p_3} \, dx dy,$$

where, p_1, p_2 and p_3 are nonnegative integers. Take λ_1 and λ_2 as the independent variables and $\lambda_3 = 1 - \lambda_1 - \lambda_2$. By (4.6),

$$\lambda_i(x, y) = \frac{1}{2\Delta}(\eta_i x - \xi_i y + \omega_i), \quad i = 1, 2.$$

This equation transforms an arbitrary triangle $\beta = (\delta_1, \delta_2, \delta_3)$ in the $(x - y)$-plane into the right triangle on the $(\lambda_1 - \lambda_2)$-plane

$$\beta' : 0 \le \lambda_2 \le 1, \ 0 \le \lambda_1 \le 1 - \lambda_2.$$

The Jacobian of this transformation is

$$\frac{\partial(\lambda_1, \lambda_2)}{\partial(x, y)} = \begin{vmatrix} \dfrac{\partial \lambda_1}{\partial x} & \dfrac{\partial \lambda_1}{\partial y} \\[2mm] \dfrac{\partial \lambda_2}{\partial x} & \dfrac{\partial \lambda_2}{\partial y} \end{vmatrix} = \frac{1}{(2\Delta)^2} \begin{vmatrix} \eta_1 - \xi_1 \\ \eta_2 - \xi_2 \end{vmatrix} = \frac{1}{2\Delta},$$

$$\frac{\partial(x, y)}{\partial(\lambda_1, \lambda_2)} = 2\Delta.$$

Thus, the integration over the triangle β can be carried out over the right triangle β' by transforming to centroid coordinates:

$$\iint_\beta \lambda_1^{p_1} \lambda_2^{p_2} \lambda_3^{p_3} \, dxdy = \iint_{\beta'} \lambda_1^{p} \lambda_2^{p}(1 - \lambda_1 - \lambda_2)^{p_3} \left| \frac{\partial(x, y)}{\partial(\lambda_1, \lambda_2)} \right| d\lambda_1 d\lambda_2$$

$$= 2\Delta \int_0^1 d\lambda_2 \int_0^{1-\lambda_2} \lambda_1^{p_1} \lambda_2^{p_2} (1 - \lambda_1 - \lambda_2)^{p_3} d\lambda_1$$

$$= 2\Delta \int_0^1 \lambda_2^{p_2}(1 - \lambda_2)^{p_1+p_3+1} d\lambda_2 \int_0^1 t^{p_1}(1 - t)^{p_3} dt,$$

then by the use of Euler's formula (2.10), we obtain

$$\iint_\beta \lambda_1^{p_1} \lambda_2^{p_2} \lambda_3^{p_3} \, dxdy = 2\Delta \frac{p_1! p_2! p_3!}{(p_1 + p_2 + p_3 + 2)!}. \tag{4.11}$$

We can compute the following Table from this equation and (4.6).

Numerical quadrature on triangles can also be expressed in terms of the centroid coordinates. The general form is

$$\iint_\beta F(\lambda_1, \lambda_2, \lambda_3) dxdy \approx \Delta \sum_{k=1}^m \rho^{(k)} F(\lambda_1^{(k)}, \lambda_2^{(k)}, \lambda_3^{(k)}), \tag{4.12}$$

where $(\lambda_1^{(k)}, \lambda_2^{(k)}, \lambda_3^{(k)})$, $k = 1, \cdots, m$, are nodes specified in terms of centroid coordinates and $\rho^{(k)}$ are the corresponding weights. Several frequently used formulas for numerical quadrature are listed in Table 4, where accuracy n implies that the formula is exact for polynomials of order n.

Table 3 $\iint_\beta \varphi\psi\, dx\, dy$

$\diagdown \begin{matrix}\varphi\\\psi\end{matrix}$	1	λ_j	$\dfrac{\partial\lambda_f}{\partial x}$	$\dfrac{\partial\lambda_j}{\partial y}$
1	Δ	$\dfrac{\Delta}{3}$	$\dfrac{\eta_j}{2}$	$\dfrac{-\xi_j}{2}$
λ_i		$\dfrac{(1+\delta_{ij})\Delta}{12}$	$\dfrac{\eta_j}{6}$	$\dfrac{-\xi_j}{6}$
$\dfrac{\partial\lambda_i}{\partial x}$			$\dfrac{\eta_i\eta_j}{4\Delta}$	$\dfrac{-\xi_j\eta_i}{4\Delta}$
$\dfrac{\partial\lambda_i}{\partial y}$				$\dfrac{\xi_i\eta_j}{4\Delta}$

Like the chord length s, (λ_1, λ_2) can also be taken as the coordinates for the line element $\gamma = (\delta_1, \delta_2)$. Here (λ_1, λ_2) are determined from (2.6) and (2.7),

$$\lambda_1 = 1 - \frac{s}{l}, \quad \lambda_2 = \frac{s}{l}$$

$$\lambda_1 + \lambda_2 = 1.$$

Let

l_1 = the length of the line segment $\delta\delta_2 = l - s$,

l_2 = the length of the line segment $\delta_1\delta = s$.

Then the centroid coordinates (λ_1, λ_2) corresponding to the point $\delta = \delta(s)$ are just the ratios of the line segment lengths l_1/l and l_2/l.

Numerical quadrature over line elements can also be expressed in terms of centroid coordinates. Its general form is

$$\int_\gamma F(\lambda_1, \lambda_2)ds = l\sum_{k=1}^{m} \rho^{(k)} F(\lambda_1^{(k)}, \lambda_2^{(k)}). \qquad (4.13)$$

For the nodes $(\lambda_1^{(k)}, \lambda_2^{(k)})$ and weights $\rho^{(k)}$, see Table 5.

Table 4   Formulas for numerical quadrature on a triangle

number of nodes m	coordinates of nodes $(\lambda_1, \lambda_2, \lambda_3)$	weights ρ	accuracy n
1	$\left(\dfrac{1}{3}, \dfrac{1}{3}, \dfrac{1}{3}\right)$	1	1
3	$\left(0, \dfrac{1}{2}, \dfrac{1}{2}\right)$ $\left(\dfrac{1}{2}, 0, \dfrac{1}{2}\right)$ $\left(\dfrac{1}{2}, \dfrac{1}{2}, 0\right)$	$\dfrac{1}{3}$	2
7	$\left(\dfrac{1}{3}, \dfrac{1}{3}, \dfrac{1}{3}\right)$	$\dfrac{27}{60}$	
7	$\left(0, \dfrac{1}{2}, \dfrac{1}{2}\right)$ $\left(\dfrac{1}{2}, 0, \dfrac{1}{2}\right)$ $\left(\dfrac{1}{2}, \dfrac{1}{2}, 0\right)$	$\dfrac{8}{60}$	3
	$(1, 0, 0)$ $(0, 1, 0)$ $(0, 0, 1)$	$\dfrac{3}{60}$	
7	$\left(\dfrac{1}{3}, \dfrac{1}{3}, \dfrac{1}{3}\right)$	0.225	
	$(\alpha_1, \beta_1, \beta_1)$ $(\beta_1, \alpha_1, \beta_1)$ $(\beta_1, \beta_1, \alpha_1)$	0.13239415	5
$\alpha_1 = 0.05961587$ $\beta_1 = 0.47014206$ $\alpha_2 = 0.79742699$ $\beta_2 = 0.10128651$	$(\alpha_2, \beta_2, \beta_2)$ $(\beta_2, \alpha_2, \beta_2)$ $(\beta_2, \beta_2, \alpha_2)$	0.12593918	

Table 5 Formulas for numerical quadrature over line element

number of nodes m	coordinates of nodes (λ_1, λ_2)	wieghts ρ	accuracy n
1	$\left(\dfrac{1}{2}, \dfrac{1}{2}\right)$	1	1
2 $\alpha_1 = 0.2113248654$ $1 - \alpha_1 = 0.7886751346$	$\left.\begin{array}{c} \alpha_1(1 - \alpha_1) \\[1em] (1 - \alpha_1, \alpha_1) \end{array}\right\}$	$\dfrac{1}{2}$	3
3	$\left.\begin{array}{c} \left(\dfrac{1}{2}, \dfrac{1}{2}\right) \\[1em] (0, 1) \\[1em] (1, 0) \end{array}\right\}$	$\dfrac{4}{6}$ $\dfrac{1}{6}$	3
3 $\alpha_2 = 0.1127016654$ $1 - \alpha_2 = 0.8872983346$	$\left.\begin{array}{c} \left(\dfrac{1}{2}, \dfrac{1}{2}\right) \\[1em] (\alpha_2, 1 - \alpha_2) \\[1em] (1 - \alpha_2, \alpha_2) \end{array}\right\}$	$\dfrac{8}{18}$ $\dfrac{5}{18}$	5

(4) Quadratic interpolation on triangles

As with quadratic interpolation on line elements, we can, in order to raise the accuracy, construct quadratic interpolations on triangular area elements. The first case is the six-point quadratic interpolation.

Use the three vertices $\delta_1, \delta_2, \delta_3$ of the triangle $\beta = (\delta_1, \delta, \delta_3)$ and the midpoints $\delta_4, \delta_5, \delta_6$ of their opposite edges as interpolation points. Adopt the convention that, δ_1 is opposite δ_4, δ_2 is opposite δ_5, and δ_3 is opposite δ_6 (Fig. 69). Form a complete quadratic polynomial

$$Q(x, y) = a_1 + a_2 x + a_3 y + a_4 x^2 + a_5 xy + a_6 y^2$$

such that

$$Q(x_i, y_i) = u_i, \quad i = 1, 2, \cdots, 6.$$

From these equations the coefficients a_i $(i = 1, 2, \cdots, 6)$ can be determined. The interpolation polynomial can be written as

$$Q(x, y) = \sum_{i=1}^{6} u_i \mu_i(x, y), \tag{4.14}$$

where, $\mu_i(x, y)$ are quadratic polynomials

$$\mu_i(x_j, y_j) = \delta_{ij}, \quad i, j = 1, 2, \cdots, 6. \tag{4.15}$$

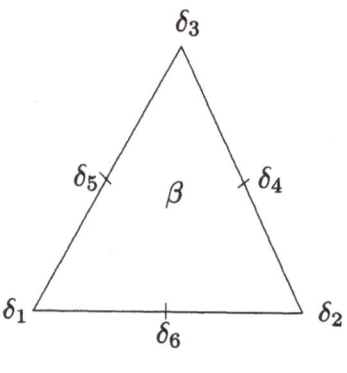

Fig. 69

These functions can be expressed in terms of the centroid coordinates λ_1, λ_2 and λ_3 as

$$\begin{cases} \mu_i(x, y) = 2\lambda_i^2 - \lambda_i, \quad i = 1, 2, 3, \\ \mu_4(x, y) = 4\lambda_2\lambda_3, \quad \mu_5(x, y) = 4\lambda_3\lambda_1, \quad \mu_6(x, y) = 4\lambda_1\lambda_2. \end{cases} \tag{4.16}$$

$\mu_i(x, y)(i = 1, 2, \cdots, 6)$ are the basis functions of the quadratic interpolation on triangles. They satisfy

$$\frac{\partial \mu_i}{\partial x} = \eta_i(4\lambda_i - 1), \quad i = 1, 2, 3,$$

$$\frac{\partial \mu_4}{\partial x} = 4(\eta_3\lambda_2 + \eta_2\lambda_3), \quad \frac{\partial \mu_5}{\partial x} = 4(\eta_1\lambda_3 + \eta_3\lambda_1),$$

$$\frac{\partial \mu_6}{\partial x} = 4(\eta_2\lambda_1 + \eta_1\lambda_2),$$

$$\frac{\partial \mu_i}{\partial y} = -\xi_i(4\lambda_i - 1), \quad i = 1, 2, 3,$$

$$\frac{\partial \mu_4}{\partial y} = -4(\xi_3\lambda_2 + \xi_2\lambda_3), \quad \frac{\partial \mu_5}{\partial y} = -4(\xi_1\lambda_3 + \xi_3\lambda_1),$$

$$\frac{\partial \mu_6}{\partial y} = -4(\xi_2\lambda_1 + \xi_1\lambda_2).$$

Table 6 $\displaystyle\iint_\beta \varphi\psi\,dxdy$

ψ \ φ	1	μ_j	μ_{3+j}
1	Δ	0	$\Delta/3$
μ_i		$(7\delta_{ij}-1)\Delta/180$	$-\delta_{ij}\Delta/45$
μ_{3+i}		$4(1+\delta_{ij})\Delta/45$	

Table 7 $\displaystyle\iint_\beta \varphi\psi\,dxdy$

ψ \ φ	1	$\dfrac{\partial\mu_1}{\partial y}$	$\dfrac{\partial\mu_2}{\partial y}$	$\dfrac{\partial\mu_3}{\partial y}$	$\dfrac{\partial\mu_4}{\partial y}$	$\dfrac{\partial\mu_5}{\partial y}$	$\dfrac{\partial\mu_6}{\partial y}$
1	Δ	$\dfrac{-\xi_1}{6}$	$\dfrac{-\xi_2}{6}$	$\dfrac{-\xi_3}{6}$	$\dfrac{2\xi_1}{3}$	$\dfrac{2\xi_2}{3}$	$\dfrac{2\xi_3}{3}$
$\dfrac{\partial\mu_1}{\partial x}$	$\eta_1/6$	$\dfrac{\eta_2\xi_1}{12\Delta}$	$\dfrac{\eta_1\xi_2}{12\Delta}$	$\dfrac{\eta_1\xi_3}{12\Delta}$	0	$\dfrac{-\eta_1\xi_3}{3\Delta}$	$\dfrac{-\eta_1\xi_2}{3\Delta}$
$\dfrac{\partial\mu_2}{\partial x}$	$\eta_2/6$	$\dfrac{\eta_3\xi_1}{12\Delta}$	$\dfrac{-\eta_2\xi_2}{4\Delta}$	$\dfrac{-\eta_3\xi_2}{3\Delta}$	$\dfrac{-\eta_3\xi_1}{3\Delta}$	0	
$\dfrac{\partial\mu_3}{\partial x}$	$\eta_3/6$	$\dfrac{\eta_3\xi_1}{12\Delta}$	$\dfrac{\eta_3\xi_2}{12\Delta}$	$\dfrac{-\eta_3\xi_3}{4\Delta}$	$\dfrac{-\eta_3\xi_2}{3\Delta}$	$\dfrac{-\eta_3\xi_1}{3\Delta}$	0
$\dfrac{\partial\mu_4}{\partial x}$	$\dfrac{-2\eta_1}{3}$	0	$\dfrac{-\eta_3\xi_3}{3\Delta}$	$\dfrac{-\eta_2\xi_3}{3\Delta}$	$\dfrac{-\sum_{k=1}^{3}\eta_k\xi_k}{3\Delta}$	$\dfrac{-(\eta_1\xi_2+\eta_2\xi_1)}{3\Delta}$	$\dfrac{-(\eta_1\xi_3+\eta_3\xi_1)}{3\Delta}$
$\dfrac{\partial\mu_5}{\partial x}$	$\dfrac{-2\eta_2}{3}$	$\dfrac{-\eta_3\xi_1}{3\Delta}$	0	$\dfrac{-\eta_1\xi_3}{3\Delta}$	$\dfrac{-(\eta_2\xi_1+\eta_1\xi_2)}{3\Delta}$	$\dfrac{-\sum_{k=1}^{3}\eta_k\xi_k}{3\Delta}$	$\dfrac{-(\eta_2\xi_3+\eta_3\xi_2)}{3\Delta}$
$\dfrac{\partial\mu_4}{\partial x}$	$\dfrac{-2\eta_3}{3}$	$\dfrac{-\eta_2\xi_1}{3\Delta}$	$\dfrac{-\eta_1\xi_2}{3\Delta}$	0	$\dfrac{-(\eta_3\xi_1+\eta_1\xi_3)}{3\Delta}$	$\dfrac{-(\eta_3\xi_2+\eta_2\xi_3)}{3\Delta}$	$\dfrac{-\sum_{k=1}^{3}\eta_k\xi_k}{3\Delta}$

The integration formulas for the basis functions μ_i and their partial derivatives on the triangle β are given in Table 6 and 7. The value of the integral of $\dfrac{\partial\mu_i}{\partial x}\dfrac{\partial\mu_j}{\partial x}$ (or $\dfrac{\partial\mu_i}{\partial y}\dfrac{\partial\mu_j}{\partial y}$), can be obtained immediately by substituting η for ξ (or ξ for η) in the tables and reversing the sign.

4.3   Analysis of elements
(linear elements and quadratic elements)

As described above, triangulate the domain Ω, and assume that the boundaries Γ_1 and Γ_3 are decomposed into sums of line elements:

$$\Gamma_1 : \{\gamma_k'\}, \ k = 1, 2, \cdots, M_1;$$

$$\Gamma_3 : \{\gamma_k''\}, \ k = 1, 2, \cdots, M_3.$$

Then the total energy $J(u)$ of the variational problem (4.3) can be into approximated by the sum of the related elements

$$J(u) \approx \sum_{i=1}^{N_2} J_{\beta_i}(u) + \sum_{k=1}^{M_3} J_{\gamma_k''}(u), \tag{4.18}$$

where

$$J_\beta(u) = \iint_\beta \left\{ \frac{1}{2} p \left[\left(\frac{\partial u}{\partial x} \right)^2 + \left(\frac{\partial u}{\partial y} \right)^2 \right] - fu \right\} dx dy, \tag{4.19}$$

$$J_{\gamma_k''}(u) = \int_{\gamma_k''} \left[\frac{1}{2} cu^2 - gu \right] ds. \tag{4.20}$$

Since point elements contribute nothing to $J(u)$, we have

$$J_\delta(u) = 0.$$

When the domain Ω is not a polygonal, errors will be induced by substituting the straight for curved on the boundary. That is why the energy decomposition (4.18) is only approximate.

Substituting a quadratic interpolation function for u on each element will discretize the energies of the individual elements (4.19) and (4.20) into quadratic functions. We will now carry out the analysis of elements with respect to linear interpolation (linear elements) and quadratic interpolation (quadratic elements) on triangles.

(1) The linear element

1. Analysis of area elements

Consider the linear interpolation on the triangle $\beta = (\delta_1, \delta_2, \delta_3)$,

$$Q(x, y) = \sum_{i=1}^{3} u_i \lambda_i,$$

as an approximation of $u(x, y)$. Let

$$U_\beta = (u_1, u_2, u_3)^T, \quad N(x, y) = (\lambda_1, \lambda_2, \lambda_3)^T,$$

then

$$Q = U_\beta^T N.$$

The energy integral on β is discretized to

$$\begin{aligned} J_\beta(u) \approx J_\beta(Q) &= \iint_\beta \left\{ \frac{1}{2} p \left[\left(\frac{\partial Q}{\partial x} \right)^2 + \left(\frac{\partial Q}{\partial y} \right)^2 \right] - fQ \right\} dxdyz \\ &= \iint_\beta \left\{ \frac{1}{2} p \left[U_\beta^T \left(\frac{\partial N}{\partial x} \right) \left(\frac{\partial N}{\partial x} \right)^T U_\beta \right. \right. \\ &\quad \left. \left. + U_\beta^T \left(\frac{\partial N}{\partial y} \right) \left(\frac{\partial N}{\partial y} \right)^T U_\beta \right] - f U_\beta^T N \right\} dxdy \\ &= \frac{1}{2} U_\beta^T K_\beta U_\beta - U_\beta^T F_\beta. \end{aligned} \tag{4.21}$$

The element stiffness matrix is

$$\begin{aligned} K_\beta &= \iint_\beta p \left[\left(\frac{\partial N}{\partial x} \right) \left(\frac{\partial N}{\partial x} \right)^T + \left(\frac{\partial N}{\partial y} \right) \left(\frac{\partial N}{\partial y} \right)^T \right] dxdy \\ &= \left[\iint_\beta p \left(\frac{\partial \lambda_i}{\partial x} \frac{\partial \lambda_j}{\partial x} + \frac{\partial \lambda_i}{\partial y} \frac{\partial \lambda_j}{\partial y} \right) dxdy \right]_{,i,1=1,2,3} \end{aligned} \tag{4.22}$$

$$F_\beta = \iint_\beta fN dxdy = \left(\iint_\beta f\lambda_1 dxdy, \iint_\beta f\lambda_2 dxdy, \iint_\beta f\lambda_3 dxdy \right)^T. \tag{4.23}$$

When p and f are constant on β, the entries of K_β and F_β can be evaluated using Table 3 as follows:

$$\begin{aligned} [K_\beta]_{ij} &= \iint_p p \left(\frac{\partial \lambda_i}{\partial x} \frac{\partial \lambda_j}{\partial x} + \frac{\partial \lambda_i}{\partial y} \frac{\partial \lambda_i}{\partial y} \right) dxdy \\ &= \frac{p}{4\Delta} (\eta_i \eta_j + \xi_i \xi_j), \quad i, j = 1, 2, 3, \end{aligned} \tag{4.24}$$

$$[F_\beta]_i = \iint_\beta f\lambda_i dxdy = \frac{f\Delta}{3}, \quad i = 1, 2, 3. \tag{4.25}$$

For general functions p and f, K_β and F_β can be evaluated numerically using Table 4.

2. Analysis of the line element

Consider a line element γ. It is an edge of some triangle β, say (δ_1, δ_2). Considering the values on (δ_1, δ_2) of the linear interpolation Q on β, since the centroid coordinate $\lambda_3 = 0$ on the boundary (δ_1, δ_2), we have

$$Q(s) = u_1\lambda_1(s) + u_2\lambda_2(s),$$

where, s indicates the chord length pointing to δ_2 from δ_1, i.e.,

$$s = \sqrt{(x - x_1)^2 + (y - y_1)^2}.$$

On the other hand, we can see from the identities (4.10) for the centroid coordiantes that, when $\lambda_3 = 0$, we have

$$\lambda_1 + \lambda_2 = 1,$$
$$x = x_1\lambda_1 + x_2\lambda_2, \ y = y_1\lambda_1 + y_2\lambda_2.$$

It follows that

$$x = x_1(1 - \lambda_2) + x_2\lambda_2 = x_1 + \lambda_2(x_2 - x_1),$$
$$y = y_1(1 - \lambda_2) + y_2\lambda_2 = y_1 + \lambda_2(y_2 - y_1).$$

Hence

$$s = \sqrt{(x - x_1)^2 + (y - y_1)^2} = \lambda_2\sqrt{(x_2 - x_1)^2 + (y_2 - y_1)^2} = \lambda_2 l,$$

where, $l = \sqrt{(x_2 - x_1)^2 + (y_2 - y_1)^2}$ is just the length of the line segment (δ_1, δ_2). Then

$$\lambda_1(s) = 1 - \frac{s}{l}, \quad \lambda_2(s) = \frac{s}{l}.$$

This implies that $\lambda_1(s)$ and $\lambda_2(s)$ are basis functions for the linear interpolation on (δ_1, δ_2). Therefore, the values taken on the edge γ by the three-vertex linear interpolation function Q on the area element β are the same as those of the two-point linear interpolation done independently on γ. Consequently, the linear interpolation function (2.5) in §2 is:

$$Q(s) = U_\gamma^T N(s),$$
$$U_\gamma = (u_1, u_2)^T, \quad N(s) = (\lambda_1, \lambda_2)^T$$

which can be directly used for the linear element γ. The energy integral on the linear element is discretized to

$$
\begin{aligned}
J_\gamma(u) &= \int_\gamma \left(\frac{1}{2}cu^2 - gu\right)ds \approx J_\gamma(Q) \\
&= \int_\gamma \left(\frac{1}{2}cU_\gamma^T NN^T U_\gamma - gU_\gamma^T N\right)ds \\
&= \frac{1}{2}U_\gamma^T K_\gamma U_\gamma - U_\gamma^T F_\gamma,
\end{aligned}
\tag{4.26}
$$

where

$$
K_\gamma = \int_\gamma cNN^T ds = \left[\int_\gamma c\lambda_i\lambda_j ds\right]_{i,j=1,2,}
\tag{4.27}
$$

$$
F_\gamma = \int_\gamma gN ds = \left(\int_\gamma g\lambda_1 ds, \int_\gamma g\lambda_2 ds\right)^T.
\tag{4.28}
$$

If c and g are constant on γ, then according to Table 1, we have

$$
K_\gamma = \frac{cl}{6}\begin{bmatrix} 2 & 1 \\ 1 & 2 \end{bmatrix}, \quad F_\gamma = \frac{gl}{2}\begin{bmatrix} 1 \\ 1 \end{bmatrix}.
\tag{4.29}
$$

To sum up, for the function $u(x,y)$ on Ω with respect to the triangular subdivision, the three-vertex linear interpolations performed on each area element individually have the same values on the common edges and at the common vertices of the area elements. Therefore, the piecewise linear function $Q(x,y)$ obtained by piecing together all the individual ones on Ω is determined uniquely by the values $u_1, \cdots, u_N$ of $u(x,y)$ at the nodal points, and is the linear interpolation of the nodal points on each element (whether an area element or a line element). $Q(x,y)$ is everywhere continuous on Ω, while its first derivative has discontinuity on the line elements.

(2) The quadratic element

1. Analysis of the area element

Consider an arbitrary area element $\beta = (\delta_1, \delta_2, \delta_3)$, and let the midpoints of its three edges be δ_4, δ_5, and δ_6 (Fig. 69). Quadratically interpolate μ on β according to formula (4.14):

$$
Q(x,y) = \sum_{i=1}^{6} u_i \mu_i(x,y).
$$

Let

$$U_\beta = (u_1, \cdots, u_6)^T, N(x, y) = (\mu_1, \cdots, \mu_6)^T,$$

then

$$Q = U_\beta^T N.$$

Therefore,

$$J_\beta(u) \approx J_\beta(Q) = \frac{1}{2} U_\beta^T K_\beta U_\beta - U_\beta^T F_\beta,$$

where

$$K_\beta = \iint_\beta p \left[\left(\frac{\partial N}{\partial x} \right) \left(\frac{\partial N}{\partial x} \right)^T + \left(\frac{\partial N}{\partial y} \right) \left(\frac{\partial N}{\partial y} \right)^T \right] dx dy$$

$$= \left[\iint_\beta p \left(\frac{\partial \mu_i}{\partial x} \frac{\partial \mu_j}{\partial x} + \frac{\partial \mu_i}{\partial y} \frac{\partial \mu_j}{\partial y} \right) dx dy \right]_{i,j=1,\cdots,6}, \qquad (4.30)$$

$$F_\beta = \iint_\beta f N dx dy = \left[\iint_\beta f \mu_i dx dy \right]_{i=1,\cdots,6}. \qquad (4.31)$$

If p and f are constants on β, then according to Table 7, we immediately obtain

$$K_\beta = \frac{p}{12\Delta}$$

$$\begin{bmatrix} 3(\eta_1^2 + \xi_1^2) & -(\eta_1\eta_2 + \xi_1\xi_2) & -(\eta_3\eta_1 + \xi_3\xi_1) & 0 & 4(\eta_3\eta_1 + \xi_3\xi_1) & 4(\eta_1\eta_2 + \xi_1\xi_2) \\ & 3(\eta_2^2 + \xi_2^2) & -(\eta_2\eta_3 + \xi_2\xi_3) & 4(\eta_2\eta_3 + \xi_2\xi_3) & 0 & 4(\eta_1\eta_2 + \xi_1\xi_2) \\ & & 3(\eta_3^2 + \xi_3^2) & 4(\eta_2\eta_3 + \xi_2\xi_3) & 4(\eta_3\eta_1 + \xi_3\xi_1) & 0 \\ & & & 4\sum_{k=1}^{3}(\eta_k^2 + \xi_k^2) & 8(\eta_1\eta_2 + \xi_1\xi_2) & 8(\eta_3\eta_1 + \xi_3\xi_1) \\ & \text{sym.} & & & 4\sum_{k=1}^{3}(\eta_k^2 + \xi_k^2) & 8(\eta_2\eta_3 + \xi_2\xi_3) \\ & & & & & 4\sum_{k=1}^{3}(\eta_k^2 + \xi_k^2) \end{bmatrix}$$

$$(4.32)$$

$$F_\beta = \frac{f\Delta}{3}(0, 0, 0, 1, 1, 1)^T. \qquad (4.33)$$

For general p and f, K_β and F_β can be evaluated numerically method according to Table 4.

2. Analysis of the line element

Consider a linear element γ that is an edge (δ_1, δ_2) of an area element β. We have $\lambda_3 = 0$ on (δ_1, δ_2), hence by formula (4.16), we obtain

$$\mu_3 = \mu_4 = \mu_5 = 0.$$

Hence the values of Q on (δ_1, δ_2) are given by

$$Q(s) = u_1\mu_1(s) + u_2\mu_2(s) + u_6\mu_6(s),$$

where point 6 is the midpoint of (δ_1, δ_2), and

$$\mu_1(s) = 2\lambda_1^2(s) - \lambda_1(s) = 2\lambda_2^2(s) - \lambda_2(s),$$
$$\mu_6(s) = 4\lambda_1(s)\lambda_2(s).$$

We have seen before, however, that $\lambda_1(s)$ and $\lambda_2(s)$ are just the basis functions of the linear interpolation on the line element (δ_1, δ_2) when $\lambda_3 = 0$. Thus $Q(s)$ is a quadratic interpolation polynomial on (δ_1, δ_2). Hence, on the line element we can directly use formulas (2.8) and (2.9) in §2 for quadratic interpolation on a straight line:

$$Q(s) = \sum_{i=1}^{3} u_i\mu_i(s),$$

where, the subscript of the midpoint of the line element (δ_1, δ_2) has been changed to 3.

The energy integral on the line element γ is discretized to

$$J_\gamma(u) \approx J_\gamma(Q) = \frac{1}{2}U_\gamma^T K_\gamma U_\gamma - U_\gamma^T F_\gamma,$$

where,

$$K_\gamma = \int_\gamma cNN^T ds = \left[\int_\gamma c\mu_i\mu_j ds\right]_{i,j=1,2,3}, \tag{4.34}$$

$$F_\gamma = \int_\gamma gNds = \left[\int_\gamma g\mu_i ds\right]_{i=1,2,3}, \tag{4.35}$$

$$U_\gamma = (u_1, u_2, u_3)^T, N(s) = (\mu_1(s), \mu_2(s), \mu_3(s))^T.$$

If c and g are constant on γ, then according to Table 2, we obtain

$$K_\gamma = \frac{cl}{30}\begin{bmatrix} 4 & -1 & 2 \\ & 4 & 2 \\ \text{sym.} & & 16 \end{bmatrix}, \tag{4.36}$$

$$F_\gamma = \frac{gl}{6}(1, 1, 4)^T. \tag{4.37}$$

Piecing together all the quadratic interpolation functions on the various area elements, we obtain a piecewise quadratic function $Q(x, y)$ on Ω, which is determined by the values of u at the point elements and at the midpoints of the line elements. As with linear interpolation, $Q(x, y)$ is everywhere continuous on Ω, while its first derivative is still discontinuous on the line elements, i.e., its smoothness has not been improved.

4.4   Assembly and the others

Whether linear or quadratic elements are used, the total energy $J(u)$ is discretized to

$$J(u) \approx J(Q) = \sum_{i=1}^{N_2} J_{\beta_i}(Q) + \sum_{k=1}^{M_3} J_{\gamma_k''}(Q)$$

$$= \sum_{\beta} \frac{1}{2} U_\beta^T K_\beta U_\beta + \sum_{\gamma} \frac{1}{2} U_\gamma^T K_\gamma U_\gamma - \sum_{\beta} U_\beta^T F_\beta - \sum_{\gamma} U_\gamma^T F_\gamma$$

$$= \frac{1}{2} U^T K U - U^T F.$$

For linear elements, the interpolation points are the same as the point elements, so that

$$U = (u_1, \cdots, u_{N_0})^T.$$

For quadratic elements, in addition to all the point elements, the interpolation points include the midpoints of all the line elements. Consequently, we need to count all the interpolation points. The number of interpolation points is the number of the point elements plus the number of the line elements, that is, $N_0 + N_1$. Hence

$$U = (u_1, \cdots, u_{N_0+N_1})^T.$$

The coefficient matrix K (or the vector F) is obtained by superposing K_β and K_γ (or F_β and F_γ) of all the individual elements. Superposition is carried out just as in §2 so long as we note that the location in K of every entry in K_β and K_γ as well as the location in F of every entry in F_β and F_γ is determined by the number of the nodal point, i.e., the geometric information of the element. Like the analysis of elements, the process of assembly is standardized. Note that the coefficient matrix is still symmetric, positive definite, and sparse.

Essential boundary conditions

The variational problem (4.3) is attached with essential boundary conditions, i.e., on the function u on a set of specified line elements γ'_k $(k = 1, \cdots, M_1)$ takes the prescribed values:

$$\gamma'_k : u = \bar{u}, \quad k = 1, \cdots, M_1.$$

For linear elements, $\bar{u}(s)$ can be discretized into the linear interpolation of the values of $\bar{u}$ at its two end points, which is compatible with the piecewise linear interpolation on Ω. Hence, provided that u_{h_k} has been prescribed at the end points of all line elements in $\{\gamma'_k\}$ by

$$u_{h_k} = \bar{u}_k, \quad k = 1, \cdots, M_1,$$

where the numberings of the end points are $h_1, \cdots, h_M$, then the conditional variational problem (4.3) is discretized into a conditional extremum problem of quadratic functions

$$
\begin{cases}
J(u_1, \cdots, u_{N_0}) = \dfrac{1}{2} U^T K U - U^T F = \min, \\[2mm]
u_{h_k} = \bar{u}_k, \quad k = 1, \cdots, M_1.
\end{cases}
\tag{4.38}
$$

Hence, after we delete the M_1 equations, $i = h_1, \cdots, h_{M_1}$ from the system of linear equations

$$KU = F$$

and the M_1 prescribed values, $u_{h_k} = \bar{u}_k$ have been substituted into the remaining equations, we finally obtain $N_0 - M_1$ equations in $N_0 - M_1$ variables. The alternate approach for dealing with essential boundary conditions introduced in §2, in which the number of equations is kept unchanged, is also applicable.

For quadratic elements, the situation is exact the same as for line elements.

We will now discuss relationship between the number of the point, line, and area elements for the triangular subdivision and the number of equations in the linear system.

As stated before, the numbers N_0, N_1, and N_2 of the point, line, and area elements satisfy Euler's formula

$$N_0 - N_1 + N_2 = 1 - q,$$

where q denotes the number of holes in the domain Ω. For simply-connected domains, $q = 0$; while for multiply-connected domains, $q > 0$. In practical problems, we generally have

$$N_0, N_1, N_2 \gg 1 - q.$$

Hence

$$\frac{1}{N_0}(N_0 - N_1 + N_2) \approx 0. \tag{4.39}$$

Furthermore, each triangular area element has three linear elements, and each line element is adjacent to two area elements (when the line element is in the interior of Ω) or adjacent to one area element (when the line element is on the boundary $\partial\Omega$). Hence, we have

$$3N_2 = 2N_1 - N_1',$$

where N_1' denotes the number of the boundary line elements. In practice, we generally have

$$N_1 \gg N_1',$$

hence,

$$3N_2 \approx 2N_1.$$

Relating this to the approximate Euler formula (4.39), we obtain the approximate ratio of the numbers of the three kinds of elements as follows

$$N_0 : N_2 : N_1 \approx 1 : 2 : 3. \tag{4.40}$$

The number of the variables related to each area element is 3 (the values of u at the three vertices) for linear interpolation and 6 (the values of u at the three vertices and at the midpoints of the three edges) for quadratic interpolation, but this does not mean that the number of the variables is doubled for quadratic interpolation. In fact, the number of variables for linear interpolation is N_0, while for quadratic interpolation it is

$$N_0 + N_1 \approx N_0 + 3N_0 = 4N_0.$$

i.e., the number of variables has quadrupled. The above comparison, however, is based on using the same subdivision, whereas the subdivision can be made coarser because of the greater accuracy of quadratic interpolation. Practice shows that, to achieve the same accuracy, the number of unknown variables required for the quadratic interpolation with a coarse subdivision is often far fewer than for the linear interpolation with a fine subdivision. Hence, quadratic interpolation is worth using.

We will not discuss here the problem of convergence of the finite element solution to the exact solution. We merely point out that, in constructing the triangular subdivision, if the maximum value of the lengths of the line elements is h, and θ and θ' are the maximum value and the minimum value respectively of the interior angles of the triangle, then the solutions of linear, quadratic, and other higher order elements (as long as the global continuity of the piecewise interpolation function on Ω is ensured) all converge to the exact solution as long as the interior angles have a positive lower bound:

$$\theta' \geq \theta_0' > 0, \text{ as } h \to 0. \tag{4.41}$$

That is, as the subdivision becomes finer, no sequence of triangles becomes flatter and flatter in such a way that $\theta' \to 0$. Consequently, the convergence of the approach can be ensured within a quite wide range.

Note that, for linear elements, the minimum interior angle condition (4.41) can be replaced by

$$\theta \leq \theta_0 < \pi, \text{ when } h \to 0. \tag{4.42}$$

That is, no sequence of triangles becomes obtuser and obtuser so that $\theta \to \pi$. Condition (4.42) is somewhat weaker than condition (4.41), because the triangle must be quite flat when it is quite obtuse but may not be obtuse when it is quite flat. For instance, if two interior angles both approach $90°$, then the minimum interior angle approaches $0°$ but the maximum interior angle does not approach $180°$. That is the minimum interior angle may approach $0°$ when (4.42) is satisfied. In such a case however, it is very likely that the coefficient matrix of the system of discretized algebraic equations will be ill-conditioned, leading to difficulties in solving it. Hence, the minimum interior angle condition for the subdivision (4.41) should be used in practice.

§5   Problems of Plane Elasticity

5.1   Variational problems

According to §1 of Chapter 3, there are two kinds of problems in the plane elasticity, i.e. the plane strain problem and the plane stress problem, that have a unified mathematical form.

Taking the x and the y directional displacement distributions $u(x, y)$ and $v(x, y)$ as the fundamental variables, in both problems the relations between the strains and the displacements are

$$\varepsilon_{11} = \frac{\partial u}{\partial x}, \quad \varepsilon_{22} = \frac{\partial v}{\partial y}, \quad \varepsilon_{12} = \frac{1}{2}\left(\frac{\partial v}{\partial x} + \frac{\partial u}{\partial y}\right). \tag{5.1}$$

The relations between the stress tensor $(\sigma_{11}, \sigma_{22}, \sigma_{12})$ and the strain tensor $(\varepsilon_{11}, \varepsilon_{22}, \varepsilon_{12})$ for the two problems can both be written as

$$\begin{cases} \sigma_{11} = \dfrac{E_1}{1 - \nu_1^2}(\varepsilon_{11} + \nu_1\varepsilon_{22}), \\[2mm] \sigma_{22} = \dfrac{E_1}{1 - \nu_1^2}(\nu_1\varepsilon_{11} + \varepsilon_{22}), \\[2mm] \sigma_{12} = \dfrac{E_1}{1 + \nu_1}\varepsilon_{12}. \end{cases} \tag{5.2}$$

For the plane strain problem,

$$E_1 = \frac{Eh}{1 - \nu^2}, \quad \nu_1 = \frac{\nu}{1 - \nu},$$

while for the plane stress problem,

$$E_1 = Eh, \quad \nu_1 = \nu,$$

where h denotes the thickness of the plate.

The equilibrium equations are

$$\Omega: \begin{cases} -\left(\dfrac{\partial\sigma_{11}}{\partial x} + \dfrac{\partial\sigma_{12}}{\partial y}\right) = f_1, \\[3mm] -\left(\dfrac{\partial\sigma_{21}}{\partial x} + \dfrac{\partial\sigma_{22}}{\partial y}\right) = f_2. \end{cases} \tag{5.3}$$

Assume that the boundary conditions have the following general form

$$\partial\Omega = \Gamma_1 + \Gamma_2 + \Gamma_3,$$

Γ_1: the prescribed displacements

$$u = \bar{u}, \ v = \bar{v}; \tag{5.4}$$

Γ_2: the prescribed loads

$$\begin{cases} \sigma_{11}n_1 + \sigma_{12}n_2 = g_1, \\ \sigma_{21}n_1 + \sigma_{22}n_2 = g_2; \end{cases} \tag{5.5}$$

Γ_3: the elastic supports

$$\begin{cases} \sigma_{11}n_1 + \sigma_{12}n_2 = -(c_{11}u + c_{12}v) + g_1, \\ \sigma_{21}n_1 + \sigma_{22}n_2 = -(c_{21}u + c_{22}v) + g_2; \end{cases} \tag{5.6}$$

where (n_1, n_2) stands for the direction cosines of the outward normal to the boundary $\partial\Omega$, and the matrix $C = [c_{ij}]$ is symmetric positive definite. For the sake of simplicity, the second kind of boundary condition can be considered as the particular case of the third kind when $c_{ij} = 0$.

When eigher of the the medium coefficients E_1 or ν_1 is discontinuous, the displacements must be continuous on the line L where E_1 or ν_1 is discontinuous

$$[u]_-^+ = 0, \ [v]_-^+ = 0. \tag{5.7}$$

The following two interface conditions must also be satisfied

$$\left[\sum_{j=1}^{2} \sigma_{ij}\nu_j\right]_-^+ = 0, \quad i = 1, 2, \tag{5.8}$$

where (ν_1, ν_2) are the direction cosines of the outward normal to the discontinuity line L.

The equilibrium equations (5.2), together with all boundary (including all interface) conditions, are equivalent to the following variational problem

$$\begin{cases} J(u,v) = \iint_\Omega \frac{1}{2} \sum_{i,j=1}^{2} \sigma_{ij}\varepsilon_{ij}dxdy + \int_{\Gamma_3} \frac{1}{2}(c_{11}u^2 + 2c_{12}uv + c_{22}v^2)ds \\ \qquad - \iint_\Omega (f_1u + f_2v)dxdy - \int_{\Gamma_3}(g_1u + g_2v)ds = \min, \\ \Gamma_1 : u = \bar{u}, \ v = \bar{v}. \end{cases} \tag{5.9}$$

The continuity conditions (5.7) of the displacements on the discontinuity line need not to be listed out as constraint conditions so long as the displacements are prescribed to be single-valued on L.

5.2   Bilinear rectangular elements
(1) Interpolation polynomials

Triangular linear element and the quadratic elements were introduced in §4. Both of these kinds of elements are also applicable to the problems of plane elasticity. It is possible to use rectangles as the fundamental

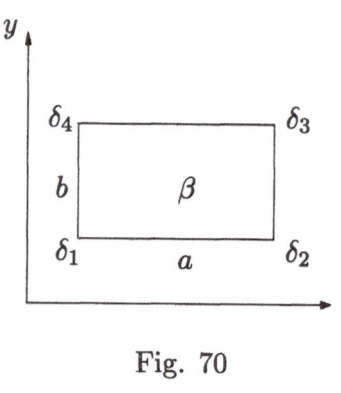

Fig. 70

elements of the subdivision when the geometry of the domain is sufficiently regular. For this purpose, we now introduce bilinear interpolation on the rectangles.

Consider a rectangle β. Suppose the four vertices of β are denoted $\delta_1, \delta_2, \delta_3, \delta_4$, taken in the anticlockwise direction, $\delta_i = (x_i, y_i)$. For simplicity, suppose the two edges of the rectangle are respectively parallel to the x and y coordinate axes (Fig. 70). Construct a bilinear interpolation for u on β,

$$P(x,y) = a_1 + a_2 x + a_3 y + a_4 xy,$$

such that

$$P(x_i, y_i) = u(x_i, y_i) = u_i, \quad i = 1, 2, 3, 4,$$

The conditions above determine the coefficients $a_i, i = 1, 2, 3, 4$ can be determined. One can write $P(x, y)$ in the form

$$P(x,y) = \sum_{i=1}^{4} u_i \varphi_i(x, y), \tag{5.10}$$

where φ_i are bilinear expressions of x and y satisfying

$$\varphi_i(x_j, y_j) = \delta_{ij}, \quad i, j = 1, 2, 3, 4.$$

Making the coordinate transformation

$$\begin{cases} \xi = \dfrac{2}{a}(x - x_0), \ x_0 = \dfrac{1}{2}(x_1 + x_2), \\[2mm] \eta = \dfrac{2}{b}(y - y_0), \ y_0 = \dfrac{1}{2}(y_1 + y_4), \end{cases} \tag{5.11}$$

the rectangle β is transformed into a square of edge length of 2 in the (ξ, η)-coordinates,

$$\beta' : -1 \leq \xi, \ \eta \leq 1.$$

The four vertices $\delta_1, \delta_2, \delta_3$, and δ_4 of β are, respectively, transformed into the four vertices of β':

$$(\xi_1, \eta_1) = (-1, -1), (\xi_2, \eta_2) = (1, -1),$$
$$(\xi_3, \eta_3) = (1, 1), (\xi_4, \eta_4) = (-1, 1). \tag{5.12}$$

Because ξ, x and η as well as y are related linearly, the original bilinear expressions of x and y are transformed into the bilinear expressions of ξ and η, i.e.,

$$\varphi_i(x, y) = \varphi_i(\xi, \eta) = b_{i1} + b_{i2}\xi + b_{i3}\eta + b_{i4}\xi\eta,$$
$$\varphi_i(\xi_j, \eta_j) = \delta_{ij}, \quad i, j = 1, 2, 3, 4.$$

From above we get

$$\varphi_i(\xi, \eta) = \frac{1}{4}(1 + \xi_i\xi)(1 + \eta_i\eta), \tag{5.13}$$

where $\varphi_i, i = 1, 2, 3, 4$ are the basis functions of the bilinear interpolation.

For the same reason, the bilinear interpolation of the displacement v on β is

$$Q(x, y) = \sum_{i=1}^{4} v_i\varphi_i(\xi, \eta). \tag{5.14}$$

The integrals of the products of the φ_i's and their partial derivatives on β are listed in the following table.

Table 8 $\displaystyle\iint_\beta \varphi\psi \, dx dy$

$\begin{array}{c}\varphi \\ \hline \psi\end{array}$	1	φ_j	$\dfrac{\partial \varphi_j}{\partial x}$	$\dfrac{\partial \varphi_j}{\partial y}$
1	ab	$\dfrac{ab}{4}$	$\dfrac{b}{2}\xi_j$	$\dfrac{a}{2}\eta_j$
φ_i	$\dfrac{b}{16}\left(1 + \dfrac{1}{3}\xi_i\xi_j\right) \times \left(1 + \dfrac{1}{3}\eta_i\eta_j\right)$	$\dfrac{b}{8}\xi_j \times \left(1 + \dfrac{1}{3}\eta_i\eta_j\right)$	$\dfrac{a}{8}\eta_j \times \left(1 + \dfrac{1}{3}\xi_i\xi_j\right)$
$\dfrac{\partial \varphi_i}{\partial x}$			$\dfrac{1}{4}\dfrac{b}{a}\xi_i\xi_j \times \left(1 + \dfrac{1}{3}\eta_i\eta_j\right)$	$\dfrac{1}{4}\xi_i\eta_j$
$\dfrac{\partial \varphi_i}{\partial y}$				$\dfrac{1}{4}\dfrac{a}{b}\eta_i\eta_j \times \left(1 + \dfrac{1}{3}\xi_i\xi_j\right)$

(2) Analysis of elements

Divide the domain Ω into rectangles, and then decompose the total energy into the sum of energies of all the individual elements. Except for the analysis of elements, the remaining steps are all the same as for the triangular subdivision and will not be repeated.

1. Analysis of the area element

Consider the energy integral on the rectangle β

$$J_\beta(u,v) = \frac{1}{2} \iint_\beta \sum_{i,j=1}^{2} \sigma_{ij}\varepsilon_{ij}\,dxdy - \iint_\beta (f_1 u + f_2 v)\,dxdy.$$

Let

$$\sigma = (\sigma_{11}, \sigma_{22}, \sigma_{12})^T, \quad \varepsilon = (\varepsilon_{11}, \varepsilon_{22}, 2\varepsilon_{12})^T,$$

and rewrite (5.2) in matrix form

$$\sigma = R\varepsilon, \tag{5.15}$$

where

$$R = \frac{E_1}{1 - \nu_1^2} \begin{bmatrix} 1 & \nu_1 & 0 \\ \nu_1 & 1 & 0 \\ 0 & 0 & \dfrac{1 - \nu_1}{2} \end{bmatrix} \tag{5.16}$$

is a symmetric matrix, whose entries depend only on the elasticity constants E_1 and ν_1.

On the other hand, let

$$U_\beta = (u_1, v_1, u_2, v_2, u_3, v_3, u_4, v_4)^T, \tag{5.17}$$

$$N = (N_1, N_2, N_3, N_4), \quad N_i = \varphi_i \begin{bmatrix} 1 & 0 \\ 0 & 1 \end{bmatrix}. \tag{5.18}$$

Then the interpolation formulas (5.10) and (5.14) can be written in the matrix form as

$$\begin{bmatrix} P \\ Q \end{bmatrix} = N U_\beta. \tag{5.19}$$

Substituting the bilinear interpolations P and Q for u and v in the strain ε of (5.1), we get

$$
\varepsilon = \begin{bmatrix} \dfrac{\partial u}{\partial x} \\[2mm] \dfrac{\partial v}{\partial y} \\[2mm] \dfrac{\partial u}{\partial y} + \dfrac{\partial v}{\partial x} \end{bmatrix} = \begin{bmatrix} \dfrac{\partial}{\partial x} & 0 \\[2mm] 0 & \dfrac{\partial}{\partial y} \\[2mm] \dfrac{\partial}{\partial y} & \dfrac{\partial}{\partial x} \end{bmatrix} \begin{bmatrix} u \\ v \end{bmatrix} \approx \begin{bmatrix} \dfrac{\partial}{\partial x} & 0 \\[2mm] 0 & \dfrac{\partial}{\partial y} \\[2mm] \dfrac{\partial}{\partial y} & \dfrac{\partial}{\partial x} \end{bmatrix} \begin{bmatrix} P \\ Q \end{bmatrix}
$$

$$
= \begin{bmatrix} \dfrac{\partial}{\partial x} & 0 \\[2mm] 0 & \dfrac{\partial}{\partial y} \\[2mm] \dfrac{\partial}{\partial y} & \dfrac{\partial}{\partial x} \end{bmatrix} N U_\beta = B U_\beta,
\tag{5.20}
$$

where

$$
B = (B_1, B_2, B_3, B_4), \quad B_i = \begin{bmatrix} \dfrac{\partial \varphi_i}{\partial x} & 0 \\[2mm] 0 & \dfrac{\partial \varphi_i}{\partial y} \\[2mm] \dfrac{\partial \varphi_i}{\partial y} & \dfrac{\partial \varphi_i}{\partial x} \end{bmatrix}, \quad i = 1, 2, 3, 4.
$$

Consequently, from (5.15), (5.19) and (5.20), we obtain

$$
\begin{aligned}
J_\beta(u, v) &= \frac{1}{2} \iint_\beta \sigma^T \varepsilon \, dx dy - \iint_\beta (f_1 u + f_2 v) dx dy \\
&\approx \frac{1}{2} \iint_\beta U_\beta^T B^T R B U_\beta dx dy - \iint_\beta U_\beta^T N^T f dx dy \\
&= \frac{1}{2} U_\beta^T K_\beta U_\beta - U_\beta^T F_\beta,
\end{aligned}
$$

where

$$
K_\beta = \iint_\beta B^T R B \, dx dy,
\tag{5.22}
$$

$$
F_\beta = \iint_\beta N^T f \, dx dy, \quad f = (f_1, f_2)^T.
\tag{5.23}
$$

The element stiffness matrix K_β is a 8×8 matrix, which can be partitioned according to the four vertices δ_1, δ_2, δ_3, and δ_4 of β into

$$K_\beta = \begin{bmatrix} K_{11} & K_{12} & K_{13} & K_{14} \\ & K_{22} & K_{23} & K_{24} \\ & & K_{33} & K_{34} \\ & & & K_{44} \end{bmatrix}, \tag{5.24}$$

where,

$$K_{ij} = \iint_\beta B_i^T R B_j \, dx dy \qquad (i, j = 1, 2, 3, 4)$$

is a 2×2 matrix. From Table 8, we can evaluate

$$K_{ij} = \frac{E_1}{1 - \nu_1^2} \begin{bmatrix} \iint_\beta \left(\frac{\partial \varphi_i}{\partial x} \frac{\partial \varphi_j}{\partial x} + \frac{1 - \nu_1}{2} \frac{\partial \varphi_i}{\partial y} \frac{\partial \varphi_j}{\partial y} \right) dx dy \\ \iint_\beta \left(\nu_1 \frac{\partial \varphi_i}{\partial y} \frac{\partial \varphi_j}{\partial x} + \frac{1 - \nu_1}{2} \frac{\partial \varphi_i}{\partial x} \frac{\partial \varphi_j}{\partial y} \right) dx dy \end{bmatrix}$$

$$\begin{matrix} \iint_\beta \left(\nu_1 \frac{\partial \varphi_i}{\partial x} \frac{\partial \varphi_j}{\partial y} + \frac{1 - \nu_1}{2} \frac{\partial \varphi_i}{\partial y} \frac{\partial \varphi_j}{\partial x} \right) dx dy \\ \iint_\beta \left(\frac{\partial \varphi_i}{\partial y} \frac{\partial \varphi_j}{\partial y} + \frac{1 - \nu_1}{2} \frac{\partial \varphi_i}{\partial x} \frac{\partial \varphi_j}{\partial x} \right) dx dy \end{matrix}$$

$$= \frac{E_1}{4(1 - \nu_1^2)} \begin{bmatrix} \frac{b}{a} \xi_i \xi_j \left(1 + \frac{1}{3} \eta_i \eta_j \right) + \frac{1 - \nu_1}{2} \frac{a}{b} \eta_i \eta_j \left(1 + \frac{1}{3} \xi_i \xi_j \right) \\ \nu_1 \xi_i \eta_i + \frac{1 - \nu_1}{2} \xi_i \eta_j \end{bmatrix}$$

$$\begin{matrix} \nu_1 \xi_i \eta_j + \frac{1 - \nu_1}{2} \xi_j \eta_i \\ \frac{a}{b} \eta_i \eta_j \left(1 + \frac{1}{3} \xi_i \xi_j \right) + \frac{1 - \nu_1}{2} \frac{b}{a} \xi_i \xi_j \left(1 + \frac{1}{3} \eta_i \eta_j \right) \end{matrix} \Bigg].$$

$$\tag{5.25}$$

When f_1 and f_2 are constants on β,

$$F_\beta = \frac{ab}{4} (f_1, f_2, f_1, f_2, f_1, f_2, f_1, f_2)^T. \tag{5.26}$$

2. Analysis of the line elements.

Consider a line element γ which is an edge of the rectangle β, say $\gamma = (\delta_1, \delta_2)$. Because $\eta = -1$ on (δ_1, δ_2) and, from (5.12), $\eta_3 = \eta_4 = 1$, the values of the basis functions φ_3 and φ_4 on (δ_1, δ_2) are

$$\bar{\varphi}_3 = \frac{1}{4}(1 + \xi_3\xi)(1 + \eta_3\eta) = 0,$$

$$\bar{\varphi}_4 = \frac{1}{4}(1 + \xi_4\xi)(1 + \eta_4\eta) = 0.$$

Hence of the interpolation polynomial $P(x, y)$ is given on (δ_1, δ_2) by

$$P(\xi) = u_1\bar{\varphi}_1(\xi) + u_2\bar{\varphi}_2(\xi), \tag{5.27}$$

where

$$\bar{\varphi}_1(\xi) = \frac{1}{2}(1 - \xi), \quad \bar{\varphi}_2(\xi) = \frac{1}{2}(1 + \xi).$$

This means that the value of the rectanglar bilinear interpolation polnomial on a boundary line element is equal to the value of the linear interpolation yielded by the function values at the two end points of this line element. It follows that the interpolation function pieced together from the bilinear interpolation polynomials is everywhere continuous on Ω, although its first derivative may be discontinuous on the boundaries of the rectangles.

The energy integral on the line element can be discretized using (5.27). That is, let

$$\begin{bmatrix} u \\ v \end{bmatrix} \approx \begin{bmatrix} P \\ \Omega \end{bmatrix} = \bar{N}U_\gamma, \ U_\gamma = (u_1, v_1, u_2, v_2)^T,$$

$$\bar{N} = (\bar{N}_1, \bar{N}_2),$$

$$\bar{N}_i = \bar{\varphi}_i \begin{bmatrix} 1 & 0 \\ 0 & 1 \end{bmatrix}, \quad \bar{\varphi}_1(\xi) = \frac{1}{2}(1 - \xi),$$

$$\bar{\varphi}_2(\xi) = \frac{1}{2}(1 + \xi), \ g = (g_1, g_2)^T,$$

then

$$J_\gamma(u, v) = \int_\gamma \frac{1}{2}(c_{11}u^2 + 2c_{12}uv + c_{22}v^2)ds - \int_\gamma (g_1u + g_2v)ds$$

$$\approx J_\gamma(P, Q) = \frac{1}{2}\int_\gamma U_\gamma^T \bar{N}^T C\bar{N}U_\gamma ds - \int_\gamma U_\gamma^T \bar{N}^T g ds$$

$$= \frac{1}{2}U_\gamma^T K_\gamma U_\gamma - U_\gamma^T F_\gamma,$$

where

$$K_\gamma = \int_\gamma \bar{N}^T C \bar{N} ds = \begin{bmatrix} \left(\int_\gamma \bar{\varphi}_1^2 ds\right) C & \left(\int_\gamma \bar{\varphi}_1 \bar{\varphi}_2 ds\right) C \\ \left(\int_\gamma \bar{\varphi}_1 \bar{\varphi}_2 ds\right) C & \left(\int_\gamma \bar{\varphi}_2^2 ds\right) C \end{bmatrix}$$

$$= \frac{1}{6} \begin{bmatrix} 2C & C \\ C & 2C \end{bmatrix}, C = [c_{ij}]_{i,j=1,2}, \tag{5.28}$$

$$F_\gamma = \left(\int_\gamma g_1 \bar{\varphi}_1 ds, \int_\gamma g_2 \bar{\varphi}_1 ds, \int_\gamma g_1 \bar{\varphi}_2 ds, \int_\gamma g_2 \bar{\varphi}_2 ds\right)^T. \tag{5.29}$$

If g_1 and g_2 are constant on γ, then

$$F_\gamma = \frac{l}{2}(g_1, g_2, g_1, g_2)^T. \tag{5.30}$$

Note that bilinear rectangular elements and linear triangular elements can be used together, i.e. one can partition the domain Ω into some rectangular and some triangular elements. For example, we can subdivide the domain into rectangles in the interior and triangles on the boundary, use bilinear interrpolation on the rectangles and linear interpolation on the triangles, evaluate the stiffness matrices of these elements individually, and then assemble them. This is because these two kinds of elements both induce linear interpolation on their boundary line elements, so an interpolation function pieced together from these two kinds of elements will still be continuous on the common boundary. The program implementation may not be unified in this way. However, if a rectangle is divided by the diagonal into two right triangles, then the original bilinear rectangular element is equivalent to the two linear triangular elements.

5.3   Essential boundary conditions

The essential boundary conditions of the variational problem (5.9) are

$$\Gamma_1 : u = \bar{u}, v = \bar{v}.$$

These can be dealt with by the two standard methods introduced before. In some practical problems, however, the two displacements u and v in the coordinate directions may not be entirely fixed, but the displacement in another direction, e.g., the normal displacement u_n on the boundary, is fixed:

$$\Gamma_1 : u_n = n_1 u + n_2 v = \bar{u}_n. \tag{5.31}$$

In that case, a boundary condition for the tangential stress needs to be added

$$\Gamma_1 : (\sigma_{11} n_1 + \sigma_{12} n_2) t_1 + (\sigma_{21} n_1 + \sigma_{22} n_2) t_2 = g_t,$$

where t_1 and t_2 are the tangential direction cosines, and (t, n) forms a righthanded system, i.e., $t_1 = -n_2$, $t_2 = n_1$. Correspondingly, the energy expression needs to be increased by a term

$$- \int_{\Gamma_1} g_t (t_1 u + t_2 v) ds.$$

This term of integral is similar to the integral over Γ_3 in form, i.e., by letting

$$g_1 = g_t t_1, g_2 = g_t t_2, c_{ij} = 0, i, j = 1, 2,$$

it can be handled unifiedly with the integral on Γ_3.

The form of the essential boundary condition (5.31) is not as simple as that previously discussed, being expressed in terms of linear combinations of the unknown variables u and v.

Suppose that the global energy is discretized by bilinear rectangular elements into

$$\frac{1}{2} U^T K U - U^T F,$$

and that the boundary Γ_1 is subdivided into a set line elements whose endpoints are numbered $h_1, \cdots, h_{M_1}$. Then the essential boundary condition (5.31) takes the following form.

$$n_{1 h_k} u_{h_k} + n_{2 h_k} v_{h_k} = \bar{u}_{h_k}, \quad k = 1, \cdots, M_1, \tag{5.32}$$

where $n_{1 h_k}$ and $n_{2 h_k}$ denote the outward normal direction cosines of the end point numbered h_k on the boundary line element.

Let

$$1\cdots 2h_1 - 1\cdots 2h_2 - 1\cdots 2h_M - 1\cdots N_0$$

$$N = \begin{bmatrix} \cdots n_{1h_1} & n_{2h_1} \cdots & & \\ & \cdots n_{1h_2} & n_{2h_2} \cdots & 0 \\ 0 & & \cdots\cdots & \\ & & \cdots n_{1h_M} & n_{2h_M} \cdots \end{bmatrix},$$

$$d_c = (\bar{u}_{h_1}, \cdots, \bar{u}_{h_{M_1}})^T.$$

Then (5.32) can be expressed as

$$NU = d_c. \tag{5.33}$$

Thus, the problem reduces to minimizing the quadratic function

$$J(u_1, \cdots, u_{N_0}) = \frac{1}{2} U^T K U - U^T F$$

under the subsidiary condition (5.33). To this end, we use the Lagrange multiplier method. Introducing new variables $\lambda_1, \cdots, \lambda_{M_1}$ and letting

$$\Lambda = (\lambda_1, \lambda_2, \cdots, \lambda_{M_1})^T,$$

the quadratic function

$$G(u_1, \cdots, u_{N_0}, \lambda_1, \cdots, \lambda_{M_1}) = \frac{1}{2} U^T K U - U^T F + (NU - d_c)^T \Lambda$$

has $N_0 + M_1$ independent variables. The conditional extremum problem for J is equivalent to the unconditional extremum problem for G. The extremum conditions for the latter are

$$\frac{\partial G}{\partial u_i} = 0, \quad i = 1, \cdots, N_0;$$

$$\frac{\partial G}{\partial \lambda_i} = 0, \quad i = 1, \cdots, M_1.$$

It follows that

$$\begin{bmatrix} K & N^T \\ N & 0 \end{bmatrix} \begin{bmatrix} U \\ \Lambda \end{bmatrix} = \begin{bmatrix} F \\ d_c \end{bmatrix}. \tag{5.34}$$

The new coefficient matrix is still symmetric, but is no longer positive definite, and the order is increased from N_0 to $N_0 + M_1$. We will not discuss here the special methods needed to solve this kind of equation.

§6 Bending Thin Plates

6.1 Variational principles

According to §4 of Chapter 4, the equilibrium equation of the problem of bending thin plates is

$$\Omega : -\left(\frac{\partial^2 M_{11}}{\partial x_1^2} + 2\frac{\partial^2 M_{12}}{\partial x_1 \partial x_2} + \frac{\partial^2 M_{22}}{\partial x_2^2}\right) = f, \tag{6.1}$$

where,

$$M_{11} = D(K_{11} + \nu K_{12}), \quad M_{22} = D(\nu K_{11} + K_{22}),$$

$$M_{12} = D(1 - \nu)K_{12}, \tag{6.2}$$

$$K_{ij} = -\frac{\partial^2 w}{\partial x_i \partial x_j}, \quad i, j = 1, 2, 3, \tag{6.3}$$

where w denotes the transverse deflection, f denotes the transverse load distribution, and $D = \dfrac{Eh^3}{12(1 - \nu^2)}$ denotes the flexural rigidity.

We assume the boundary conditions to be

$$\Gamma_1' : \ w = \frac{\partial w}{\partial n} = 0 \ \text{(fixed support)};$$

$$\Gamma'' : \ w = M_{nn} = 0 \ \text{(simply support)};$$

$$\Gamma_2 : \ Q_{3n} + \frac{\partial M_{tn}}{\partial l} = 0, \ M_{nn} = 0 \ \text{(free support)},$$

If the plate and the foundation are elastically coupled then the equilibrium equation (6.1) must be changed to

$$\Omega : -\left(\frac{\partial^2 M_{11}}{\partial x_1^2} + 2\frac{\partial^2 M_{12}}{\partial x_1 \partial x_2} + \frac{\partial^2 M_{22}}{\partial x_2^2}\right)cu = f, \tag{6.4}$$

where c is the constant of elastic coupling.

For the sake of convenience, change the coordinate notation (x_1, x_2) to (x, y). The equilibrium equation (6.4), together with the boundary conditions, is equivalent to the following variational problem:

$$
\begin{cases}
J(w) = \dfrac{1}{2} \iint_\Omega \sum_{i,j=1}^{2} M_{ij} K_{ij} dx dy + \dfrac{1}{2} \iint_\Omega cw^2 dx dt \\
\qquad\quad - \iint_\Omega f w dx dy, \\
\Gamma_1' : w = \dfrac{\partial w}{\partial n} = 0; \ \Gamma_1'' : w = 0.
\end{cases}
\tag{6.5}
$$

In the following sections, we will introduce several kinds of plate elements often used in engineering computations.

6.2   Incomplete bicubic rectangular elements (Addini–Clough–Melosh elements)

(1) Interpolation Polynomials

Consider a rectangle $\beta = (\delta_1, \delta_2, \delta_3, \delta_4)$, $\delta_i = (x_i, y_i)$, and suppose that the edges of the rectangle are parallel to the coordinate axes, as in Fig. 70. At each vertex δ_i, take $w(\delta_i) = w_i$, $\dfrac{\partial w(\delta_i)}{\partial x} = w_{xi}$, and $\dfrac{\partial w(\delta_i)}{\partial y} = w_{yi}$ as the interpolation parameters, giving a total of 12 parameters on the rectangle β, which we write in the matrix form as

$$
U_\beta = (w_1, w_{x_1}, w_{y_1}, w_2, w_{x_2}, w_{y_2}, w_3, w_{x_3}, w_{y_3}, w_4, w_{x_4}, w_{y_4})^T. \tag{6.6}
$$

We now construct an interpolation polynomial having 12 degrees of freedom. The complete bicubic polynomial

$$
\sum_{i,j=0}^{3} a_{ij} x^i y^i
$$

has 16 degrees of freedom. Deleting the four high order terms

$$
x^3 y^3, x^3 y^2, x^2 y^3, x^2 y^2,
$$

we obtain an incomplete bicubic polynomial with 12 degrees of freedom, which we take as the interpolation polynomial for w:

$$
Q(x, y) = a_1 + a_2 x + a_3 y + a_4 x^2 + a_5 xy + a_6 y^2
$$
$$
+ a_7 x^3 + a_8 x^2 y + a_9 xy^2 + a_{10} y^3 + a_{11} x^3 y + a_{12} xy^3.
$$

The interpolation conditions are

$$Q(x_i, y_i) = w_i, \quad Q_x(x_i, y_i) = w_{xi},$$
$$Q_y(x_i, y_i) = w_{yi}, \ i = 1, 2, 3, 4.$$

From these equations we can uniquely determine the coefficients a_i, $i = 1, \cdots, 12$, of Q.

Rewrite Q as

$$Q(x, y) = \sum_{i=1}^{4} \{w_i \varphi_i(x, y) + w_{xi}\varphi_i(x, y) + w_{yi}\zeta_i(x, y)\}, \qquad (6.7)$$

where, φ_i ψ_i, and ζ_i are all bicubic polynomials with 12 degrees of freedom satisfying the following relations:

$$\varphi_i(x_j, y_j) = \delta_{ij}, \quad \frac{\partial \varphi_i(x_j, y_j)}{\partial x} = 0, \quad \frac{\partial \varphi_i(x_j, y_j)}{\partial y} = 0;$$

$$\psi_i(x_j, y_j) = 0, \quad \frac{\partial \psi_i(x_j, y_j)}{\partial x} = \delta_{ij}, \quad \frac{\partial \psi_i(x_j, y_j)}{\partial y} = 0;$$

$$\zeta_i(x_j, y_i) = 0, \quad \frac{\partial \zeta_i(x_j, y_j)}{\partial x} = 0, \quad \frac{\partial \zeta_i(x_j, y_j)}{\partial y} = \delta_{ij},$$

$$i, j = 1, \cdots, 4. \qquad (6.8)$$

These are three sets of the basis functions of interpolation polynomial.

Performing the coordinate transformation

$$\begin{cases} \xi = \dfrac{2}{a}(x - x_0), & x_0 = \dfrac{1}{2}(x_1 + x_2), \ a = x_2 - x_1, \\[2mm] \eta = \dfrac{2}{b}(y - y_0), & y_0 = \dfrac{1}{2}(y_1 + y_4), \ b = y_4 - y_1, \end{cases}$$

the rectangle β is transformed into the square β' : $-1 \le \xi, \eta \le 1$, and its four vertices correspond to

$$(\xi_1, \eta_1) = (-1, -1), \ (\xi_2, \eta_2) = (1, -1),$$
$$(\xi_3, \eta_3) = (1, 1), \ (\xi_4, \eta_4) = (-1, 1), \qquad (6.9)$$

respectively. In (ξ, η)-coordinates, it is quite easy to show that the basis functions are

$$\varphi_i(\xi, \eta) = \frac{1}{8}(1 + \xi_i\xi)(1 + \eta_i\eta)(2 + \xi_i\xi + \eta_i\eta - \xi^2 - \eta^2),$$

$$\psi_i(\xi, \eta) = -\frac{1}{16} a\xi_i(1 + \xi_i\xi)^2(1 - \xi_i\xi)(1 + \eta_i\eta),$$

$$\zeta_i(\xi, \eta) = -\frac{1}{16} b\eta_i(1 + \eta_i\eta)^2(1 - \eta_i\eta)(1 + \xi_i\xi). \qquad (6.10)$$

In the following we consider the values of the interpolation polynomial Q on the edge γ of the rectangle. There is no loss in taking $\gamma = (\delta_1, \delta_2)$. On this edge we have $\eta = -1$, from (6.10) we obtain

$$\varphi_3 = \varphi_4 = 0, \quad \psi_3 = \psi_4 = 0,$$

$$\zeta_i = 0, \quad i = 1, 2, 3, 4.$$

Then we have

$$Q = w_1\bar{\varphi}_1 + w_2\bar{\varphi}_2 + w_{x_1}\bar{\psi}_1 + w_{x_2}\bar{\psi}_2,$$

where,

$$\bar{\varphi}_1 = \frac{1}{4}(1 - \xi)(2 - \xi - \xi^2),$$

$$\bar{\varphi}_2 = \frac{1}{4}(1 + \xi)(2 + \xi - \xi^2),$$

$$\bar{\psi}_1 = \frac{1}{8}a(1 - \xi)^2(1 + \xi),$$

$$\bar{\psi}_2 = -\frac{1}{8}a(1 + \xi)^2(1 - \xi).$$

We can see from the foregoing that the values taken by the interpolation polynomial Q on the boundary (δ_1, δ_2) of the rectangle are determined by the values of the function w_1, w_2, the values of the derivative w_{x_1}, w_{x_2} at the end points and the geometrical location of the boundary (δ_1, δ_2). Hence, the interpolation function pieced together by the interpolation polynomials on the individual rectangles is integrally continuous on Ω, but its first derivative is discontinuous on element boundaries.

(2) Analysis of elements

Let us consider an arbitrary rectangular area element $\beta = (\delta_1, \delta_2, \delta_3, \delta_4)$, and construct an interpolation polynomial Q for the deflection w using (6.7) and (6.10). Let

$$M = \begin{bmatrix} M_{11} \\ M_{22} \\ M_{12} \end{bmatrix}, \quad K = \begin{bmatrix} K_{11} \\ K_{22} \\ 2K_{12} \end{bmatrix}.$$

Write the generalized stress-strain relations (6.2) in matrix form

$$M = RK, \quad R = D \begin{bmatrix} 1 & \nu & 0 \\ \nu & 1 & 0 \\ 0 & 0 & \dfrac{1-\nu}{2} \end{bmatrix}, \qquad (6.11)$$

where R is a symmetric matrix depending only on the elastic constants E and ν and on the thickness h of the plate. Let

$$N = (N_1, N_2, N_3, N_4), \quad N_i(\varphi_i, \psi_i, \zeta_i), \quad i = 1, 2, 3, 4,$$

$$Q = NU_\beta. \qquad (6.12)$$

Then

$$K = - \begin{bmatrix} \dfrac{\partial^2 w}{\partial x^2} \\[2mm] \dfrac{\partial^2 w}{\partial y^2} \\[2mm] 2\dfrac{\partial^2 w}{\partial x \partial y} \end{bmatrix} \approx - \begin{bmatrix} \dfrac{\partial^2 Q}{\partial x^2} \\[2mm] \dfrac{\partial^2 Q}{\partial y^2} \\[2mm] 2\dfrac{\partial^2 Q}{\partial x \partial y} \end{bmatrix} = - \begin{bmatrix} \dfrac{\partial^2 N}{\partial x^2} \\[2mm] \dfrac{\partial^2 N}{\partial y^2} \\[2mm] 2\dfrac{\partial^2 N}{\partial x \partial y} \end{bmatrix} U_\beta = BU_\beta, \quad (6.13)$$

where,

$$B = - \begin{bmatrix} \dfrac{\partial^2 N}{\partial x^2} \\[2mm] \dfrac{\partial^2 N}{\partial y^2} \\[2mm] 2\dfrac{\partial^2 N}{\partial x \partial y} \end{bmatrix} = (B_1, B_2, B_3, B_4),$$

$$B_i = - \begin{bmatrix} \dfrac{\partial^2 N_i}{\partial x^2} \\[2mm] \dfrac{\partial^2 N_i}{\partial y^2} \\[2mm] 2\dfrac{\partial^2 N_i}{\partial x \partial y} \end{bmatrix} = - \begin{bmatrix} \dfrac{\partial^2 \varphi_i}{\partial x^2} & \dfrac{\partial^2 \psi_i}{\partial x^2} & \dfrac{\partial^2 \zeta_i}{\partial x^2} \\[2mm] \dfrac{\partial^2 \varphi_i}{\partial y^2} & \dfrac{\partial^2 \psi_i}{\partial y^2} & \dfrac{\partial^2 \zeta_i}{\partial y^2} \\[2mm] 2\dfrac{\partial^2 \varphi_i}{\partial x \partial y} & 2\dfrac{\partial^2 \psi_i}{\partial x \partial y} & 2\dfrac{\partial^2 \zeta_i}{\partial x \partial y} \end{bmatrix},$$

$$i = 1, 2, 3, 4, \qquad (6.14)$$

and we have

$$\frac{\partial^2 \varphi_i}{\partial x^2} = -\frac{3}{a^2}\xi_i\xi(1 + \eta_i\eta),$$

$$\frac{\partial^2 \varphi_i}{\partial y^2} = -\frac{3}{b^2}\eta_i\eta(1 + \xi_i\xi),$$

$$\frac{\partial^2 \varphi_i}{\partial x \partial y} = \frac{1}{2ab}\xi_i\eta_i[4 - 3(\xi^2 + \eta^2)];$$

$$\frac{\partial^2 \psi_i}{\partial x^2} = \frac{1}{2a}(1 + \eta_i\eta)(\xi_i + 3\xi),$$

$$\frac{\partial^2 \psi_i}{\partial y^2} = 0,$$

$$\frac{\partial^2 \psi_i}{\partial x \partial y} = -\frac{1}{4b}\eta_i(1 - 2\xi_i\xi - 3\xi^2);$$

$$\frac{\partial^2 \zeta_i}{\partial x^2} = 0,$$

$$\frac{\partial^2 \zeta_i}{\partial y^2} = \frac{1}{2b}(1 + \xi_i\xi)(\eta_i + 3\eta),$$

$$\frac{\partial^2 \zeta_i}{\partial x \partial y} = -\frac{1}{4a}\xi_i(1 - 2\eta_i\eta - 3\eta^2). \tag{6.15}$$

The energy integral on the rectangle β is discretized into

$$J_\beta(w) = \frac{1}{2}\iint_\beta (M^T(w)K(w)dxdy + \frac{1}{2}\iint_\beta cw^2dxdy$$

$$- \iint_\beta fwdxdy \approx \frac{1}{2}\iint_\beta M^T(Q)K(Q)dxdy$$

$$+ \frac{1}{2}\iint_\beta cQ^2dxdy - \iint_\beta fQdxdy$$

$$= \frac{1}{2}\iint_\beta U_\beta^T B^T RBU_\beta dxdy + \frac{1}{2}\iint_\beta cU_\beta^T N^T NU_\beta dxdy$$

$$- \iint_\beta fU_\beta^T N^T dxdy = \frac{1}{2}U_\beta^T(K_\beta + G_\beta)U_\beta - U_\beta^T F_\beta, \tag{6.16}$$

where the element stiffness matrix is

$$K_\beta = \iint_\beta B^T RBdxdy = \left[\iint_\beta B_i^T Rb_j dxdy\right]_{i,j=1,2,3,4}. \tag{6.17}$$

Then we have

$$\iint_\beta B_i^T RB_j dxdy = \frac{D}{ab} \begin{bmatrix} k_{11} & k_{12} & k_{13} \\ k_{21} & k_{22} & k_{23} \\ k_{31} & k_{32} & k_{33} \end{bmatrix},$$

where the values of the entries k_{ij} are given by

$$k_{11} = \frac{b^2}{a^2}\xi^*(3+\eta^*) + \frac{a^2}{b^2}\eta^*(3+\xi^*) + \frac{2(7-2\nu)}{5}\xi^*\eta^*,$$

$$k_{12} = -\frac{b^2}{2a}\xi_i(3+\eta^*) - \frac{a}{2}\eta^*\left(\nu\xi_j + \frac{2+3\nu}{5}\xi_i\right),$$

$$k_{13} = -\frac{a^2}{2b}\eta_i(3+\xi^*) - \frac{b}{2}\xi^*\left(\nu\eta_j + \frac{2+3\nu}{5}\eta_i\right),$$

$$k_{21} = -\frac{b^2}{2a}\xi_i(3+\eta^*) - \frac{a}{2}\eta^*\left(\nu\xi_j + \frac{2+3\nu}{5}\xi_i\right),$$

$$k_{22} = \frac{b^2}{12}(3+\xi^*)(3+\eta^*) + \frac{1-\nu}{30}a^2\eta^*(5\xi^*+3),$$

$$k_{23} = \frac{ab}{4}(\xi_i+\xi_j)(\eta_i+\eta_j),$$

$$k_{31} = -\frac{a^2}{2b}\eta_i(3+\xi^*) - \frac{b}{2}\xi^*\left(\nu\eta_j + \frac{2+3\nu}{5}\eta_i\right),$$

$$k_{32} = \frac{ab}{4}(\xi_i+\xi_j)(\eta_i+\eta_j),$$

$$k_{33} = \frac{a^2}{12}(3+\xi^*)(3+\eta^*) + \frac{1-\nu}{30}b^2\xi^*(5\eta^*+3).$$

The energy matrix of the elastic coupling with the foundation is

$$G_\beta = \iint_\beta cN^T N dxdy = \left[\iint_i cN_i^T N_j dxdy\right]_{i,j=1,2,3,4}. \tag{6.18}$$

If the coefficient of the coupling c is constant on β, then we have

$$\iint_\beta cN_i^T N_j dxdy = cab \begin{bmatrix} g_{11} & g_{12} & g_{13} \\ g_{21} & g_{22} & g_{23} \\ g_{31} & g_{32} & g_{33} \end{bmatrix}, \tag{6.19}$$

where the values of the entries g_{ij} are given by

$$g_{11} = \frac{1}{16}\left\{1 + \frac{17}{35}(\xi^* + \eta^*) + \frac{349}{1575}\xi^*\eta^*\right\},$$

$$g_{12} = -\frac{1}{32}a\xi_j\left\{\frac{1}{3} + \frac{3}{35}\xi^* + \frac{2}{15}\eta^* + \frac{52}{1575}\xi^*\eta^*\right\},$$

$$g_{13} = -\frac{1}{32}b\eta_j\left\{\frac{1}{3} + \frac{2}{15}\xi^* + \frac{3}{35}\eta^* + \frac{52}{1575}\xi^*\eta^*\right\},$$

$$g_{21} = -\frac{1}{32}a\xi_j\left\{\frac{1}{3} + \frac{3}{35}\xi^* + \frac{2}{15}\eta^* + \frac{52}{1575}\xi^*\eta^*\right\},$$

$$g_{22} = \frac{1}{256}a^2\xi^*\left(1 + \frac{1}{3}\eta^*\right)\left\{\frac{1}{3} - \frac{1}{15}\xi^* + \frac{1}{5}\xi^{*2} + \frac{1}{7}\xi^{*3}\right\},$$

$$g_{23} = \frac{1}{576}ab\xi_i\eta_j\left\{1 + \frac{1}{5}(\xi^* + \eta^*) + \frac{1}{25}\xi^*\eta^*\right\},$$

$$g_{31} = -\frac{1}{32}b\eta_i\left\{\frac{1}{3} + \frac{2}{15}\xi^* + \frac{3}{35}\eta^* + \frac{52}{1575}\xi^*\eta^*\right\},$$

$$g_{32} = \frac{1}{576}ab\xi_i\eta_j\left\{1 + \frac{1}{5}(\xi^* + \eta^*) + \frac{1}{25}\xi^*\eta^*\right\},$$

$$g_{33} = \frac{1}{256}b^2\eta^*\left(1 + \frac{1}{3}\eta^*\right)\left\{\frac{1}{3} - \frac{1}{15}\eta^* + \frac{1}{5}\eta^{*2} + \frac{1}{7}\eta^{*3}\right\}.$$

In all the expressions above,

$$\xi^* = \xi_i\xi_j, \quad \eta^* = \eta_i\eta_j,$$

where the values of ξ_i and η_i $(i = 1, 2, 3, 4)$ are as in (6.9).

The load vector is

$$F_\beta = \iint_\beta fN^T dxdy.$$

This equation has the following two important special cases.

1.   If f is constant on β, then

$$F_\beta = f\frac{ab}{4}\left(1, \frac{a}{6}, \frac{b}{6}, 1, -\frac{a}{6}, \frac{b}{6}, 1, -\frac{a}{6}, -\frac{b}{6}, 1, \frac{a}{6}, -\frac{b}{6}\right)^T. \tag{6.20}$$

2.   If f is a concentrated force of intensity q acting at the point $\delta_p = (x_p, y_p)$ in the rectangle β, then

$$F_\beta = q(N_1(\delta_p), N_2(\delta_p), N_3(\delta_p), N_4(\delta_p))^T.$$

In particular, if δ_p is the center of the rectangle, i.e., $\delta_p = (x_0, y_0)$, then

$$F_\beta = \frac{q}{4}\left(1, \frac{a}{4}, \frac{b}{4}, 1, -\frac{a}{4}, \frac{b}{4}, 1, -\frac{a}{4}, -\frac{b}{4}, 1, \frac{a}{4}, -\frac{b}{4}\right)^T. \tag{6.21}$$

Superpose the stiffness matrices, the energy matrix of the foundation, and the load vectors of the various elements so as to assemble the global coefficient matrix and the right-hand vector. Then consider the essential boundary conditions on Γ_1' and Γ_1''. On Γ_1',

$$\frac{\partial w}{\partial n} = w_x \cos(n, x) + w_y \cos(n, y) = 0.$$

This is a linear combination of the unknown variables w_x and w_y (several methods for dealing with the essential boundary conditions have been discussed in detail before). Finally, we obtain the system linear equations

$$KU = F.$$

Solving this equation for U, we obtain at once the deflection w, the rotation angle about the x–axis $\theta_x = w_y$ and the rotation angle about the y–axis $\theta_y = -w_x$ at the various nodal points.

If the internal forces of the plate need to be evaluated too, i.e.,

$$M = \begin{bmatrix} M_{11} \\ M_{22} \\ M_{12} \end{bmatrix},$$

then by (6.11) and (6.13), we have $M \approx RBU_\beta$, on each element. M evaluated using this equation is discontinuous at nodal points. However, we may average appropriately the values of M at the centroids of the two adjacent elements and take the average as the internal force at the nodal point.

The incomplete bicubic rectangular element with 12 degrees of freedom is generally called the ACM element (Adini-Clough-Melosh) in engineering computations.

6.3   Incomplete cubic triangular elements (Zienkiewicz elements)

(1) Interpolation Polynomials

We construct a cubic interpolation polynomial Q of w on a triangle $\beta = (\delta_1, \delta_2, \delta_3)$. As we know, the complete cubic polynomial has a total of 10 terms, i.e.,

$$1, x, y, x^2, xy, y^2, x^3, x^2 y, xy^2, y^3,$$

To determine the coefficients of these 10 terms, 10 conditions are needed. Three interpolation conditions at each vertex of the triangle are

$$Q(\delta_i) = w_i, \quad Q_x(\delta_i) = w_{xi}, Q_y(\delta_i) = w_{yi}, \quad i = 1, 2, 3. \tag{6.22}$$

So a total of 9 conditions can be given, but they are still not enough to determine Q completely. We still need one more condition. We want to choose a condition satisfied by the complete quadratic polynomial.

The complete quadratic polynomial has a total of 6 terms, which can be expressed in terms of the centroid coordinates as

$$P_2(\lambda_1, \lambda_2, \lambda_3) = a_1 \lambda_1^2 + a_2 \lambda_2^2 + a_3 \lambda_3^2 + a_4 \lambda_1 \lambda_2 + a_5 \lambda_2 \lambda_3 + a_6 \lambda_1 \lambda_3.$$

It is easy to verify that, P_2 identically satisfies the equality

$$6P(\delta_0) - 2\sum_{i=1}^{2} P(\delta_i) + \frac{2}{3} P_{\lambda_1}(\delta_1) - \frac{1}{3} P_{\lambda_1}(\delta_2) - \frac{1}{3} P_{\lambda_1}(\delta_3) - \frac{1}{3} P_{\lambda_2}(\delta_1)$$

$$+ \frac{2}{3} P_{\lambda_2}(\delta_2) - \frac{1}{3} P_{\lambda_2}(\delta_3) - \frac{1}{3} P_{\lambda_3}(\delta_1) - \frac{1}{3} P_{\lambda_3}(\delta_2) + \frac{2}{3} P_{\lambda_3}(\delta_3)$$

$$= 0, \tag{6.23}$$

where δ_0 stands for the centroid of the triangle β, whose coordinates are

$$\left(\frac{1}{3}, \frac{1}{3}, \frac{1}{3} \right).$$

We now take the complete cubic polynomial as the interpolation function Q. It can be expressed in terms of the centroid coordinates as

$$Q(\lambda_1, \lambda_2, \lambda_3) = b_1 \lambda_1^3 + b_2 \lambda_2^3 + b_3 \lambda_3^3 + b_4 \lambda_1^2 \lambda_2 + b_5 \lambda_1 \lambda_2^2$$

$$+ b_6 \lambda_2^2 \lambda_3 + b_7 \lambda_2 \lambda_3^2 + b_8 \lambda_3^2 \lambda_1 + b_9 \lambda_3 \lambda_1^2$$

$$+ b_{10} \lambda_1 \lambda_2 \lambda_3. \tag{6.24}$$

We would like to attach a condition to Q so that it satisfies the equality (6.23) that the complete quadratic polynomial satisfies. Thus, the complete quadratic polynomial is included into this kind of interpolation polynomial. Substituting (6.24) into (6.23), we obtain

$$2(b_1 + b_2 + b_3 + b_{10}) - (b_4 + b_5 + b_6 + b_7 + b_i + b_9) = 0. \qquad (6.25)$$

It is this condition that we will require the bicubic polynomial Q to satisfy.

The other 9 interpolation conditions (6.22) can be expressed in terms of the centroid coordinates as

$$Q(\delta_i) = w(\delta_i), \quad Q_{\lambda_1}(\delta_i) = w_{\lambda_1}(\delta_i), Q_{\lambda_2}(\delta_i) = w_{\lambda_2}(\delta_i), \quad i = 1, 2, 3. \qquad (6.26)$$

Because

$$\frac{\partial Q}{\partial \lambda_1} = \frac{\partial Q}{\partial x}\frac{\partial x}{\partial \lambda_1} + \frac{\partial Q}{\partial y}\frac{\partial y}{\partial \lambda_1} = -\xi_2\frac{\partial Q}{\partial x} - \eta_2\frac{\partial Q}{\partial y},$$

$$\frac{\partial Q}{\partial \lambda_2} = \frac{\partial Q}{\partial x}\frac{\partial x}{\partial \lambda_2} + \frac{\partial Q}{\partial y}\frac{\partial y}{\partial \lambda_2} = \xi_1\frac{\partial Q}{\partial x} + \eta_1\frac{\partial Q}{\partial y}, \qquad (6.27)$$

and the determinant

$$\begin{vmatrix} -\xi_2 & -\eta_2 \\ \xi_1 & \eta_1 \end{vmatrix} = \xi_1\eta_2 - \xi_2\eta_1 = 2\Delta > 0,$$

$Q_x(\delta_i)$ and $Q_y(\delta_i)$ taken together are equivalent to $Q_{\lambda_1}(\delta_i)$ and $Q_{\lambda_2}(Q_i)$. Rewrite the interpolation function Q as

$$Q(\lambda_1, \lambda_2, \lambda_3) = \sum_{i=1}^{3}\{\bar{\varphi}_i w(\delta_i) + \bar{\psi}_i w_{\lambda_1}(\delta_i) + \xi_i w_{\lambda_2}(\delta_i)\}, \qquad (6.28)$$

where $\bar{\varphi}_i$, $\bar{\psi}_i$ and ξ_i are all complete cubic polynomials. Based on the conditions (6.26), we have

$$\bar{\varphi}_i(\delta_j) = \delta_{ij}, \quad \frac{\partial\bar{\varphi}_i(\delta_j)}{\partial\lambda_1} = \frac{\partial\bar{\varphi}_i(\delta_j)}{\partial\lambda_2} = 0,$$

$$\bar{\psi}_i(\delta_j) = 0, \quad \frac{\partial\bar{\psi}_i(\delta_j)}{\partial\lambda_1} = \delta_{ij}, \quad \frac{\partial\bar{\psi}_i(\delta_j)}{\partial\lambda_2} = 0,$$

$$\xi_i(\delta_j) = 0, \quad \frac{\partial\xi_i(\delta_j)}{\partial\lambda_1} = 0, \quad \frac{\partial\xi_i(\delta_j)}{\partial\lambda_2} = \delta_{ij},$$

$$i, j = 1, 2, 3. \qquad (6.29)$$

These three sets of functions of the interpolation polynomials.

Each basis function should satisfy the constraint condition (6.25) attached to Q. For example, if $w = \xi_i$, then by (6.28) and (6.29),

$$Q = \xi_i = w,$$

so that ξ_i must satisfy condition (6.25). From the interpolation conditions (6.29) and the additional constraint (6.25), one can uniquely determine the basis functions. After simple computations, we get the following formulas:

$$\bar{\varphi}_1 = \lambda_1^2(3 - 2\lambda_1) + 2\lambda_1\lambda_2\lambda_3,$$

$$\bar{\varphi}_2 = \lambda_2^2(3 - 2\lambda_2) + 2\lambda_1\lambda_2\lambda_3,$$

$$\bar{\varphi}_3 = \lambda_3^2(3 - 2\lambda_3) + 2\lambda_1\lambda_2\lambda_3,$$

$$\bar{\psi}_1 = \lambda_1^2(\lambda_1 - 1) - \lambda_1\lambda_2\lambda_3,$$

$$\bar{\psi}_2 = \lambda_1\lambda_2^2 + \frac{1}{2}\lambda_1\lambda_2\lambda_3,$$

$$\bar{\psi}_3 = \lambda_1\lambda_3^2 + \frac{1}{2}\lambda_1\lambda_2\lambda_3,$$

$$\xi_1 = \lambda_1^2\lambda_2 + \frac{1}{2}\lambda_1\lambda_2\lambda_3,$$

$$\xi_2 = \lambda_2^2(\lambda_2 - 1) - \lambda_1\lambda_2\lambda_3,$$

$$\xi_3 = \lambda_2\lambda_3^2 + \frac{1}{2}\lambda_1\lambda_2\lambda_3.$$

Using the transformation relations (6.27) between the derivatives, the interpolation polynomial (6.28) can be written in rectangular coordinate form:

$$
\begin{aligned}
Q(x, y) &= \sum_{i=1}^{3}\{\bar{\varphi}_i(\lambda_1(x, y), \lambda_2(x, y), \lambda_3(x, y))w_i \\
&\quad + \bar{\psi}_i(\lambda_1(x, y), \lambda_2(x, y), \lambda_3(x, y))(-\xi_2 w_{xi} - \eta_2 w_{yi}) \\
&\quad + \xi_i(\lambda_1(x, y), \lambda_2(x, y), \lambda_3(x, y))(\xi_1 w_{xi} + \eta_1 w_{yi})\} \\
&= \sum_{i=1}^{3}\{\varphi_i w_i + \psi_i w_{xi} + \zeta_i w_{yi}\},
\end{aligned}
\tag{6.31}
$$

where,

$$\varphi_i = \bar{\varphi}_i, \psi_i = -\xi_2\bar{\psi}_i + \xi_1\xi_i, \zeta_i = -\eta_2\bar{\psi}_i + \eta_1\xi_i, \ i = 1, 2, 3. \tag{6.32}$$

The concrete forms of ψ_i and ζ_i are as follows:

$$
\begin{cases}
\psi_1 = \lambda_1^2(\xi_1\lambda_2 - \xi_2\lambda_1 - \xi_2) + \left(\frac{1}{2}\xi_1 + \xi_2\right)\lambda_1\lambda_2\lambda_3, \\[2mm]
\psi_2 = \lambda_2^2(\xi_1\lambda_2 - \xi_2\lambda_1 - \xi_2) - \left(\frac{1}{2}\xi_2 + \xi_1\right)\lambda_1\lambda_2\lambda_3, \\[2mm]
\psi_3 = \lambda_3^2(\xi_1\lambda_2 - \xi_2\lambda_1) + \left(\frac{1}{2}\xi_1 - \xi_2\right)\lambda_1\lambda_2\lambda_3,
\end{cases}
\tag{6.33}
$$

$$
\begin{cases}
\zeta_1 = \lambda_1^2(\eta_1\lambda_2 - \eta_2\lambda_1 + \eta_2) + \left(\frac{1}{2}\eta_1 + \eta_2\right)\lambda_1\lambda_2\lambda_3, \\[2mm]
\zeta_2 = \lambda_2^2(\eta_1\lambda_2 - \eta_2\lambda_1 - \eta_1) - \left(\frac{1}{2}\eta_2 + \eta_1\right)\lambda_1\lambda_2\lambda_3, \\[2mm]
\zeta_3 = \lambda_3^2(\eta_1\lambda_2 - \eta_2\lambda_1) + \frac{1}{2}(\eta_1 - \eta_2)\lambda_1\lambda_2\lambda_3.
\end{cases}
\tag{6.34}
$$

Consider now the values of the interpolation polynomial Q on the boundary γ of the triangle. Take $\gamma = (\delta_1, \delta_2)$. On this edge we have $\lambda_3 = 0$. Hence from (6.33) and (6.34) we obtain

$$
\varphi_3 = \psi_3 = \zeta_3 = 0,
$$

so that

$$
Q(s) = w_1\varphi_1(s) + w_2\varphi_2(s) + w_{x_1}\psi_1(s) + w_{x_2}\psi_2(s)
$$
$$
+ w_{y_1}\zeta_1(s) + w_{y_2}\zeta_2(s),
\tag{6.35}
$$

where s still denotes the chord length pointing to δ_2 from δ_1 on γ, while

$$
\begin{aligned}
\varphi_1(s) &= \lambda_1^2(s)(3 - 2\lambda_1(s)), \\
\varphi_2(s) &= \lambda_2^2(s)(3 - 2\lambda_2(s)), \\
\psi_1(s) &= \lambda_1^2(s)(\xi_1\lambda_2(s) - \xi_2\lambda_1(s) + \xi_2) = \lambda_1^2(s)(\xi_1\lambda_2(s) + \xi_2\lambda_2(s)) \\
&= \lambda_1^2(s)\lambda_2(s)(\xi_1 + \xi_2) = \lambda_1^2(s)\lambda_2(s)(x_2 - x_1), \\
\psi_2(s) &= \lambda_2^2(s)(\xi_1\lambda_2(s) - \xi_2\lambda_1(s) - \xi_1) = -\lambda_1(s)\lambda_2^2(s)(x_2 - x_1), \\
\zeta_1(s) &= \lambda_1^2(s)(\eta_1\lambda_2(s) - \eta_2\lambda_1(s) + \eta_2) = \lambda_1^2(s)\lambda_2(s)(y_2 - y_1), \\
\zeta_2(s) &= \lambda_2^2(s)(\eta_1\lambda_2(s) - \eta_2\lambda_1(s) - \eta_2) = -\lambda_1(s)\lambda_2^2(s)(y_2 - y_1), \\
\lambda_1(s) &= 1 - \frac{s}{l}, \quad \lambda_2(s) = \frac{s}{l}, \\
l &= \sqrt{(x_2 - x_1)^2 + (y_2 - y_1)^2}.
\end{aligned}
\tag{6.36}
$$

We can see from the foregoing that $Q(s)$ depends only on the values w_1, w_{x_1}, x_{y_1} and w_2, w_{x_2}, w_{y_2} at the two end points as well as the geometric location of these two end points. Since $Q(s)$ is independent of the third vertex δ_3, it is a cubic Hermite interpolation polynomial with respect to s. Hence the interpolation function pieced together by the interpolation polynomials Q on the individual triangles is everywhere continuous on Ω, although its first derivatives are discontinuous on the boundaries of triangles.

(2) Analysis of elements

The interpolation polynomial Q in (6.31) and that introduced in (6.7) for the rectangular element are very similar in that they all have three parameters w, w_x, and w_y per nodal point. We can directly use the process of element analysis for rectangular elements so long as the subscript variables i and j are changed from 1, 2, 3 and 4 to 1, 2 and 3. Hence we have the element stiffness matrix

$$K_\beta = \iint_\beta B^T R B dx dy = \left[\iint_\beta B_i^T R B_j dx dy \right]_{i,j=1,2,3},$$

the energy matrix of the foundation

$$G_\beta = \iint_\beta c N^T N dx dy = \left[\iint_\beta c N_i^T N_j dx dy \right]_{i,j=1,2,3},$$

and the load vector

$$F_\beta = \iint_\beta f N^T dx dy.$$

Using the formulas for the basis functions φ_i, ψ_i, and ζ_i, Table 3 in §4 for integrals and Euler's formula (4.11), we can compute the entries of K_β, G_β, and F_β. Although they are too long to give explicity here, readers are recommended to carry out the derivation by themselves.

For engineering computations, there is still another kind of incomplete cubic triangular element determined by a total of ten parameters, nine determined at the three vertices and one determined by the value of the function at the centroid of the triangle. However, for this kind of element the theory cannot guarantee convergence, so we will not introduce it here.

6.4   Complete quadratic triangular elements
       (Morley elements)

In this section, we will introduce a rather simple but useful triangular element.

(1) Interpolation polynomials

Consider a triangle $\beta = (\delta_1, \delta_2, \delta_3)$, and denote the midpoints of the respective opposite edges by δ_4, δ_5, and δ_6, numbered in the anticlockwise direction (see Fig. 69). Taking these 6 points as the interpolation points, we construct a complete quadratic polynomial

$$Q(x, y) = a_1 + a_2 x + a_3 y + a_4 x^2 + a_5 xy + a_6 y^2,$$

which is required to have the same values at the vertices as the function w:

$$Q(\delta_i) = w_i, i = 1, 2, 3. \tag{6.37}$$

It is, however, different from the quadratic interpolation polynomial in §4 such as to have the same normal derivative values at the midpoints of the three edges:

$$Q_n(\delta_i) = w_{mi}, \quad i = 4, 5, 6. \tag{6.38}$$

These 6 interpolation conditions uniquely determine the 6 coefficients of the polynomial Q.

Write Q as

$$Q(x, y) = \sum_{i=1}^{3} w_i \varphi_i(x, y) + \sum_{i=4}^{6} w_{mi} \varphi_i(x, y), \tag{6.39}$$

where all φ_i, $i = 1, \cdots, 6$ are complete quadratic polynomials satisfying the interpolation conditions:

$$\begin{cases} \varphi_i(\delta_j) = \delta_{ij}, & i, j = 1, 2, 3 \\ \dfrac{\partial \varphi_i(\delta_j)}{\partial n} = 0, & i = 1, 2, 3; \ j = 4, 5, 6; \end{cases} \tag{6.40}$$

$$\begin{cases} \varphi_i(\delta_j) = 0, \ i = 4, 5, 6; \ j = 1, 2, 3, \\ \dfrac{\partial \varphi_i(\delta_j)}{\partial n} = \delta_{ij}, \ i, j = 4, 5, 6. \end{cases} \tag{6.41}$$

We again use the centroid coordinates to construct the basis functions φ_i. Consider φ_1 first, which is a homogeneous quadratic expression of λ_1, λ_2 and λ_3:

$$\varphi_1(x, y) = \varphi_1(\lambda_1, \lambda_2, \lambda_3) = b_1 \lambda_1^2 + b_2 \lambda_2^2 + b_3 \lambda_3^2 + b_4 \lambda_2 \lambda_3$$

$$b_5 \lambda_3 \lambda_1 + b_6 \lambda_1 \lambda_2.$$

From the first expression of condition (6.40), we have

$$\varphi_1(\delta_1) = \varphi_1(1,0,0) = b_1 = 1,$$
$$\varphi_1(\delta_2) = \varphi_1(0,1,0) = b_2 = 0,$$
$$\varphi_1(\delta_3) = \varphi_1(0,0,1) = b_3 = 0,$$

so that

$$\varphi_1 = \lambda_1^2 + b_4\lambda_2\lambda_3 + b_5\lambda_3\lambda_1 + b_6\lambda_1\lambda_2.$$

Moreover, from the second expression of the condition (6.40), we have

$$\frac{\partial\varphi_1(\delta_j)}{\partial n} = \frac{\partial\varphi_1(\delta_j)}{\partial x}\cos(n_j,x) + \frac{\partial\varphi_1(\delta_j)}{\partial y}\cos(n_j,y)$$

$$= \frac{1}{2\Delta}\sum_{i=1}^{3}\left\{\frac{\partial\varphi_1(\delta_j)}{\partial\lambda_i}r_i\cos(n_j,x)\right.$$

$$\left.-\frac{\partial\varphi_1(\delta_j)}{\partial\lambda_i}\xi_i\cos(n_j,y)\right\}$$

$$= \frac{1}{2\Delta}\sum_{i=1}^{3}\frac{\partial\varphi_1(\delta_j)}{\partial\lambda_i}(\eta_i\cos(n_j,x) - \xi_i\cos(n_j,y))$$

$$= 0, \quad j = 4,5,6,$$

where $\cos(n_j,x)$ and $\cos(n_j,y)$ denote the normal direction cosines of the edge with midpoint δ_j lies. Clearly, we have

$$\cos(n_{3+i},x) = -\frac{\eta_i}{l_i}, \quad \cos(n_{3+i},y) = \frac{\xi_i}{l_i}, \quad i = 1,2,3,$$

where l_1, l_2 and l_3 denote, respectively, the lengths of the three edges $\delta_2\delta_3$, $\delta_3\delta_1$ and $\delta_1\delta_2$ of the triangle. We now obtain the system of equations

$$\begin{cases} -b_4 + b_5 + b_6 = 0, \\ b_4 - b_5 + b_6 = \dfrac{2s_{12}}{l_2^2}, \\ b_4 + b_5 - b_6 = \dfrac{2s_{13}}{l_3^2}, \\ s_{ij} = -(\xi_i\xi_j + \eta_i\eta_j). \end{cases} \tag{6.42}$$

From (6.42), we get

$$b_4 = \frac{s_{12}}{l_2^2} + \frac{s_{31}}{l_3^2}, \quad b_5 = \frac{s_{31}}{l_3^2}, \quad b_6 = \frac{s_{12}}{l_2^2}.$$

Hence

$$\varphi_1 = \lambda_1^2 + \left(\frac{s_{12}}{l_2^2} + \frac{s_{31}}{l_3^2}\right)\lambda_2\lambda_3 + \frac{s_{31}}{l_3^2}\lambda_3\lambda_1 + \frac{s_{12}}{l_2^2}\lambda_1\lambda_2.$$

For the same reason, we obtain

$$\varphi_2 = \lambda_2^2 + \frac{s_{23}}{l_3^2}\lambda_2\lambda_3 + \left(\frac{s_{23}}{l_3^2} + \frac{s_{12}}{l_1^2}\right) + \frac{s_{12}}{l_1^2}\lambda_1\lambda_2,$$

$$\varphi_3 = \lambda_3^2 + \frac{s_{23}}{l_2^2}\lambda_2\lambda_3 + \frac{s_{31}}{l_1^2}\lambda_3\lambda_1 + \left(\frac{s_{31}}{l_1^2} + \frac{s_{23}}{l_2^2}\right)\lambda_1\lambda_2,$$

$$\varphi_4 = \frac{2\Delta}{l_1}\lambda_1(\lambda_1 - 1), \quad \varphi_5 = \frac{2\Delta}{l_2}\lambda_2(\lambda_2 - 1),$$ (6.43)

$$\varphi_6 = \frac{2\Delta}{l_3}\lambda_3(\lambda_3 - 1).$$

(2) Analysis of elements
Let

$$U_\beta = (w_1, w_2, w_3, w_{n1}, w_{n2}, w_{n3})^T,$$ (6.44)

$$N = (\varphi_1, \varphi_2, \varphi_3, \varphi_4, \varphi_5, \varphi_6),$$ (6.45)

$$B = (B_1, B_2, B_4, B_5, B_6),$$

$$B_i = - \begin{bmatrix} \dfrac{\partial^2 \varphi_i}{\partial x^2} \\[2mm] \dfrac{\partial^2 \varphi_i}{\partial y^2} \\[2mm] 2\dfrac{\partial^2 \varphi_i}{\partial x \partial y} \end{bmatrix}, \quad i = 1, \cdots, 6.$$ (6.46)

The whole analysis process can be carried out in the same manner as for the rectangular element in 6.2.

The element stiffness matrix is

$$K_\beta = \iint_\beta B^T RB dx dy = \left[\iint_\beta B_i^T RB_j dx dy\right]_{i,j=1,\cdots,6},$$

where B is a constant matrix, because φ_i is a quadratic polynomial of x and y, whose second derivatives are constant. Hence,

$$K_\beta = \Delta[B_i^T RB_j]_{i,j=1,\cdots,6} = \Delta D[k_{ij}]_{i,j=1,\cdots,6}.$$ (6.47)

where

$$k_{ij} = \frac{\partial^2 \varphi_i}{\partial x^2} \frac{\partial^2 \varphi_j}{\partial x^2} + \frac{\partial^2 \varphi_i}{\partial y^2} \frac{\partial^2 \varphi_j}{\partial y^2} + \nu \left(\frac{\partial^2 \varphi_i}{\partial x^2} \frac{\partial^2 \varphi_j}{\partial y^2} \right.$$
$$\left. + \frac{\partial^2 \varphi_i}{\partial y^2} \frac{\partial^2 \varphi_j}{\partial x^2} \right) + 2(1 - \nu) \frac{\partial^2 \varphi_i}{\partial x \partial y} \frac{\partial^2 \varphi_j}{\partial x \partial y}.$$

From (6.43), it follows that

$$\frac{\partial^2 \varphi_1}{\partial x^2} = \frac{1}{2\Delta^2} \left(\eta_1^2 - \frac{s_{12}}{l_2^2} \eta_2^2 - \frac{s_{31}}{l_3^2} \eta_3^2 \right),$$

$$\frac{\partial^2 \varphi_1}{\partial y^2} = \frac{1}{2\Delta^2} \left(\xi_1^2 - \frac{s_{12}}{l_2^2} \xi_2^2 - \frac{s_{31}}{l_3^2} \xi_3^2 \right),$$

$$\frac{\partial^2 \varphi_1}{\partial x \partial y} = -\frac{1}{2\Delta^2} \left(\xi_1 \eta_1 - \frac{s_{12}}{l_2^2} \xi_2 \eta_2 - \frac{s_{31}}{l_3^2} \xi_3 \eta_3 \right),$$

$$\frac{\partial^2 \varphi_2}{\partial x^2} = \frac{1}{2\Delta^2} \left(\eta_2^2 - \frac{s_{23}}{l_3^2} \eta_3^2 - \frac{s_{12}}{l_1^2} \eta_1^2 \right),$$

$$\frac{\partial^2 \varphi_2}{\partial y^2} = \frac{1}{2\Delta^2} \left(\xi_2^2 - \frac{s_{23}}{l_3^2} \xi_3^2 - \frac{s_{12}}{l_1^2} \xi_1^2 \right),$$

$$\frac{\partial^2 \varphi_2}{\partial x \partial y} = -\frac{1}{2\Delta^2} \left(\xi_2 \eta_2 - \frac{s_{23}}{l_3^2} \xi_3 \eta_3 - \frac{s_{12}}{l_1^2} \xi_1 \eta_1 \right),$$

$$\frac{\partial^2 \varphi_3}{\partial x^2} = \frac{1}{2\Delta^2} \left(\eta_3^2 - \frac{s_{31}}{l_1^2} \eta_1^2 - \frac{s_{23}}{l_2^2} \eta_2^2 \right),$$

$$\frac{\partial^2 \varphi_3}{\partial y^2} = \frac{1}{2\Delta^2} \left(\xi_3^2 - \frac{s_{31}}{l_1^2} \xi_1^2 - \frac{s_{23}}{l_2^2} \xi_2^2 \right),$$

$$\frac{\partial^2 \varphi_3}{\partial x \partial y} = -\frac{1}{2\Delta^2} \left(\xi_3 \eta_3 - \frac{s_{31}}{l_1^2} \xi_1 \eta_1 - \frac{s_{23}}{l_2^2} \xi_2 \eta_2 \right),$$

$$\frac{\partial^2 \varphi_{3+i}}{\partial x^2} = \frac{1}{\Delta l_i} \eta_i^2, \quad \frac{\partial^2 \varphi_{3+i}}{\partial y^2} = \frac{1}{\Delta l_i} \xi_i^2,$$

$$\frac{\partial^2 \varphi_{3+i}}{\partial x \partial y} = \frac{-1}{\Delta l_i} \xi_i \eta_i, \quad i = 1, 2, 3.$$

Thus we obtain K_β.

The energy matrix of the elastic support is

$$G_\beta = \iint_\beta c N^T N \, dx dy = \left[\iint_\beta c \varphi_i \varphi_j dx dy \right]_{i,j=1,\cdots 6}. \qquad (6.48)$$

When c is constant on β, it is quite easy to compute the various entries of G_β using Euler's formula (4.11) in §4.

The load vector is

$$F_\beta = \iint_\beta f N^T \, dx dy.$$

If f is constant on β, then

$$F_\beta = f \frac{\Delta}{6} \Big(1 + \frac{s_{12}}{l_2^2} + \frac{s_{31}}{l_3^2}, 1 + \frac{s_{23}}{l_3^2} + \frac{s_{12}}{l_1^2}, 1 + \frac{s_{31}}{l_1^2} + \frac{s_{23}}{l_2^2},$$
$$-\frac{2\Delta}{l_1}, -\frac{2\Delta}{l_2}, -\frac{2\Delta}{l_3} \Big)^T. \tag{6.49}$$

6.5   On nonconforming elements

The energy integral $J(w)$ of the plate bending problem contains the second derivatives of the deflection w. Therefore, an admissible function for the variational problem (6.5) must itself be continuous and must have continuous first-order partial derivatives as well as piecewise continuous second-order partial derivatives on the domain Ω. The three kinds of plate elements introduced so far do not satisfy those continuity conditions. Because the first derivatives of the interpolation functions of both the ACM rectangular element and the Zienkiewicz triangular element are discontinuous on element boundaries, the first order partial derivatives are only piecewise continuous. The Morely triangular element, is much worse, since even the interpolation functions themselves are discontinuous on element boundaries.

An element or interpolation function which satisfies the continuity requirement for admissible functions in the variational problem is called a conforming element; otherwise it is called a nonconforming element. The three kinds of plane elements stated above, therefore, are all nonconforming when they are used as plate elements.

The following theoretical problem arises from the use of nonconforming elements. When the energy integral on the whole domain is decomposed into the sum of the energy integrals of the individual elements and the interpolation polynomials are substituted for the undetermined function, since the interpolation functions do not satisfy the continuity requirement of the variational problem, some derivative terms in the energy integral present singular distribution of the δ-function. Hence, we cannot ensure that it is correct to use Green's formula piecewise. The analysis of nonconforming elements is therefore theoretically much more difficult than that of conforming elements.

Recently, the theory of the finite element method (conforming elements), which has already been established in the early sixties by the first author of the book, has developed into a complete mathematical theory. Research on convergence of nonconforming elements has made much progress recently, and some criteria for verifying convergence have been presented. From those criteria we can prove that the ACM element is convergent. As for the Zienkiewisz triangular element, so long as the element subdivision satisfies the so-called parallel condition, i.e., all three edges of the triangle are required to be parallel to three fixed directions, or the whole domain Ω is partitioned into a finite number of domains and on each subdomain the parallel condition is satisfied, this kind of triangular element is convergent, too. It is interesting that convergence of the Morely triangular element can, on the contrary, be proved without any geometric restriction.

We can see from the foregoing analysis that, it is much more difficult to construct a conforming plate element than a conforming element for plane elasticity. Generally, to construct a conforming element for plane elasticity, we do not need to use interpolation polynomials of high order, which cause many problems in practical computations. Because of the great difficulty in constructing conforming elements, nonconforming elements are used rather widely platebending problems.

§7 Composite Structures

The mechanical model and the corresponding variational principle for composite structure, have been illustrated in detail in Chapter 4. The implementation of the finite element method for this theory will be described in the present section.

7.1 Plane composite structures

A special case of plane composite structures is the plane system of rods in tension, discussed in §3 of Chapter 1. Its mechanical model has only finitely many degrees of freedom, and the equilibrium equations of the system form a system of linear equations. It corresponds to a system of equations of a finite element discretization obtained by taking each rod as an element and by using the linear interpolation mode. If a rod is not uniform, then it must be divided into several segments, i.e. several elements. Each element is assumed to be uniform and is analyzed by the method in §2 of Chapter 2 or by the equivalent finite element method.

The fundamental members of the general plane composite structure are the one-dimensional rod in tension and bending and the two-dimensional plate in tension (i.e. the plane stress plate) within the plane. The total energy of the structure is obtained by summing three kinds energies of the members: the tensile energy of the rod; the flexural energy of the rod; and the tensile energy of the plate.

Generally speaking, the finite element discretization of the composite structure is accomplished by the following steps.

1. Settle upon the global coordinates and the local coordinates for the various members.

2. Subdivide the various members in geometry, and give out the geometric information and the mechanical parameters of the various elements.

3. Employ some interpolation formula, write out the element stiffness matrices and the load vectors of the various members in local coordinates, and then transform them into the global coordinates if necessary.

4. Couple and superimpose the stiffness metrices and load vectors of the various members according to the connection conditions among members, i.e. the elastic coupling relationships.

5. Deal with the essential boundary condition, and finally obtain a system of linear equations.

Compared with the finite element method for a single elastic body discussed before, here we have two more steps: the transformation of the element stiffness matrices and of the load vectors from local coordinates to global coordinates; and the coupling among the stiffness matrices of the members. These are the main features of the finite element method for composite structures. These two problems will be analyzed and discussed in the following.

(1) Coordinate transformation

Choose an orthogonal right hand system $(\bar{x}, \bar{y})$ as the global coordinates in the plane of the structure. Fix local coordinates for each member. Write the local coordinates as (x^β, y^β) for a plane element and (x^γ, y^γ) for a rod element, and adopt the convention that x^γ denotes the longitudinal direction of the rod and y^γ denotes the transverse direction.

The coordinate transformation for rod elements.

Let the angle of rotation from the $\bar{x}$-axis of the global coordinates to the x^γ-axis in the anticlockwise direction be θ_γ, then the transformation relations between the local components (u^γ, v^γ) and the global components $(\bar{u}, \bar{v})$ of the displacement vector on the rod element are

$$
\begin{bmatrix} u^\gamma \\ v^\gamma \end{bmatrix} = A_\gamma \begin{bmatrix} \bar{u} \\ \bar{v} \end{bmatrix}, \quad A_\gamma = \begin{bmatrix} \cos\theta_\gamma & \sin\theta_\gamma \\ -\sin\theta_\gamma & \cos\theta_\gamma \end{bmatrix}, \tag{7.1}
$$

while the value of the rotation angle $\theta^{1)}$ in the $(\bar{x}, \bar{y})$–plane is kept unchanged after the rotation of the coordinates.

Suppose that both the tensile rigidity EA and the flexural rigidity El_x on the rod are constant. Superposing the tensile (using the linear element) and the flexural (using the cubic Hermite element) element stiffness matrices and the load vectors, all in local coordinates, according to (2.17), (2.20), (3.15) and (3.17), we obtain the element stiffness matrix and the load vector of the plane rod element in tension and bending as follows:

$$
K_\gamma = \begin{bmatrix}
\dfrac{EA}{l} & 0 & 0 & \dfrac{-EA}{l} & 0 & 0 \\[2mm]
 & \dfrac{12El_x}{l^3} & \dfrac{6El_x}{l^2} & 0 & \dfrac{-12El_x}{l^3} & \dfrac{6El_x}{l^2} \\[2mm]
 & & \dfrac{4El_x}{l} & 0 & \dfrac{-6El_x}{l^2} & \dfrac{2EL_x}{l} \\[2mm]
 & & & \dfrac{EA}{l} & 0 & 0 \\[2mm]
 & \text{sym.} & & & \dfrac{12El_x}{l^3} & \dfrac{-6El_x}{l^2} \\[2mm]
 & & & & & \dfrac{4El_x}{l}
\end{bmatrix}, \tag{7.2}
$$

$$
F_\gamma = \left(\int_\gamma f_s\lambda_1 ds, \int_\gamma f_b\varphi_1 ds, \int_\gamma f_b\psi_1 ds, \int_\gamma f_s\lambda_2 ds, \int_\gamma f_b\varphi_2 ds, \int_\gamma f_b\psi_2 ds \right)^T, \tag{7.3}
$$

where f_s and f_b denote the longitudinal and the transverse load distributions, respectively. The element nodal vector in local coordinates is

$$
U_\gamma = (u_1^\gamma, v_1^\gamma, \theta_1, u_2^\gamma, v_2^\gamma, \theta_2)^T, \tag{7.4}
$$

---

[1] In the present section, we unitedly use θ_x, θ_y, and θ_z but not the notation ω to denote the angles of rotation about the x–, y– and x–axes, respectively, in order to avoid being confused with the z–directional displacement w. In the plane structure, θ_z is abbreviated as θ.

and the element energy (including both tensile and flexural parts) is

$$\frac{1}{2}U_\gamma^T K_\gamma U_\gamma - U_\gamma^T F_\gamma. \tag{7.5}$$

We now transform the element nodal vector U_γ in local coordinates into the global coordinates in order to couple and superimpose it with the adjacent elements. Because

$$\begin{bmatrix} u_1^\gamma \\ v_1^\gamma \end{bmatrix} = A_\gamma \begin{bmatrix} \bar{u}_1 \\ \bar{v}_1 \end{bmatrix}, \quad \begin{bmatrix} u_2^\gamma \\ v_2^\gamma \end{bmatrix} = A_\gamma \begin{bmatrix} \bar{u}_2 \\ \bar{v}_2 \end{bmatrix},$$

in the element nodal vector in global coordinates is

$$\bar{U}_\gamma = (\bar{u}_1, \bar{v}_1, \theta_1, \bar{u}_2, \bar{v}_2, \theta_2)^T, \tag{7.6}$$

then we have the transformation relation

$$U_\gamma = S_\gamma \bar{U}_\gamma, \tag{7.7}$$

where the transformation matrix is

$$S_\gamma = \begin{bmatrix} A_\gamma & & & 0 \\ & 1 & & \\ & & A_\gamma & \\ 0 & & & 1 \end{bmatrix}. \tag{7.8}$$

Hence the element energy is transformed into

$$\frac{1}{2}U_\gamma^T K_\gamma U_\gamma - U_\gamma^T F_\gamma = \frac{1}{2}\bar{U}_\gamma^T S_\gamma^T K_\gamma S_\gamma \bar{U}_\gamma - \bar{U}_\gamma^T S_\gamma^T F_\gamma$$

$$= \frac{1}{2}\bar{U}_\gamma^T \bar{K}_\gamma \bar{U}_\gamma - \bar{U}_\gamma^T \bar{F}_\gamma,$$

where $\bar{K}_\gamma$ and $\bar{F}_\gamma$ are the element stiffness matrix and load vector in the global coordinates:

$$\bar{K}_\gamma = S_\gamma^T K_\gamma S_\gamma, \quad \bar{F}_\gamma = S_\gamma^T F_\gamma. \tag{7.9}$$

Coordinate transformation for a plate element.

The element stiffness matrix and the load vector of a plate can be transformed in a similar way. Suppose the transformation relation between the local coordinates and global coordinates of the plate is

$$\begin{bmatrix} u^\beta \\ v^\beta \end{bmatrix} = A_\beta \begin{bmatrix} \bar{u} \\ \bar{v} \end{bmatrix}, \quad A_\beta = \begin{bmatrix} \cos\theta_\beta & \sin\theta_\beta \\ -\sin\theta_\beta & \cos\theta_\beta \end{bmatrix}. \tag{7.10}$$

For example, given a bilinear rectangular element, whose the element nodal vector in local coordinates is

$$U_\beta = (u_1^\beta, v_1^\beta, u_2^\beta, v_2^\beta, u_3^\beta, v_3^\beta, u_4^\beta, v_4^\beta)^T,$$

the element stiffness matrix and the load vector in the global coordinates are

$$\bar{K}_\beta = S_\beta^T K_\beta S_\beta, \quad \bar{F}_\beta = S_\beta^T F_\beta. \tag{7.11}$$

The transformation matrix is

$$S_\beta = \begin{bmatrix} A_\beta & & & \\ & A_\beta & 0 & \\ & 0 & A_\beta & \\ & & & A_\beta \end{bmatrix}, \tag{7.12}$$

and the element nodal vector in global coordinates is

$$\bar{U}_\beta = (\bar{u}_1, \bar{v}_1, \ \bar{u}_2, \ \bar{v}_2, \ \bar{u}_3, \ \bar{v}_3, \ \bar{u}_4, \ \bar{v}_4)^T,$$

$$U_\beta = S_\beta \bar{U}_\beta. \tag{7.13}$$

(2) The realization of the connection condition

We consider two kinds of connections among members: rigid connections and pinned connections.

Rigid connections.

According to §2 of Chapter 4, in the case of plane structures, the continuity properties of a rigid connection are:

1. The displacement at a joint point of rods is continuous;
2. The displacement on a joint line of plates and plates is continuous;
3. The displacement on a joint line of plates and rods is continuous;
4. The rotation angle at a joint point of rods and rods is continuous.

These four continuity condition are equivalent in the finite element method to the following requirements:

1. The values of the global displacement components $\bar{u}$, $\bar{v}$ and θ of the rod elements joined at a point rigid connection are respectively equal.

2. The values of the global displacement components $\bar{u}$ and $\bar{v}$ of the plate elements joined at a line rigid connection are respectively equal at the notal points on that line;

3. The values of the global displacement components $\bar{u}$ and $\bar{v}$ of the plate element and the rod element rigidly connected at a point are respectively equal.

In what follows, we will refer to nodal points on a line of rigid connection as points of rigid connection.

These three requirements, tell us how to couple and superpose the element stiffness matrices and load vectors of the elements at a rigid-connected point. Note that, there is no rotation angle component among the nodal parameters of a plate element, but there are three parameters $\bar{u}$, $\bar{v}$ and θ at a point rigid connection of a plate element and a rod element, where θ belongs to the rod element.

Pinned connections.

The displacement continuous at a point pin connection of rods, but the rotation angles are indepenedent.

According to this condition, the values of the two global displacements $\bar{u}$ and $\bar{v}$ at point pin connection of rod elements, are respectively equal, while the rotation angle components θ are independent parameters. Thus, at a pin-connected point of γ rod elements, there are two displacement components $\bar{u}$ and $\bar{v}$ as well as γ independent components of the rotation angle. Such a simple approach for dealing with pin-connected points is quite inconvenient for practical computations, since it leads not only to an increase in the number of unknown variables at the nodal point but also to nonuniformity in the form of the stiffness matrix form.

Another method of condensation is as follows.

Consider a rod element $\gamma = (\delta_1, \delta_2)$, and suppose that one of its ends δ_2 is pin-connected. According to the foregoing, in the element nodal vector (local)

$$U_\gamma = (u_1^\gamma, v_1^\gamma, \theta_1, u_2^\gamma, v_2^\gamma, \theta_2)^T,$$

the rotation angle component θ_2 is an independent variable and should not be coupled with the rotation angle components of the adjacent elements.

This implies that, except for that element r, the unknown variable θ_2 is not related with the other elements, i.e., the adjacent elements contribute nothing to θ_2. We can thus eliminate θ_2 a priori in the element γ without waiting to assemble the system of linear equations and solve them all together. By (7.2) and (7.3), the equation corresponding to θ_2 in the element γ is

$$\frac{6El_x}{l^2}v_1^\gamma + \frac{2El_x}{l}\theta_1 - \frac{6El_x}{l^2}v_2^\gamma + \frac{4El_x}{l}\theta_2 = \int_\gamma f_b\psi_2 ds;$$

hence

$$\theta_2 = -\frac{3}{2l}v_1^\gamma - \frac{1}{2}\theta_1 + \frac{3}{2l}v_2^\gamma + \frac{l}{4El_x}\int_\gamma f_b\psi_2 ds. \qquad (7.14)$$

Substituting this equation into the element energy (7.5), we can eliminate θ_2 and obtain a quadratic function in the five unknown variables u_1^γ, v_1^γ, θ_1, u_2^γ and v_2^γ. The concrete computation goes as follows: let the new element nodal vector (local) be

$$U_\gamma' = (u_1^\gamma, v_1^\gamma, \theta_1, u_2^\gamma, v_2^\gamma)^T, \qquad (7.15)$$

Then by (7.14),

$$U_\gamma = CU_\gamma' + e,$$

where

$$C = \begin{bmatrix} 1 & & & & \\ & 1 & 0 & & \\ & & 1 & & \\ & 0 & & 1 & \\ & & & & 1 \\ 0 & \dfrac{-3}{2l} & \dfrac{-1}{2} & 0 & \dfrac{3}{2l} \end{bmatrix},$$

and

$$e = \left(0, 0, 0, 0, 0, \frac{l}{4El_x}\int_\gamma f_b\psi_2 ds\right)^T.$$

Substituting U_γ into the energy formula (7.5), we obtain

$$\frac{1}{2}U_\gamma^T K_\gamma U_\gamma - U_\gamma^T F_\gamma = \frac{1}{2}(U_\gamma'^T G^T + e^T)K_\gamma(CU_\gamma' + e)$$

$$-(U_\gamma'^T G^T + e^T)F_\gamma = dfrac12 U_\gamma'^T G^T K_\gamma C U_\gamma' - U_\gamma'^T(C^T F_\gamma$$

$$-C^T K_\gamma e) + \left(\frac{1}{2}e^T K_\gamma e - e^T F_\gamma\right).$$

The last term in the above expression,

$$\frac{1}{2}e^T K_\gamma e - e^T F_\gamma,$$

is a constant and hence does not affect the solution of the global energy extremum problem. Therefore we can delete it and obtain a new quadratic energy function

$$\frac{1}{2}U_\gamma'^T K_\gamma' U_\gamma' - U_\gamma'^T F_\gamma', \tag{7.16}$$

where

$$K_\gamma' = C^T K_\gamma C = \begin{bmatrix} \frac{EA}{l} & 0 & 0 & \frac{-EA}{l} & 0 \\ & \frac{3El_x}{l^3} & \frac{3El_x}{l^2} & 0 & \frac{-3El_x}{l^3} \\ & & \frac{3El_x}{l} & 0 & \frac{-3El_x}{l^2} \\ & & & \frac{EA}{l} & 0 \\ \text{sym.} & & & & \frac{3El_x}{l^3} \end{bmatrix}, \tag{7.17}$$

$$F_\gamma' = C^T(F_\gamma - K_\gamma e) = \Big(\int_\gamma f_s\lambda_1 ds, \int_\gamma f_b\Big(\varphi_1 - \frac{3}{2l}\psi_2\Big)ds,$$

$$\int_\gamma f_b\Big(\psi_1 - \frac{1}{2}\psi_2\Big)ds, \int_\gamma f_s\lambda_2 ds, \int_\gamma f_b\Big(\varphi_2 + \frac{3}{2l}\psi_2\Big)ds\Big)^T. \tag{7.18}$$

There are just the element stiffness matrix and the load vector in local coordinates for a element in tension and bending which is rigid-connected at the left end and pin-connected at the right end. Transforming into global coordinates, we have

$$\bar{K}_\gamma' = S_\gamma'^T K_\gamma' S_\gamma', \quad \bar{F}_\gamma' = S_\gamma'^T F_\gamma', \tag{7.19}$$

where the transformation matrix is

$$
S'_\gamma = \begin{bmatrix} A_\gamma & & 0 \\ & 1 & \\ 0 & & A_\gamma \end{bmatrix}, \tag{7.20}
$$

and the element nodal vector (global) is

$$
\bar{U}'_\gamma = (\bar{u}_1, \bar{v}_1, \theta_1, \bar{u}_2, \bar{v}_2)^T.
$$

Using the forms of the element stiffness matrix and of the load vector shown in (7.19), we have only two displacement variables $\bar{u}$ and $\bar{v}$ but no rotation angle variable θ at the pin-connected point. Therefore, a pinned connection of rod elements can be solved immediately as long as we respectively equate the values of $\bar{u}$ and $\bar{v}$ of the adjacent rod elements and superpose the stiffness matrices. Consequently, the independent rotation component of the angle need not be considered. This method is quite convenient and helpful in computations.

The stiffness matrix (7.17) and the load vector (7.18) can be derived formally by another intuitive method. Consider the system of equations

$$
K_\gamma U_\gamma = F_\gamma, \tag{7.21}
$$

where K_γ is the coefficient matrix, F_γ is the right-hand member, and U_γ is the vector of unknown variables. Taking the diagonal entry $\dfrac{4EI_x}{l}$ at the crossing of the 6-th row and the 6-th column as the pivot, and performing one step of Gaussian elimination so as to eliminate the variable θ_2, we immediately obtain the required matrix K'_γ and vector F'_γ. Note that this derivation is only formal, because we have only the element energy (7.5) but no system of equations (7.21) on the individual elements.

7.2   Space composite structures

The fundamental members of a space composite structure are of three kinds: one dimensional rod members, i.e. rods in tension, in bending and in torsion; two dimensional plate members, i.e. plates in stretching and bending; and three dimensional body elements, i.e. general three dimensional elastic bodies. For the sake of simplicity, we will not consider three dimensional body elements but discuss only space plate-rod structures in the following.

The process of finite element discretization for space composite structures is similar to that for plane structures. It, too, consists mainly of
the coordinate transformation and the realization of the conditions on the
connections of the elements. In particular, the connection conditions are
more complicated and many-sided in space structures. Consequently, the
treatment of the finite element method is also much more difficult.

(1) Coordinate transformation

Fix a global coordinate system $(\bar{x}, \bar{y}, \bar{z})$ and individual local coordinates for each element. The local coordinates of the plate element are
$(x^\beta, y^\beta, z^\beta)$ under the convention that, x^β and y^β denote the longitudinal
directions of the plate, while z^β denotes the direction transverse to the
plate; the local coordinates of the rod element are $(x^\gamma, y^\gamma, z^\gamma)$ with the
convention that x^γ denotes the longitudinal direction of the rod, while y^γ
and z^γ denote the transverse directions.

Coordinate transformation of a rod element.

A rod member in space may have three kinds of deformations: x^γ-
directional stretching, $y^\gamma-$ and $z^\gamma-$bidirectional bending; and torsion about
the x_γ-axis.

In local coordinates, the three stiffness matrices and the three load
vectors for the tensile (linear element), the flexural (cubic Hermite element
used in both directions) and the torsional (linear element) deformations
are superposed to form the overall stiffness matrix and the load vector
for the rod element. Suppose that the tensile rigidity (EA), the flexural
rigidity $(El_z$ for the y-directional bending, and El_y for the z-directional
bending), and the torsional rigidity (GJ) are all constant. Then there are
6 parameters at each nodal point

$$u^\gamma, v^\gamma, w^\gamma, \theta_x^\gamma, \theta_y^\gamma, \theta_z^\gamma,$$

where u^γ denotes the displacement of the rod in the x^γ-direction; v^τ and
$\theta_x^\gamma \left(= \dfrac{dv^\tau}{dx^\gamma} \right)$ denote the flexural deflection and the rotation angle of the
beam in the (x^γ, y^γ)-plane; w^γ and $\theta_y^\gamma \left(= -\dfrac{dw^\gamma}{dx^\gamma} \right)$ denote the flexural deflection the rotation angle of the beam in the (x^γ, z^γ)-plane and θ_x^γ denotes
the angle of twist of the rod in torsion in the x^γ-direction. The reader
should note the signs of the parenthetical expressions for the rotation
angles θ_y^τ and θ_z^τ.

The element nodal vector (local) is

$$U_\gamma = (U_1, U_2)^T, \, U_i = (u_i^\gamma, \nu_i^\gamma, w_i^\gamma, \theta_{xi}^\gamma, \theta_{yi}^\gamma, \theta_{xj}^\gamma)^T, \, i = 1, 2. \qquad (7.22)$$

The element stiffness matrix for the rod in space is

$$K_\gamma = \begin{bmatrix}
\frac{EA}{l} & 0 & 0 & 0 & 0 & 0 & \frac{-EA}{l} & 0 & 0 & 0 & 0 & 0 \\[2mm]
 & \frac{12EI_x}{l^3} & 0 & 0 & 0 & \frac{6EI_x}{l^2} & 0 & \frac{-12EI_x}{l^3} & 0 & 0 & 0 & \frac{6EI_x}{l^2} \\[2mm]
 & & \frac{12EI_y}{l^3} & 0 & \frac{-6EI_y}{l^2} & 0 & 0 & 0 & \frac{-12EI_y}{l^3} & 0 & \frac{-6EI_y}{l^2} & 0 \\[2mm]
 & & & \frac{GJ}{l} & 0 & 0 & 0 & 0 & 0 & \frac{-GJ}{l} & 0 & 0 \\[2mm]
 & & & & \frac{4EI_y}{l} & 0 & 0 & 0 & \frac{6EI_y}{l^2} & 0 & \frac{2EI_y}{l} & 0 \\[2mm]
 & & & & & \frac{4EI_x}{l} & 0 & \frac{-6EI_x}{l^2} & 0 & 0 & 0 & \frac{2EI_x}{l} \\[2mm]
 & & & & & & \frac{EA}{l} & 0 & 0 & 0 & 0 & 0 \\[2mm]
 & & & & & & & \frac{12EI_x}{l^3} & 0 & 0 & 0 & \frac{-6EI_x}{l^2} \\[2mm]
 & & & & & & & & \frac{12EI_y}{l^3} & 0 & \frac{6EI_y}{l^2} & 0 \\[2mm]
 & & & & & & & & & \frac{GJ}{l} & 0 & 0 \\[2mm]
 & & & & & & & & & & \frac{4EI_y}{l} & 0 \\[2mm]
 & & & & & & & & & & & \frac{4EI_x}{l}
\end{bmatrix},$$

$$(7.23)$$

The load vector is

$$F_\gamma = (F_1, F_2)^T,$$

where

$$F_i = \left(\int_\gamma f_s \lambda_i ds, \int_\gamma f_{yb} \varphi_i ds, \int_\gamma f_{zb} \varphi_i ds, \int_\gamma f_s \lambda_i ds, \right.$$
$$\left. - \int_\gamma f_{xb} \psi_i ds, \int_\gamma f_{yb} \psi_i ds \right)^T, \qquad (7.24)$$

where f_i denotes the x^γ–longitudinal load distribution, f_{yb} and f_{xb} denote the y^γ– and z^γ–transverse load distributions, respectively, and f_i denotes the x^γ–longitudinal torque distribution.

The element stiffness matrix (7.23) and the load vector (7.24) in the local coordinates must now be transformed into that in the global coordinates.

The transformation relations between local components and global components of the displacement on rod are

$$
\begin{bmatrix} u^\gamma \\ v^\gamma \\ w^\gamma \end{bmatrix} = A_\gamma \begin{bmatrix} \bar{u} \\ \bar{v} \\ \bar{w} \end{bmatrix},
$$

where the transformation matrix A_γ is

$$
A_\gamma = \begin{bmatrix} \cos(x^\gamma, \bar{x}) & \cos(x^\gamma, \bar{y}) & \cos(x^\gamma, \bar{z}) \\ \cos(y^\gamma, \bar{x}) & \cos(y^\gamma, \bar{y}) & \cos(y^\gamma, \bar{z}) \\ \cos(z^\gamma, \bar{x}) & \cos(z^\gamma, \bar{y}) & \cos(z^\gamma, \bar{z}) \end{bmatrix}. \tag{7.25}
$$

Similarly, the transformation relations between the local components and the global components of the rotation angle are

$$
\begin{bmatrix} \theta_x^\gamma \\ \theta_y^\gamma \\ \theta_x^\gamma \end{bmatrix} = A_t \begin{bmatrix} \bar{\theta}_x \\ \bar{\theta}_y \\ \bar{\theta}_z \end{bmatrix}.
$$

Therefore, the transformation relations between the local components and the global components of the element nodal vector are

$$
U_\gamma = S_\gamma \bar{U}_\gamma, \quad S_\gamma = \begin{bmatrix} A_\gamma & 0 \\ 0 & A_\gamma \end{bmatrix}, \tag{7.26}
$$

and the transformation relations between the local coordinates and the global coordinates of the stiffness matrix and the load vector of the space rod element are

$$
\bar{K}_\gamma = S_\gamma^T K_\gamma S_\gamma, \quad \bar{F}_\gamma = S_\gamma^T F_\gamma. \tag{7.27}
$$

Now we consider coordinate transformation of a plate element.

A space plate member may have two kinds of deformations, i.e., the longitudinal stretching in the plane of the plate and the transverse bending perpendicular to the plane of the plate.

The element stiffness matrices and the load vectors of the plate in stretching and of the plate in bending in local coordinates are superposed to construct the stiffness matrix and the load vector of the space plate member. For the sake of definiteness, assume the rectangular (also may be the triangular) subdivision is adopted, the bilinear element being used for the plate in stretching, and the ACM element for the plate in bending. Thus, there are 5 parameters at each nodal point:

$$u^\beta, v^\beta, w^\beta, \theta_x^\beta \left(= \frac{\partial w^\beta}{\partial y^\beta} \right), \ \theta_y^\beta \left(= -\frac{\partial w^\beta}{\partial x^\beta} \right),$$

where u^β and v^β denote the longitudinal displacements and w^β, θ_x^β and θ_y^β respectively denote the transverse displacement and the angles of rotation about the x^β-and the y^β-axes of the plate in bending. Note that, the rotation angle θ_z^β about the z^β-direction does not appear in this finite element model.

The element nodal vector (local) is

$$U_\beta = (U_1, U_2, U_3, U_4)^T,$$
$$U_i = (u_i^\beta, v_i^\beta, w_i^\beta, \theta_{xi}^\beta, \theta_{yi}^\beta)^T. \tag{7.28}$$

Suppose that the element stiffness matrix and the load vector of the plate in stretching are

$$K_\beta^{(i)} = [K_{ij}^{(s)}]_{i,j=1,\cdots,4}, \quad F_\beta^{(s)} = (F_1^{(s)}, F_2^{(s)}, F_3^{(s)}, F_4^{(s)})^T,$$

and the element stiffness matrix and the load vector of the plate in bending are

$$K_\beta^{(b)} = [K_{ij}^{(b)}]_{i,j=1,\cdots,4}, \quad F_\beta^{(b)} = (F_1^{(b)}, F_2^{(b)}, F_3^{(b)}, F_4^{(b)})^T.$$

Then the element stiffness matrix and the load vector of the space plate in stretching and bending are

$$K_\beta = [K_{ij}]_{i,j=1,\cdots,4}, \ F_\beta = (F_1, F_2, F_3, F_4)^T, \tag{7.29}$$

where

$$K_{ij} = \begin{bmatrix} K_{ij}^{(s)} & 0 & 0 & 0 \\ & 0 & 0 & 0 \\ 0 & 0 & & \\ 0 & 0 & K_{ij}^{(b)} \\ 0 & 0 & \end{bmatrix}, \ F_i = (F_i^{(s)}, F_i^{(b)})^T. \tag{7.30}$$

Similar to coordinate transformation for the space rod element, the transformation relations between the local components and the global components of the displacement on the space plate element are

$$
\begin{bmatrix} u^\beta \\ v^\beta \\ w^\beta \end{bmatrix} = A_\beta \begin{bmatrix} \bar{u} \\ \bar{v} \\ \bar{w} \end{bmatrix},
$$

where the transformation matrix A_β is

$$
A_\beta = \begin{bmatrix} \cos(x^\beta, \bar{x}) & \cos(x^\beta, \bar{y}) & \cos(x^\beta, \bar{z}) \\ \cos(y^\beta, \bar{x}) & \cos(y^\beta, \bar{y}) & \cos(y^\beta, \bar{z}) \\ \cos(z^\beta, \bar{x}) & \cos(z^\beta, \bar{y}) & \cos(z^\beta, \bar{z}) \end{bmatrix}. \tag{7.31}
$$

The transformation relations among the components of the rotation angle are

$$
\begin{bmatrix} \theta_x^\gamma \\ \theta_y^\gamma \end{bmatrix} = A_\beta' \begin{bmatrix} \bar{\theta}_x \\ \bar{\theta}_y \\ \bar{\theta}_z \end{bmatrix},
$$

where the transformation matrix A_β' is derived from A_β by deleting its third row.

Suppose that the element nodal vector in global coordinates is

$$
\bar{U}_\beta = (\bar{U}_1, \bar{U}_2, \bar{U}_3, \bar{U}_4)^T, \quad \bar{U}_i = (\bar{u}_i, \bar{v}_i, \bar{w}_i, \bar{\theta}_{xi}, \bar{\theta}_{yi}, \bar{\theta}_{zi})^T, \quad i = 1, 2, 3, 4,
$$

then the transformation relation is

$$
U_\beta = S_\beta \bar{U}_\beta,
$$

where

$$
S_\beta = \begin{bmatrix} S_1 & & & 0 \\ & S_2 & & \\ & & S_3 & \\ 0 & & & S_4 \end{bmatrix}, \quad S_i = \begin{bmatrix} A_\beta & 0 \\ 0 & A_\beta' \end{bmatrix}, \quad i = 1, 2, 3, 4. \tag{7.32}
$$

The stiffness matrix and the load vector of the plate element in global coordinates are

$$\bar{K}_\beta = S_\beta^T K_\beta S_\beta, \quad \bar{F}_\beta = S_\beta^T F_\beta. \tag{7.33}$$

Note that when all plate elements related to some nodal point are on the same plane, the displacements and the rotation angles at that nodal point are just expressed in terms of local coordinates (5 parameters) but not transformed into global coordinates (6 parameters), in order to avoid singularity of the stiffness matrix.

(2) Realizing connection conditions

The connection conditions among the various members of a space composite structure are far more complex than those among the members of a plane structure, but they can also be classified into the two cases of rigid connection and pinned connection. We consider these two cases as follows.

Rigid connections

According to §3 of Chapter 4, the continuity conditions are as follows.

1. The displacement is continuous a joint of members of the same or the different dimension.

2. The rotation angle vector is continuous at a point joining of rod members to rod members.

3. The tangential rotation angle is continuous at a line joining plate members to plate members.

4. The tangential angle of rotation is continuous on a line joining plate members and rod members.

These four continuity condition are equivalent to the following requirements in the finite element method.

1. At a point rigidly connecting rods, the three global displacement components $\bar{u}$, $\bar{v}$ and $\bar{w}$ and the three global components of the rotation angle $\bar{\theta}_x$, $\bar{\theta}_y$ and $\bar{\theta}_z$ of the different rods elements are respectively equal.

2. At a point rigidly connecting plate elements and plate elements, the global displacement components $\bar{u}$, $\bar{v}$ and $\bar{w}$ and the tangential angles of rotation θ_t^β of the different plates the rotation angles θ_n^β in the normal direction are independent. The subscript t denotes the direction of the rigid-connected line, n denotes the direction perpendicular to t on the plane of the plate, and (t, n) is required to form a right-handed system.

3. At a point rigidly connecting plate members and rod members, the three global displacement components $\bar{u}$, $\bar{v}$ and $\bar{w}$ of the rod and the plate

are respectively equal, and the tangential angle of rotation θ_x^γ of the rod element and the tangential angle of rotation θ_t^β of the plate element are the same, the rotation angle θ_n^β normal the plate element and the rotation angles θ_y^γ and θ_z^γ in transverse the rod element are mutually independent.

Pinned connections

On a line joining plates and plates or plates and rods, and at a point joining rods and rods, the displacement is continuous but the rotation angles are independent.

The corresponding requirement in the finite element method is:
At any pin-connected point, the three global displacement components of the elements joined are respectively equal, while the three components of their rotation angles are independent.

We give a concrete scheme realizing these connection conditions. The basic idea is similar to that of the method of condensation which eliminates the independent rotation angle of a rod member at a pin-connected point in a plane structure. In a space structure, however, the rotation angle at the joint point is sometimes related to adjacent elements so that it cannot be eliminated immediately in the element stiffness matrix but should be eliminated in a somewhat enlarged stiffness matrix.

Consider each member (plate or rod) of the structure as an independent unit, or called a substructure. Its boundary is either just the boundary of the whole structure, or is connected with the other members.

Carry out the geometric subdivision of each member, write out the stiffness matrices and the load vectors of the individual elements in terms of the local coordinates on that member, then obtain the stiffness matrix and the load vector of the member by superposition.

For the stiffness matrix of a rod member, we transform the three local displacement components at each nodal point into the global components, but keep the components of the rotation angle in the local coordinates of the rod member. There are 6 parameters at each nodal point.

For the stiffness matrix of a plate member, we classify the nodal points into two kinds: boundary points of the plate member, and internal points. Since the internal points of the plate member contribute nothing to the stiffness matrix of any other member, all nodal parameters at internal points can be eliminated in the stiffness matrix and in the load vector of the plate member so as to obtain a stiffness matrix and a load vector both expressed in terms of the nodal parameters at the boundary points of the plate member. There are five parameters: u^β, v^β, w^β, θ_x^β and θ_y^β at each nodal point.

Let us further analyze the stiffness matrix of a plate member. The boundary points of the plate member are of two kinds: boundary points of the structure; and points shared the other members. Consider a point shared with other members. At a rigid connnection, the displacements as well as the tangential angles of rotation must be equal but the rotation angles in the normal direction are independent. At a pinned connection the displacements must be equal, but the rotation angles in both directions are independent. In either case, the externally imposed boundary conditions only involve the displacement and the tangential angle of rotation. Thus, the normal angle of rotation at the boundary point of the plate member is independent of the other member and the boundary condition. Hence we can try to eliminate it in the stiffness matrix and the load vector of the plate member. Therefore, the two components of the rotation angle θ_x^β and θ_y^β at the boundary point (except for the corner point) of the plate member are compounded into the tangential angle of rotation θ_t^β and the normal angle of rotation θ_n^β:

$$\theta_t^\beta = \theta_x^\beta \cos(t, x) + \theta_y^\beta \cos(t, y),$$
$$\theta_n^\beta = \theta_x^\beta \cos(n, x) + \theta_y^\beta \cos(n, y), \tag{7.34}$$

and at the corner point of the plate member, they are compounded into two tangential angles of rotation $\theta_{t_1}^\beta$ and $\theta_{t_2}^\beta$ through that point:

$$\theta_{t_1}^\beta = \theta_x^\beta \cos(t_1, x) + \theta_y^\beta \cos(t_1, y),$$
$$\theta_{t_2}^\beta = \theta_x^\beta \cos(t_2, x) + \theta_y^\beta (t_2, y). \tag{7.35}$$

The normal angle of rotation at each boundary point (except for the corner point) can thus be eliminated from the stiffness matrix and from the load vector of the plate member. At the same time, transform the three local displacement components at every boundary nodal point into global components. Therefore, at every non-corner boundary nodal point, there are four parameters: three displacements and one tangential angle of rotation. At a corner point, there are five parameters: three displacements and two tangential angles of rotation. In what follows we shall assume that both the stiffness matrix and the load vector of each plate member have already been expressed in such a standard form.

Their stiffness matrices and load vectors can now be coupled and super-posed according to the connection relations among the various members.

1. Rigid connection of plates and plates.

Since the stiffness matrices of the plates β_1 and β_2 have the same nodal parameters, i.e., three global displacements and one tangential angle of rotation at the joint point δ, the rigid connection at δ can be realized immediately by letting their values be equal. At the rigid-connecting point of these two plates, there are four parameters: three displace-ments and one tangential angle of rotation along the rigid-connecting line. If at δ there is still an-other plate member which is rigid-connected with it (see Fig. 71), then at the common nodal point δ there will be three

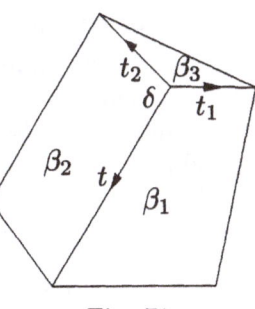

Fig. 71

displacement components as well as three tangential angles of rotation along the line of triple rigid-connection after the matrices of these three plate members β_1, β_2 and β_3 have been coupled. Namely, for each rigid-connected line through δ, there is always a tangential angle of rotation presented as the nodal parameter in the stiffness matrix correspondingly.

2. Rigid connection of plates and rods.

At the joint point δ, the three displacements and the tangential angle of rotation in the stiffness matrix of the plate equal the three displacements and the tangential angle of rotation in the stiffness matrix of the rod. This tells us immediately how to couple the two stiffness matrices. At δ there are 6 nodal parameters: three displacements and three rotation angles, the tangential angle of rotation of the joint line and the other two components of the rotation angle vector of the rod member. The latter two must not to be eliminated, because it is sometimes necessary to use them to couple with other member or to deal with boundary conditions.

3. Rigid connection of rods and rods.

The coupling is realized by setting the 6 parameters in the stiffness matrices of both members to be respectively equal at the joint point.

4. Pinned connection of plates and plates or plates and rods.

Let the three displacements in the stiffness matrices of both members be equal at the joint point, and let all the rotation angles be independent.

5. Pinned connection of rods and rods.

As we did for pinned connection of rods in the plane, corsider a space rod element, one end of which is rigid-connected and the other is pin-connected. There are 6 nodal parameters at the rigid-connected end but only three displacements and no rotation angle at the pin-connected end. When a space rod member is pin-connected, such kind of rod element can be used to eliminate the three independent rotation angles on each rod member at the common nodal point.

The stiffness matrix and the load vector of the space rod element whose left end is rigid-connected and whose right end is pin-connected are as follows:

$$
K_\gamma = \begin{bmatrix}
\dfrac{EA}{l} & 0 & 0 & 0 & 0 & 0 & \dfrac{-EA}{l} & 0 & 0 \\[2mm]
 & \dfrac{3EL_x}{l^3} & 0 & 0 & 0 & \dfrac{3El_x}{l^2} & 0 & \dfrac{-3El_x}{l^3} & 0 \\[2mm]
 & & \dfrac{3El_y}{l^3} & 0 & \dfrac{-3El_y}{l^2} & 0 & 0 & 0 & \dfrac{-3EL_y}{l^3} \\[2mm]
 & & & 0 & 0 & 0 & 0 & 0 & 0 \\[2mm]
 & & & & \dfrac{3El_y}{l} & 0 & 0 & 0 & \dfrac{3El_y}{l^2} \\[2mm]
 & & & & & \dfrac{3El_x}{l} & 0 & \dfrac{-3El_x}{l^2} & 0 \\[2mm]
 & & \text{sym.} & & & & \dfrac{EA_x}{l} & 0 & 0 \\[2mm]
 & & & & & & & \dfrac{3El_y}{l^3} & 0 \\[2mm]
 & & & & & & & & \dfrac{3El_y}{l^3}
\end{bmatrix},
$$

$$\tag{7.36}$$

$$
F_\gamma = \Big(f_1, \; f_2 - \frac{3}{2}\frac{1}{l}f_{12}, \; f_3 + \frac{3}{2}\frac{1}{l}f_{11}, \; 0, \; f_5 - \frac{1}{2}f_{11},
$$
$$
f_6 - \frac{1}{2}f_{12}, \; f_7, \; f_8 + \frac{3}{2}\frac{1}{l}f_{12}, \; f_9 - \frac{3}{2}\frac{1}{l}f_{11}\Big)^T, \tag{7.37}
$$

where $f_1, \cdots, f_{12}$ denote the 12 components of the load vector (7.24) of the rod element.

Having coupled and superposed the stiffness matrices and the load vectors of the individual members at each joint point according to the rules stated above, we obtain the stiffness matrix and the load vector of the whole structure. Its nodal points are just the joint points of the individual members and the boundary points of the structure. The number of the parameters at each nodal point is determined by the behaviour of the joint point. Finally, solving this system of linear algebraic equations according to the boundary conditions of the problem, we obtain the displacements and the rotation angles at the boundary nodal points (including the joint points and the boundary points of the structure) of the individual members. Then solving the stiffness matrices of the individual members, we obtain the displacements and the rotation angles at the internal nodal points. In engineering computations, this is called the substructure method, which we use here in combination with the treatment of the connection condition.

7.3   Nonstandard connections and the treatment
of the offset distance

The mechanical model of composite structures given in Chapter 4 requires the connections of the various members to be standard (§1 of Chapter 4). That is, the difference between the geometric dimension at the joint of members and the dimension of the its own member should not be larger than 1. However, the mechanical model is quite a idealized after all. A number of structures are often encountered of which the connection among various member in engineering practice is nonstandard. For instance, in the case of plate-rod structure, the plate and the rod are simply connected at a point. In the following, we give a practical method for treating this kind of nonstandard connection.

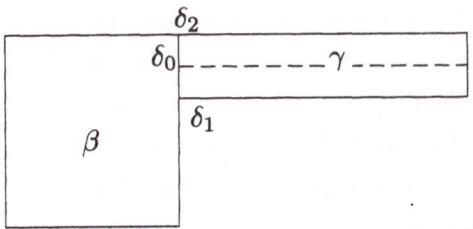

Fig. 72

Example 1. The nonstandard connection of a plate and a rod in the plane (Fig. 72). According to the mechanical model, the one-dimensional rod member γ and the two-dimensional plate member β are connected at the point δ_0, where δ_0 is a point on the neutral line of the rod (the dotted line in Fig. 72). This is one kind of nonstandard connection. The actual contacting condition is, however, different from that. If the thickness of the plate is neglected, then the joint should be a line segment $\delta_1\delta_2$, not just a point δ_0.

In the finite element method, the element stiffness matrix of the rod r has three parameters: two displacements $\bar{u}(\delta_0)$, $\bar{v}(\delta_0)$ and one rotation angle $\theta(\delta_0)$ at the nodal point δ_0. For the plate member β, by taking δ_1 and δ_2 as two nodal points of the element adjacent to the rod, the corresponding matrix has two individual displacements at δ_1 and δ_2: $\bar{u}(\delta_1)$, $\bar{v}(\delta_1)$ and $\bar{u}(\delta_2)$, $\bar{v}(\delta_2)$. In the case of a small deformation, assume there exists a rigid relationship between δ_0 and δ_1, δ_2. Then the displacement on δ_1, δ_2 can be expressed in terms of the displacement and the rotation angle on δ_0 according to the formula for rigid motion. That is,

$$\bar{u}(\delta_1) = \bar{u}(\delta_0) + \frac{l_1}{2}\theta(\delta_0), \bar{v}(\delta_1) = \bar{v}(\delta_0);$$

$$\bar{u}(\delta_2) = \bar{u}(\delta_0) - \frac{l_1}{2}\theta(\delta_0), \ \bar{v}(\delta_2) = \bar{v}(\delta_0),$$

$$(7.38)$$

where l_1 denotes the width of the rod, i.e., the distance between δ_1 and δ_2. Then the stiffness matrices the plate element and the rod element can be coupled and superimposed according to (7.38). Thus we model the nonstandard connection of plates and rods.

Example 2. The most often encountered case of nonstandard connection of plates and rods in a space structure is that the rod and the plate are rigid-connected perpendicularly or obliquely to the plane of the plate (Fig. 73). The rectangle $\delta_1\delta_2\delta_3\delta_4$ shown in Fig. 73 is the cross section at the joint of the rod and the plate. Assume that the rod has rectangular cross section, and that δ_0 is the point of intersection of the plate and the neutral line of the rod. According to the mechanical model, the one-dimensional rod and the two-

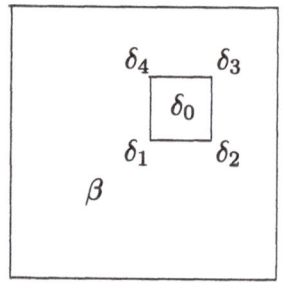

Fig. 73

dimensional plate intersect at the point δ_0 which is also a nonstandard connection. But what is actually connected is an area element, not just a point element.

In the finite element method, the space rod element has 6 parameters: three displacment $\bar{u}_0$, $\bar{v}_0$ and $\bar{w}_0$ and three rotation angles $\bar{\theta}_{x_0}$, $\bar{\theta}_{y_0}$ and

$\bar{\theta}_{x_0}$ at δ_0. By considering the interface $\delta_1\delta_2\delta_3\delta_4$ of the plate and the rod as an element of the plate, the stiffness matrix of the plate element has 6 parameters (global): three displacements $\bar{u}_i$, $\bar{v}_i$ and $\bar{w}_i$ as well as three rotation angles $\bar{\theta}_{xi}$, $\bar{\theta}_{yi}$, and $\bar{\theta}_{xi}$ at the nodal point δ_i. By approximating the relationship between δ_i and δ_0 by a rigid connection, the displacements and the rotation angles at δ_i can be expressed in terms of the displacements and the rotation angles at δ_0 according to the formula of rigid motion as follows:

$$U_i = U_0 + \Theta_0 \wedge \overrightarrow{\delta_0\delta_i},$$

$$\Theta_i = \Theta_0, i = 1, 2, 3, 4,$$

(7.39)

where the displacement vector and the rotation angle vector of are, respectively,

$$U_i = (\bar{u}_i, \bar{v}_i, \bar{w}_i)^T, \quad \Theta_i = (\bar{\Theta}_{xi}, \bar{\theta}_{yi}, \bar{\theta}_{zi})^T,$$

and the distance vector is

$$\overrightarrow{\delta_0\delta_i} = (dx_i, dy_i, dz_i)^T.$$

Rewriting (7.39) in the general matrix form,

$$
\begin{bmatrix} \bar{u}_i \\ \bar{v}_i \\ \bar{w}_i \\ \bar{\theta}_{xi} \\ \bar{\theta}_{yi} \\ \bar{\theta}_{zi} \end{bmatrix}
=
\begin{bmatrix}
1 & 0 & 0 & 0 & dz_i & -dy_i \\
 & 1 & 0 & -dz_i & 0 & dx_i \\
 & & 1 & dy_i & -dx_i & 0 \\
 & & & 1 & 0 & 0 \\
 & & 0 & & 1 & 0 \\
 & & & & & 1
\end{bmatrix}
\begin{bmatrix} \bar{u}_0 \\ \bar{v}_0 \\ \bar{w}_0 \\ \bar{\theta}_{x0} \\ \bar{\theta}_{y0} \\ \bar{\theta}_{z0} \end{bmatrix}.
$$

(7.40)

The stiffness matrices of the plate element and of the rod element can then be coupled accordingly.

One can see from the two examples above that, at the nonstandard joint point, one can couple the stiffness matrices of the members by assuming a rigid relationship between the idealized joint point and the actual joint line or surface. Thus the connection of members is realized by using the formula for rigid motion. From the practical point of view, this method is reasonable for small deformations, but this kind of nonstandard connection is not permitted in the mathematical theory of the composite structure, so that it is not rigorously justified.

The treatment of offset distance.

Another aspect of the difference between the mechanical model of a composite structure and the actual structure is that, there is often a certain distance between the actual joint line (point) of the members and the idealized joint line (point), called the offset distance. This problem was discussed and a method for treating it given in Section 3.9 of Chapter 4. The basic idea here is also to consider the relationship between the actual joint line (point) and the idealized joint line (point) as an approximately rigid relationship and then to modify the mathematical formulas related to the idealized joint line (point) using the formula for rigid motion. This kind of method. We can be used directly in the finite element method need only deal with the offset distance at the relevant nodal points. An example follows.

Suppose the plate β is rigidly connected with the rod γ (Fig. 74). Since there is a certain distance, the offset distance, between the neutral surface of the plate and the neutral line of the rod, in the finite element method the nodal point of the rod element is line and the nodal point of the plate element is taken on the neutral taken on the neutral surface. We must transform

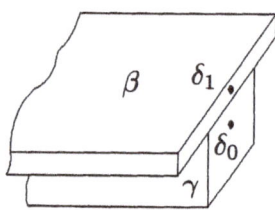

Fig. 74

the parameters at the nodal point δ_1 of the plate element into the parameters at the nodal point δ_0 of the rod element or vice versa before we can couple and superpose the stiffness matrices of the plate element and of the rod element. In general we express the displacements and the rotation angle of the member with the higher dimensions in terms of the corresponding quantitites of the member with the lower dimensions.

Assuming there exists a rigid relationship between δ_1 and δ_0, and letting $i = 1$ according to (7.40), we immediately obtain a formula expressing the nodal parameters of the plate element in terms of the nodal parameters of the rod element. This is a way of dealing with the offset distance at a nodal point in the finite element method.

Finally, we want to point out that the mechanical model and the variational principle as well as the corresponding finite element method of the composite elastic structure presented in the book are all founded on a rigorous mathematical theory, and that the convergence of the finite element method can be ensured[4].

References

[1] Argyris J.H., Energy theorems and structural analysis, *Aircraft Eng.*,26, 347–356, 383–387, 394, 1954; 27, 42–58, 80–94, 125–134, 1955.

[2] Argyris J.H., Haase M. and Mlejnek H.P., On an unconventional but natural formation of a stiffness matrix, *Comput. Methods. Appl. Mech. Eng.*, **22** (1980), 1–22.

[3] Babuska I., Aziz A.K., Survey lectures on the mathematical foundation of the finite element method, in A.K. Aziz, ed.: *The Mathematical Foundations of the Finite Element Method with Applications to Partial Differential Equations*, Academic Press, N.Y., 5–359, 1972.

[4] Bazeley G.P., Cheung Y.K., Irons B.M. and Zienkiewicz O.C., Triangular elements in bending-conforming and nonconforming solutions, in *Proc. Conf. Matrix Methods in Structural Mechanics*, Air Force Ins. Tech., Wright-Patterson A.F. Base, Ohio, 1965.

[5] Bergan P.G. and Hansen L., A new approach for deriving "good" finite element stiffness matrices, in Whiteman J.R. ed.: *The Mathematics of Finite Elements and Applications*, Vol II, 1976, 483–498.

[6] Bergan P.G., Finite elements based on energy orthogonal functions, *Int. J. Numer. Methods. Eng.*, **15**(1980), 1541–1555.

[7] Bergan P.G., and Nygard M.K., Finite elements with increased freedom in choosing shape functions, *Int. J. Numer. Methods. Eng.*, **20**(1984), 634–664.

[8] Brezzi F., On the existence, uniqueness and approximation of saddle-point problems arising from Lagrangian multipliers, *RAIRO, Anal. Numer.*, **R-2**(1984), 129–151.

[9] Carey G.F., An analysis of finite element equations and mesh subdivisions, *Comput. Methods. Appl. Mech. Engrg.*, **9**(1976), 165–179.

[10] Chen Shaochun and Shi Zhong-ci, On the free formulation scheme for construction of stiffness matrices, in *Proc. International Conference on Scientific Computation*, (T.Chan & Z-C Shi, eds.), World Scientific, 1992, 18–27.

[11] Chen Shaochun and Shi Zhong-ci, Double set parameter method of constructing stiffness matrices, *Numerica Mathematica Sinica*, **13**, 3(1991), 286–296.

[12] Chien Weizhang, Yee Kaiyuan, *Theory of Elasticity*, Science Press, Beijing, 1956.

[13] P.G.Ciarlet, *The Finite Element Method for Elliptic Problems*, North-Holland, Amsterdam, New York, 1978.

[14] Clough R.W., The finite element method in plane stress analysis, in: *Proceedings 2nd ASCE Conference on Electronic Computation*, Pittsburgh, P.A., 1960.

[15] Courant R., Variational methods for the solution of problems of equilibrium and vibration, *Bull. Amer. Math. Soc.*, **49**(1943), 1–23.

[16] Courant R., Hilbert D., *Methods of Mathematical Physics*, I. Interscience, N.Y., 1953.

[17] Felippa C.A. and Bergan P.G., A triangular bending element based on energy-orthogonal free formulation, *Comput. Meths. Appl. Mech. Eng.*, **61**(1987), 129–160.

[18] Feng Kang, On the theory of discontinuous finite elements, *Numerica Mathematica Sinica*, **1, 4**(1979), 278–385.

[19] Feng Kang, A difference formulation based on the variational principle, *Appl. Math. Comput. Math.*, **2**(1965), 237–261.

[20] Feng Kang, The finite element method, *Math. Knowledge Appl.*, **4**(1974); **1, 2,**(1975).

[21] Feng Kang, Elliptic equations and composite elastic structures on the composite manifold, *Math. Numer. Sinica*, **1**(1979), 199–208.

[22] Fichera G., *Existence Theorems in Elasticity*, Handbuch der Physik, Encyclopedia of Physics, Band VI, 12, 347–389, Springer-Verlag, Berlin 1972.

[23] Irons B.M. and Razzaque A., Experience with the patch test, in *Proc. Symp. on Mathematical Foundations of the Finite Element Method*, (Aziz A. R. ed.), Academic Press, 1972, 557–587.

[24] Landau L., Lifchitz E., *Theory of Elasticity*, Nauka, Moscow, 1965.

[25] Lascaux P., Lesaint P., Some nonconforming finite elements for the plate bending problem, *RAIRO, Anal. Numer.*, **R-1**(1985), 9–53.

[26] Long Yu-qiu and Xin Ke-qui, Generalized conforming elements, *J. Civil Engineering*, **1**(1987), 1–14.

[27] Love, A.E.H., *Mathematical Theory of Elasticity*, 4th ed., Dover, N.Y., 1944.

[28] Morley L.S.D., The triangular equilibrium element in the solution of plate bending problems, *Aero. Quart.*, **19**(1968), 149–169.

[29] Novozhilov V.V., *Theory of Elasticity*, Pergamon Press, Oxford, 1961.

[30] Oden J.T. and Reddy J.N., *Variational Methods in Theoretical Mechanics*, Springer-Verlag, Berlin, 1976.

[31] Oden J.T. and Reedy J.N., *An introduction to the Mathematical Theory of Finite Elements*, Wiley Interscience, New York, 1986.

[32] Quarteroni A., Primal hybrid finite element methods for 4th order elliptic equations, *Calcolo*, **16**(1979), 1, 21–58.

[33] Raviart P.A. and Thomas J.M., Primal hybrid finite element methods for 2nd order elliptic equations, *Math. Comp.*, **31**(1977), 138, 391–413.

[34] Sander G. and Beckers P., The influence of the choice of connectors in the finite element methods, *Int. J. Numer. Methods Eng.* **11**(1977), 1491–1505.

[35] Shi Zhong-ci, An explicit analysis of Stummel's patch test examples, *Int. J. Numer. Methods Eng.*, **20**(1984), 1233–1246.

[36] Shi Zhong-ci, A convergence condition for the quadrilateral Wilson element, *Numer. Math.*, **44**(1984), 349–361.

[37] Shi Zhong-ci, On the convergence properties of quadrilateral elements of Sander and Beckers, *Math. Comp.*, **42**(1984), 493–504.

[38] Shi Zhong-ci, The generalized patch test for Zienkiewicz's triangles, *J. Computational Mathematics*, **2**(1984), 279–286.

[39] Shi Zhong-ci, Convergence properties of two nonconforming finite elements, *Comput. Meths. Appl. Mech. Eng.*, **48**(1985), 123–137.

[40] Shi Zhong-ci, A united formulation of shape functions for two kinds of nonconforming plate bending elements, *Numerica Mathematica Sinica*, **8**(1986), 428–434.

[41] Shi Zhong-ci, Convergence of the TRUNC plate element, *Comput. Meths. Appl. Mech. Eng.*, **62**(1987), 71–88.

[42] Shi Zhong-ci, The F-E-M-Test for nonconforming finite elements. *Math. Comput.*, **49**(1987), 391–405.

[43] Shi Zhong-ci, On the nine degree quasi-conforming plate element, *Numerica Mathematica Sinica*, **10**(1988), 100–106.

[44] Shi Zhong-ci, On Stummel's examples to the patch test, *Computational Mechanics*, **5**(1989), 81–87.

[45] Shi Zhong-ci, On the accuracy of the quasi-conforming and generalized conforming finite elements, *Chin. Ann. Math.*, **11B**, 2(1990), 148–156.

[46] Shi Zhong-ci, On the error estimates of Morley element, *Numerica Mathematica Sinica*, **12**(1990), 113–118.

[47] Shi Zhong-ci and Chen Shaochun, An analysis of a nine parameter plate element of specht, *Numerica Mathematica Sinica*, **11**(1989), 312–318.

[48] Shi Zhong-ci and Chen Shaochun, Direct analysis of a nine parameter quasi-conforming plate element, *Numerica Mathematica Sinica*, **12**(1990), 76–84.

[49] Shi Zhong-ci and Chen Shaochun, Convergence of a nine degree generalized conforming element, *Numerica Mathematica Sinica*, **13**(1991), 193–203.

[50] Shi Zhong-ci and Zhang Fei, Construction and analysis of a new energy-orthogonal unconventional plate element, *JCM*, **8**(1990), 75–91.

[51] Strang G. and Fix G.J., *An Analysis of the Finite Element Method*, Prentice-Hall, 1973.

[52] Stummel F., The generalized patch test, *SIAM J. Numer. Analysis*, **16**(1979), 449–471.

[53] Stummel F., The limitation of the patch test, *Int. J. Numer. Methods Eng.*, **15**(1980), 177–188.

[54] Stummel F., Basic compactness properties of nonconforming and hybrid finite element spaces, *RAIRO, Anal. Numer.*, **4**(1980), 81–115.

[55] Taylor R.L., Simo T.C., Zienkiewica O.C. and Chan A.H.C., The patch test – a condition for assessing FEM convergence, *Int. J. Numer. Methods Eng.*, **22**(1986), 39–62.

[56] Washizu K., *Variational Methods in Elasticity and Plasticity*, **2nd ed.**, Pergamon Press, Oxford, 1975.

[57] Zienkiewicz O.C., *The Finite Element Method in Engineering Science*, 3rd ed., McGraw-Hill, N.Y., 1977.

[58] Zlamal M., On the finite element method, *Numer. Math.*, **12**(1968) 394–409.

Index